Looking Forward

Looking Forward

PREDICTION AND UNCERTAINTY IN MODERN AMERICA

Jamie L. Pietruska

The University of Chicago Press CHICAGO & LONDON

The University of Chicago Press, Chicago 60637
The University of Chicago Press, Ltd., London

Published 2017
Printed in the United States of America

33 32 31 30 29 28 27 26 25 2 3 4 5 6

ISBN-13: 978-0-226-47500-4 (cloth)
ISBN-13: 978-0-226-50915-0 (e-book)
DOI: https://doi.org/10.7208/chicago/9780226509150.001.0001

Publication of this book was generously supported with a grant from the Rutgers University Research Council.

Library of Congress Cataloging-in-Publication Data

Names: Pietruska, Jamie L., author.
Title: Looking forward : prediction and uncertainty in modern America / Jamie L. Pietruska.
Description: Chicago ; London : The University of Chicago Press, 2017. | Includes bibliographical references and index.
Identifiers: LCCN 2017018833 | ISBN 9780226475004 (cloth : alk. paper) | ISBN 9780226509150 (e-book)
Subjects: LCSH: Forecasting—Social aspects—United States. | Economic forecasting—United States. | Risk—United States. | Prophecy.
Classification: LCC CB158 .P54 2017 | DDC 330.973/00112—dc23
LC record available at https://lccn.loc.gov/2017018833

♾ This paper meets the requirements of ANSI/NISO Z39.48-1992 (Permanence of Paper).

To Jason

The present age is in the attitude of looking forward. It stands in an unaccustomed glare, with hand shading the eye, peering into the future. On all lips are the questions, What is coming? What have we to hope, and what to fear? What will our children live to see?

ANDOVER REVIEW, 1891

Many persons talk as if the minutest dose of disconnectedness of one part with another, the smallest modicum of independence, the faintest tremor of ambiguity about the future . . . would ruin everything, and turn this goodly universe into a sort of insane sand-heap or nulliverse, no universe at all.

WILLIAM JAMES, "The Dilemma of Determinism," 1884

CONTENTS

ACKNOWLEDGMENTS

When I began this book, I certainly did not foresee how much help I would need in order to finish it. It is gratifying to thank all the people and institutions that have supported this project over the course of many years.

I am indebted to the archivists and librarians who helped me access materials in their collections at the Boston Public Library; the Edward Bellamy House in Chicopee, Massachusetts; the Houghton Library at Harvard University; the Massachusetts Historical Society; the Milton (Massachusetts) Historical Society; the Missouri Historical Society; the National Archives in College Park, Maryland (especially Marjorie Ciarlante); the National Archives in Washington, DC; the Smithsonian Institution Archives; the New York City Municipal Archives (especially Kenneth Cobb and Joseph Van Nostrand); the Tecumseh (Michigan) District Library (especially Mary Beth Reasoner); the Tulane University Libraries; the University of Georgia Libraries; and the US Department of Agriculture's National Statistics Service (especially Amanda Pomicter). Stephen Jendrysik, president of the Edward Bellamy Memorial Association, graciously opened the Bellamy House to me on the weekends and allowed me the pleasure of roaming freely through the house and its archives. Jim Powell, descendant of British cotton forecaster Henry M. Neill, contacted me to share his fascinating family history, and Karen Christino, biographer of astrologist Evangeline Adams, kindly sent me materials from her personal collection. The Rutgers Libraries, and especially history librarian Tom Glynn, helped me track down sources quickly. I thank my research assistant, Paul Blake Nowacek, for expertly digitizing voluminous archival materials on several occasions.

Material support for research, writing, and travel was generously provided by MIT's Program in History, Anthropology, and STS; the MIT Kelly-Douglas Fund; the Society for the History of Technology; and Rutgers University. I have also been the fortunate recipient of fellowships from the American Association of University Women, the American Academy of Arts and Sciences, and the American Council of Learned Societies. A Rutgers University Research Council Subvention Grant supported publication of this book.

This book and I have had several academic homes, the first of which was the Program in History, Anthropology, and STS at the Massachusetts Institute of Technology, where I benefited from the intellectual and personal guidance of Merritt Roe Smith (my advisor), Deborah Fitzgerald, Meg Jacobs, and Chris Capozzola. Each tended to this project in its incipient stages and mentored me with exactly the right combination of rigor and encouragement. Together they taught me how to think like a historian, how to research and write a book, how to be a professor, and how to navigate the academic world. Roe deserves special thanks for his remarkable generosity, which made my journey through graduate school such a smooth and rewarding one. Leo Marx and Rosalind Williams were extraordinary guides in our explorations of cultural history and literary history. I learned much from my graduate student colleagues, especially Etienne Benson, Nick Buchanan, Kieran Downes, Brendan Foley, Xaq Frohlich, Chihyung Jeon, Dave Lucsko, Rob Martello, Lisa Messeri, Esra Ozkan, Anne Pollock, Michael Rossi, Peter Shulman, David Singerman, Jenny Smith, Bill Turkel, Rebecca Woods, and Sara Wylie. And I was especially grateful for the friendship of Candis Callison, Anita Chan, Richa Kumar, Natasha Myers, and Will Taggart during our first years of graduate school together.

I subsequently had the privilege of spending a year in the Visiting Scholars Program at the American Academy of Arts and Sciences in Cambridge, where this project was shaped by the thoughtful and incisive feedback of program director Patricia Meyer Spacks, Dan Amsterdam, Deborah Becher, Angus Burgin, Dawn Coleman, Andrew Jewett, and Jason Petrulis.

It was thanks to the American Council of Learned Societies New Faculty Fellows program, directed by Nicole Stahlmann, that I came to the History Department at Rutgers University, New Brunswick. I am grateful to deans of the School of Arts and Sciences Douglas Greenberg and Peter March and History Department chairs Jim Masschaele, Mark Wasserman, and Barbara Cooper for supporting my work and arranging sabbatical and research leaves. I thank my colleagues in the History Department who have made me feel intellectually at home in such a large institution. Jackson Lears and Ann Fabian were major sources of inspiration from the moment I first read their work, so I am incredibly lucky to now have them as mentors and colleagues. I truly appreciate all that they have done on my behalf. Special thanks to Belinda Davis, James Delbourgo, Rachel Devlin, Anthony di Battista, Marisa Fuentes, Mike Geselowitz, David Greenberg, Nancy Hewitt, Chie Ikeya, Paul Israel, Jennifer Jones, Toby Jones, Norman Markowitz, Lou Masur, Jennifer Mittelstadt, Johanna Schoen, Judith Surkis, Paola Tartakoff, Andy Urban, and Ginny Yans for conversation and kindness. I have also benefited from rich discussions at the Rutgers Center for Historical Analysis Networks of Exchange seminar directed by James Delbourgo and Toby Jones. Thanks also to Tiffany Berg, Matt Leonaggeo, Matt Steiner, and Candace Walcott-Shepherd for their logistical help and cheerful problem solving. And several classes of Rutgers undergraduates were willing and receptive audiences for some of this material.

This book has been improved by comments and questions from scholars who generously engaged with my work as fellow conference panelists, discussants, and seminar participants in various venues. I have learned so much from their work and hope they will see their influence in these pages. Many thanks to James Bergman, Dan Bouk, Eli Cook, Phaedra Daipha, Will Deringer, Jim Fleming, Erica Fretwell, Walter Friedman, Courtney Fullilove, Lisa Gitelman, Gabriel Henderson, Rebecca Herzig, Nate Holdren, Caley Horan, Sarah Igo, Richard John, Betty Ann Kevles, Dan Kevles, Josh Lauer, Jon Levy, Ken Lipartito, Rob MacDougall, Patrick McCray, David Nye, Emily Pawley, Bob Reeves, Laura Thiemann Scales, and Roger Turner.

I am grateful to those who invited me to present my research and to audiences who devoted their time and energy to some of the most rewarding discussions I have had about this topic. Their keen observations and questions reinvigorated my thinking and helped me clarify my interpretations at important junctures. Many thanks to Matthias Heymann and the workshop "Cultures of Prediction: Scientists and the Crafting of the Future" at the Centre for Science Studies, Aarhus University; Matt Stanley and the New York City History of Science Working Group at New York University; Will Deringer and the "Calculating Capitalism" conference at the Heyman Center for the Humanities at Columbia University; Seth Rockman, Lukas Rieppel, and the Nineteenth-Century US History Workshop and the Science and Capitalism Lecture Series at Brown University; and Hugh Rockoff and the participants in the Money, History and Finance seminar in the Department of Economics at Rutgers. The conferences "Uncertain Environments: Natural Hazards, Risk, and Insurance in Historical Perspective," held at the German Historical Institute, and "Understanding Markets: Information, Institutions, and History," sponsored by the Hagley Museum and Library and the German Historical Institute, shaped this book in its formative stages, and Brian Luskey and Wendy Woloson's terrific "Capitalism by Gaslight" conference and subsequent edited volume changed the way I think about nineteenth-century American capitalism.

Several colleagues at Rutgers and beyond deserve special thanks for reading and commenting on parts of the book: James Delbourgo, David Foglesong, Walter Friedman, David Greenberg, Jon Levy, Jim Livingston, and Bob Reeves. Early versions of chapters benefited from the careful readings of Andy Urban and Kyla Schuller in our nineteenth-century writing group at Rutgers. Eli Cook, Ann Fabian, Paul Israel, Toby Jones, and Jackson Lears read the entire manuscript and helped me move forward.

Portions of this book appeared previously in Jamie L. Pietruska, "'Cotton Guessers': Crop Forecasters and the Rationalizing of Uncertainty in American Cotton Markets, 1890–1905," in *The Rise of Marketing and Market Research*, edited by Hartmut Berghoff and Uwe Spiekermann (New York: Palgrave Macmillan, 2012), reproduced with permission of Palgrave Macmillan; and Jamie L. Pietruska, "US Weather Bureau Chief Willis Moore and the Reimagination of Uncertainty in Long-Range Forecasting," *Environment and History* 17 (2011): 79–105, republished with permission of the White Horse Press.

I could not have asked for a better editor than Doug Mitchell at the University of Chicago Press, whose enthusiasm (and patience) never wavered as I revised this book. Kyle Wagner and the staff at the press made the editorial and production process go smoothly from start to finish. Scott Sandage was the ideal reviewer for this manuscript—his remarkably thorough readings and constructive criticism pushed the book in the direction it needed to go. It is a much better book for his involvement in it. Two anonymous reviewers helpfully suggested additional contexts and themes to consider.

Friends old and new have provided steady encouragement and welcome diversions along the way. I will be forever grateful to Stephanie Adler for listening and always understanding me, even when I was not at my best. Jason Petrulis and Dan Amsterdam offered moral support at crucial moments. Many friends in Boston, New Jersey, and points beyond have enriched my life with joy, kindness, and delightful conversation, especially Celena Illuzzi and Ray Falke, Christine Wenc, Sue Costa Paschke, Karen Bishop and Luis Fanjul, Eman El Badawi and her wonderful family, Sarah Schwarz, Sue and Nate Domagalski, and Tim and Debi Cleary. Gina Scharoun has been a fabulous friend and impossibly fun excursionist for over twenty years. JoAnne and Richard Preiser have shared boundless love and laughter and have always made me feel at home on Newell Avenue. They have also read multiple drafts, attended my talks, and helped me in more ways than I can count or repay.

My family has supported my academic endeavors since before my first day of kindergarten. My grandparents, Michael and Irene Cetola and Stanley and Helen Pietruska, spent hours reading to me and nurturing my love for books and for them. Stan and Anne Pietruska have been so enthusiastic about every development in my career and in my life, and our discussions of history and politics always enliven my day. My parents, Elizabeth and Gerald Pietruska, did everything they could to make my life run as smoothly as possible as I became a parent and a professor. When we lived in Boston, they cheerfully commuted hours each day to provide the best childcare I could have wished for. Over the years, I have relied on the steady stream of material and moral support they have so generously provided.

Being a working mother would not have been possible without the labor, patience, and flexibility of Analicia Roeser and Tracey Flaherty, who have cared for our children and become a part of our family in the process.

No one has given more to this book and to me than Jason Burns. He has relocated for my career, adjusted his work schedule to accommodate mine, managed the home front while I traveled to conferences and archives, taken magnificent care of our children on the many weekends I spent writing this book and preparing lectures, and cooked fabulous meals—all without a word of complaint. Too often I have taken his selfless efforts for granted. His extraordinary sacrifices are matched by his extraordinary patience and love. He has always believed in me, even in my moments of self-doubt. And he read every word of this book without asking why I took so long to finish it.

Miles Henry arrived at the beginning of this project and Silas Michael arrived toward the end, each bringing the deepest love and most wondrous joy into my life. As I watch them grow, I never know what they will say or do next. From them I have learned to appreciate and even revel in the unpredictability of daily life, which is the greatest gift of all.

INTRODUCTION

Crisis of Certainty

Edward Bellamy, anointed a "Great American Prophet" by John Dewey, could never have predicted that his best-selling utopian novel would inspire a socialist reform movement.[1] *Looking Backward*, published in 1888 to worldwide acclaim, chronicles the time travel of Julian West, a wealthy Bostonian who enters a deep sleep in 1887 and wakes up in the year 2000, shocked to find himself in a world perfected by democratic Christian socialism: a harmonious and efficient cooperative commonwealth without the economic volatility, unequal distribution of wealth, and violent confrontations between labor and capital that threatened the stability of the new industrial society.[2] Bellamy's critique of the Gilded Age economy became a call to action for writers and reformers who launched a political movement to, as Bellamyite Cyrus Willard wrote, "waken the minds of the people to the fact that they are industrial slaves and that industrial slavery must be abolished."[3] The Nationalist clubs that spread across the country translated the novel's antimonopoly critique into a socialist political platform based on government ownership of major industries and

1. John Dewey, "A Great American Prophet," *Common Sense* 3, no. 4 (1934): 6–7.

2. On the novel's international reception, see Sylvia E. Bowman, *Edward Bellamy Abroad: An American Prophet's Influence* (New York: Twayne, 1962); Carl J. Guarneri, "An American Utopia and Its Global Audiences: Transnational Perspectives on *Looking Backward*," *Utopian Studies* 19, no. 2 (2008): 147–87.

3. Cyrus Field Willard to Edward Bellamy, March 22, 1889, Edward Bellamy Correspondence, 1850–1898 (MS Am 1181), Houghton Library, Harvard University.

municipal administration of utilities.[4] With this program, Nationalists sought to do for the late nineteenth century what Bellamy had imagined for the year 2000: eliminate economic uncertainties. In *Looking Backward*, when Julian West returns in a dream to 1887, he observes that everyone is haunted by "the specter of Uncertainty" whispering,

> Do your work never so well . . . rise early and toil till late, rob cunningly or serve faithfully, you shall never know security. Rich you may be now and still come to poverty at last. Leave never so much wealth to your children, you cannot buy the assurance that your son may not be servant of your servant, or that your daughter will not have to sell herself for bread.[5]

Bellamy's specter was clearly visible in soaring unemployment rates, unpredictable supply and demand, and high rates of business failure, but late nineteenth-century America was haunted by a crisis of certainty at once economic and epistemic. Americans faced not only panics and strikes in an age of boom and bust but also the crumbling foundations of traditional explanations of futurity in the contexts of science, social science, and religion. As new transcontinental railroad and telegraph networks linked producers and consumers into expanding national and global markets, and new forms of financial transactions emerged in commodity futures trading, exchanging capital in distant markets became an increasingly uncertain endeavor. The "specter of Uncertainty" also appeared in a culture in which deterministic worldviews began to accommodate the forces of chance and contingency, from Darwinian evolution to probability theory to biblical criticism. From the market to religion, Americans confronted a future more difficult but seemingly more crucial to anticipate. In that context, prediction became a ubiquitous scientific, economic, and cultural practice, and forecasts, accurate or not, offered illusions of control over one's future in what William Dean Howells recognized as the "economic chance-world" of late nineteenth-century America.[6]

4. On Nationalist clubs, see Sylvia E. Bowman, *The Year 2000: A Critical Biography of Edward Bellamy* (New York: Bookman Associates, 1958), 122–38; James J. Kopp, "Looking Backward at Edward Bellamy's Influence in Oregon, 1888–1936," *Oregon Historical Quarterly* 104, no. 1 (2003): 62–95.

5. Edward Bellamy, *Looking Backward, 2000–1887* (1888; repr., New York: Signet Classic, 1960), 212.

6. William Dean Howells, *A Hazard of New Fortunes* (1890; repr., New York: Penguin Books, 2001), 396.

This book is a history of forecasting in the United States from the 1860s to the 1920s that reveals how methods of forecasting and ideas about uncertainty changed as institutions and individuals reckoned with Howells's "world of chance."[7] It traces the production, circulation, and contestation of crop estimates, weather forecasts, economic predictions, and the predictions of fortune-tellers in order to uncover the significance of systematized forecasts as new forms of knowledge and risk management. The book's overarching argument revises historians' understanding of the late nineteenth and early twentieth centuries as a "search for order" by demonstrating that a search for predictability yielded just the opposite: acceptance of the uncertainties of economic and cultural life.[8]

The book makes four additional and interlocking arguments, the first of which is that daily forecasts became increasingly commonplace in a "culture of prediction" in late nineteenth-century America.[9] Predictions in unprecedented numbers reached millions of Americans by mail, print media, and telegraph and circulated through federal agencies, the halls of Congress, scientific institutions, commodity exchanges, and corporations. This is not a book about the unpredictability of spectacular accidents and disasters (although blizzards, panics, and train collisions occasionally appear) but rather about the routinized predictions of everyday life and how they functioned as sources of knowledge and tools for risk management in a capitalist economy.[10]

7. William Dean Howells, *The World of Chance* (New York: Harper & Brothers, 1893).

8. Robert H. Wiebe, *The Search for Order, 1877–1920* (New York: Hill and Wang, 1967).

9. For a sociological analysis of a "culture of prediction" in the National Weather Service in the early twenty-first century, see Gary Alan Fine, *Authors of the Storm: Meteorologists and the Culture of Prediction* (Chicago: University of Chicago Press, 2007). On "cultures of prediction" in late twentieth-century climate modeling, see Matthias Heymann, Gabriele Gramelsberger, and Martin Mahony, eds., *Cultures of Prediction in Atmospheric and Climate Science: Epistemic and Cultural Shifts in Computer-Based Modelling and Simulation* (New York: Routledge, 2017).

10. On the literary, legal, and cultural significance of late nineteenth-century accidents, see Jason Puskar, *Accident Society: Fiction, Collectivity, and the Production of Chance* (Palo Alto, CA: Stanford University Press, 2012); John Fabian Witt, *The Accidental Republic: Crippled Workingmen, Destitute Widows, and the Remaking of American Law* (2004; repr., Cambridge, MA: Harvard University Press, 2006); Nan Goodman, *Shifting the Blame: Literature, Law, and the Theory of Accidents in Nineteenth-Century America* (1998; repr., New York: Routledge, 2000). On the history of disasters, see Ted Steinberg, *Acts of God: The Unnatural History of Natural Disaster in America*, 2nd ed. (New York: Oxford University Press, 2006); Kevin Rozario, *The Culture of Calamity: Disaster and the Making of Modern America* (Chicago: University of Chicago Press, 2007); Steven Biel, ed., *American Disasters* (New York: NYU Press, 2001).

Second, the histories of crop and weather forecasting underscore the centrality of the countryside to late nineteenth-century American capitalism and the state.[11] The book begins with chapters on crop and weather forecasting to demonstrate that the modern bureaucratic rationality that historians have often located in the city, in the rise of corporate capitalism, has roots in the countryside, where the systematic search for predictable crop yields and weather first began.[12] The large-scale government reporting networks that produced crop and weather forecasts after the American Civil War are an important but often overlooked example of the considerable "infrastructural power" of the late nineteenth-century American state.[13] Contests over the politics of access to these information networks reveal how farmers, manufacturers, shippers, and speculators learned to operate within predictive knowledge infrastructures that linked them to the federal government on a daily basis.[14]

11. On capitalism in the countryside, see Christopher Clark, *The Roots of Rural Capitalism: Western Massachusetts, 1780–1860* (Ithaca, NY: Cornell University Press, 1990); Steven Hahn and Jonathan Prude, eds., *The Countryside in the Age of Capitalist Transformation: Essays in the Social History of Rural America* (Chapel Hill: University of North Carolina Press, 1985); William Cronon, *Nature's Metropolis: Chicago and the Great West* (New York: W. W. Norton, 1991); Christopher Clark, "The Agrarian Context of American Capitalist Development," in *Capitalism Takes Command: The Social Transformation of Nineteenth-Century America*, ed. Michael Zakim and Gary J. Kornblith (Chicago: University of Chicago Press, 2011), 13–37. On the state and the countryside, see Catherine McNicol Stock and Robert D. Johnston, eds., *The Countryside in the Age of the Modern State: Political Histories of Rural America* (Ithaca, NY: Cornell University Press, 2001).

12. Wiebe, *Search for Order*. Jonathan Levy, *Freaks of Fortune: The Emerging World of Capitalism and Risk in America* (Cambridge, MA: Harvard University Press, 2012) observes that scientific crop and weather forecasting offered farmers a new mode of risk management and signaled a "triumph for probabilistic certainty" (181–83, quotation on 183). But, as this book demonstrates, crop and weather forecasts generated widespread controversy and revealed the economic and epistemological limitations of new forms of predictive knowledge.

13. On "infrastructural power," a concept borrowed from sociologist Michael Mann, see William J. Novak, "The Myth of the 'Weak' American State," *American Historical Review* 113, no. 3 (2008): 763. On the significance of the US Department of Agriculture to late nineteenth-century federal authority, see Daniel P. Carpenter, *The Forging of Bureaucratic Autonomy: Reputations, Networks, and Policy Innovations in Executive Agencies, 1862–1928* (Princeton, NJ: Princeton University Press, 2001). For another revision of the traditional interpretation of a "weak" nineteenth-century state, see Brian Balogh, *A Government Out of Sight: The Mystery of National Authority in Nineteenth-Century America* (New York: Cambridge University Press, 2009).

14. On knowledge infrastructures, see Paul N. Edwards, *A Vast Machine: Computer Models, Climate Data, and the Politics of Global Warming* (2010; repr., Cambridge, MA: MIT Press, 2013), esp. 17–25; Paul N. Edwards, Lisa Gitelman, Gabrielle Hecht, Adrian Johns, Brian Larkin,

Third, the production of knowledge about the future was shaped by conflicts over authority and expertise within scientific institutions, the federal government, the legal system, the press, and popular discourse. The competition between government crop and weather forecasts produced as a public good and the predictions sold by commercial forecasters, predictions of future economies that drew staunch defenders and harsh critics, and fortune-tellers' prosecutions and appeals reveal ongoing struggles over which forecasters had legitimate claims to future knowledge and which did not.

And fourth, just as the production of forecasts was contested, so too was their meaning and value. Crop estimates, weather forecasts, market forecasts, and the predictions of fortune-tellers had clear economic imperatives, but they also inspired thorny epistemic debates over the possibilities and the limitations of foreknowledge. The trajectory of this book follows the shift from positivist certainty to probabilism that historians have traced from the nineteenth century to the twentieth, but it shifts the focus from philosophers and mathematicians to farmers, weather forecasters, government officials, fortune-tellers, jurists, and journalists, among others, who were keenly aware of forecasts' implications for futurity itself. *Looking Forward* is the history of epistemological conflicts over ways of knowing—and ways of *not* knowing—the future.

HISTORIES OF THE FUTURE

There is not yet extensive scholarship on forecasting, but relevant work has emerged primarily from the history and sociology of science (especially studies of meteorology), literary history (specifically utopian studies), scholarly and popular accounts of futurism, and the new history of capitalism.[15] Specific chapters explain in more detail their historiographical contributions to separate scholarly literatures; *Looking Forward* is the first synthesis of crop forecasting,

and Neil Safier, "*AHR* Conversation: Historical Perspectives on the Circulation of Information," *American Historical Review* 116, no. 5 (2011): 1393–1435.

15. For popular accounts that consider forecasting either a timeless phenomenon or a late twentieth-century development, see William A. Sherden, *The Fortune Sellers: The Big Business of Buying and Selling Predictions* (New York: John Wiley & Sons, 1998); Paul Halperin, *The Pursuit of Destiny: A History of Prediction* (Cambridge, MA: Perseus, 2000); David Orrell, *The Future of Everything: The Science of Prediction* (New York: Thunder's Mouth Press, 2007); Dan Gardner, *Future Babble: Why Expert Predictions Are Next to Worthless, and You Can Do Better* (New York: Dutton, 2011); Nate Silver, *The Signal and the Noise: Why So Many Predictions Fail—but Some Don't* (New York: Penguin Books, 2012); Philip E. Tetlock and Dan Gardner, *Superforecasting: The Art and Science of Prediction* (New York: Broadway Books, 2015).

weather forecasting, market forecasting, literary utopianism, and occult prediction into a broader cultural framework. Its periodization challenges the assumption that professional forecasting originated in the postwar contexts of big science, computational meteorology, Cold War scenario planning, and the intellectual traction of "futures studies" in the 1970s.[16] As this book demonstrates, the first systematic approaches to gathering predictive data from across the nation and translating it into knowledge about the future came a century earlier, with the federal government's crop reporting and weather observation networks. It was the late nineteenth century, not the late twentieth, that saw the emergence of professional forecasting in the United States.

This book is a history of knowledge production that emphasizes the interconnections of technoscience, capitalism, and culture.[17] In his classic account of the corporate dominance of scientific research and technical innovation in the late nineteenth-century United States, David Noble observed that "science and capitalism press forward by nature." The future orientation of both

16. For the emphasis on postwar futurity, see the recent *American Historical Review* Forum on "Histories of the Future," which includes Jenny Andersson, "The Great Future Debate and the Struggle for the World," *American Historical Review* 117, no. 5 (2012): 1411–30; Matthew Connelly et al., "'General, I Have Fought Just as Many Nuclear Wars as You Have': Forecasts, Future Scenarios, and the Politics of Armageddon," *American Historical Review* 117, no. 5 (2012): 1431–60. See also Kristine C. Harper, *Weather by the Numbers: The Genesis of Modern Meteorology* (Cambridge, MA: MIT Press, 2008); Frederik Nebeker, *Calculating the Weather: Meteorology in the Twentieth Century* (San Diego: Academic Press, 1995); Matthew Connelly, "Future Shock: The End of the World as They Knew It," in *The Shock of the Global: The 1970s in Perspective*, ed. Niall Ferguson, Charles S. Maier, Erez Manela, and Daniel J. Sargent (Cambridge, MA: Belknap Press, 2010), 337–50; Kaya Tolon, "Futures Studies: A New Social Science Rooted in Cold War Strategic Thinking," in *Cold War Social Science: Knowledge Production, Liberal Democracy, and Human Nature*, ed. Mark Solovey and Hamilton Cravens (New York: Palgrave Macmillan, 2012), 45–62.

17. For an influential formulation of *technoscience* as a practice-oriented hybrid of science and technology, see Bruno Latour, *Science in Action: How to Follow Scientists and Engineers through Society* (Cambridge, MA: Harvard University Press, 1987), chap. 4. For recent work bridging science and capitalism in nineteenth-century America, see Courtney Fullilove, *The Profit of the Earth: The Global Seeds of American Agriculture* (Chicago: University of Chicago Press, 2017); Lukas Rieppel, *Assembling the Dinosaur: Science, Museums, and American Capitalism, 1870–1930* (Cambridge, MA: Harvard University Press, forthcoming); David Roth Singerman, "Science, Commodities, and Corruption in the Gilded Age," *Journal of the Gilded Age and Progressive Era* 15, no. 3 (2016): 278–93; Emily Pawley, "Accounting with the Fields: Chemistry and Value in Nutriment in American Agricultural Improvement, 1835–1860," *Science as Culture* 19, no. 4 (2010): 461–82.

is a central theme of this book, and indeed the entanglement of technoscience and capitalism is especially apparent in the late nineteenth-century transportation systems, communications networks, and energy infrastructures that linked Wiebe's "island communities" into an unprecedented world of economic and social interdependence.[18] But the relationship between forecasting, technoscience, and capitalism in this period is only partly explained by a material account of information networks and the economic value of the forecasts they produced. Forecasts, often bought and sold, were not only tools for risk management but also new forms of predictive economic knowledge.[19] The epistemic dimensions of forecasting belong to what Jeffrey Sklansky has characterized as "the history of disciplines, genres, paradigms, and other frames of representation, in which capitalism appears as a way of seeing, a mode of organizing and conveying knowledge."[20] This book emphasizes the inseparability of ways of knowing the economy and economic action, which has been highlighted in Walter Friedman's pathbreaking history of economic forecasters in the early twentieth century as well as in a number of works on nineteenth-century American capitalism that examine insurance and risk, clerks and accounting, gambling and speculation, counterfeiting, and business

18. David F. Noble, *America by Design: Science, Technology, and the Rise of Corporate Capitalism* (New York: Knopf, 1977), 3; Wiebe, *Search for Order*, xiii. For the classic account of social interdependence and economic action at a distance in which institutional social science emerged to explain and predict human behavior, see Thomas Haskell, *The Emergence of Professional Social Science: The American Social Science Association and the Nineteenth-Century Crisis of Authority* (1977; repr., Baltimore: Johns Hopkins University Press, 2000). Historians of technology and STS scholars have used the conceptual frameworks of large technological systems, networks, and, most recently, infrastructure. See, for example, Thomas P. Hughes, "The Evolution of Large Technological Systems," in *The Social Construction of Technological Systems: New Directions in the Sociology and History of Technology*, ed. Wiebe E. Bijker, Thomas P. Hughes, and Trevor Pinch (Cambridge, MA: MIT Press, 1987), 51–82; Manuel Castells, *The Rise of the Network Society* (Malden, MA: Blackwell, 1996); Paul N. Edwards, "Infrastructure and Modernity: Force, Time, and Social Organization in the History of Sociotechnical Systems," in *Modernity and Technology*, ed. Thomas J. Misa and Philip Bray (Cambridge, MA: MIT Press, 2003), 185–225.

19. For studies of how new tools for risk management yielded new forms of predictive economic knowledge, see Josh Lauer, *Creditworthy: A History of Credit Surveillance and Financial Identity in America* (New York: Columbia University Press, 2017); Dan Bouk, *How Our Days Became Numbered: Risk and the Rise of the Statistical Individual* (Chicago: University of Chicago Press, 2015).

20. Jeffrey Sklansky, "The Elusive Sovereign: New Intellectual and Social Histories of Capitalism," *Modern Intellectual History* 9, no. 1 (2012): 241.

failure.[21] Both mechanisms for risk management and forms of knowledge, forecasts sometimes mitigated but other times exacerbated the very uncertainties they were designed to conquer. Predicting the future was a response to as well as a manifestation of the enduring uncertainties that Jonathan Levy has located in the financialization and corporatization of risk in nineteenth-century American capitalism.[22]

Looking Forward reaches beyond forecasting and futurity to integrate two major and seemingly incompatible interpretations of the late nineteenth- and early twentieth-century United States. Historians of business and technology have traditionally conceptualized a modernizing America in terms of the rise of bureaucratic order and rationality driven by large-scale technological systems and a new class of professionals.[23] More recently, cultural historians and historians of science have depicted this era as a far more unstable moment of cultural crisis in which the forces of chance and contingency held sway.[24] This

21. Levy, *Freaks of Fortune*; Michael Zakim, "Producing Capitalism: The Clerk at Work," in Zakim and Kornblith, *Capitalism Takes Command*, 223–47; Brian P. Luskey, *On the Make: Clerks and the Quest for Capital in Nineteenth-Century America* (New York: NYU Press, 2010); Ann Fabian, *Card Sharps and Bucket Shops: Gambling in Nineteenth-Century America* (1990; repr., New York: Routledge, 1999); Stephen Mihm, *A Nation of Counterfeiters: Capitalists, Con Men, and the Making of the United States* (Cambridge, MA: Harvard University Press, 2007); Edward J. Balleisen, *Navigating Failure: Bankruptcy and Commercial Society in Antebellum America* (Chapel Hill: University of North Carolina Press, 2001); Scott A. Sandage, *Born Losers: A History of Failure in America* (Cambridge, MA: Harvard University Press, 2005).

22. Levy, *Freaks of Fortune*. For a history of the lived experience and the materiality of risk in the United States from the eighteenth to the twenty-first century, see Arwen P. Mohun, *Risk: Negotiating Safety in American Society* (Baltimore: Johns Hopkins University Press, 2013).

23. Wiebe, *Search for Order*; Alfred D. Chandler, Jr., *Strategy and Structure: Chapters in the History of the American Industrial Enterprise* (Cambridge, MA: MIT Press, 1962); Alfred D. Chandler, Jr., *The Visible Hand: The Managerial Revolution in American Business* (Cambridge, MA: Harvard University Press, 1977); Louis Galambos, "The Emerging Organizational Synthesis in Modern American History," *Business History Review* 44, no. 3 (1970): 279–90; Louis Galambos, "Technology, Political Economy, and Professionalization: Central Themes of the Organizational Synthesis," *Business History Review* 57, no. 4 (1983): 471–93; Thomas P. Hughes, *Networks of Power: Electrification in Western Society, 1880–1930* (Baltimore: Johns Hopkins University Press, 1983); JoAnne Yates, *Control through Communication: The Rise of System in American Management* (Baltimore: Johns Hopkins University Press, 1989).

24. Theodore M. Porter, *The Rise of Statistical Thinking, 1820–1900* (Princeton, NJ: Princeton University Press, 1986); Stephen M. Stigler, *The History of Statistics: The Measurement of Uncertainty before 1900* (Cambridge, MA: Belknap Press, 1986); Gerd Gigerenzer et al., *The Empire of Chance: How Probability Changed Science and Everyday Life* (New York: Cambridge University Press, 1989); Ian Hacking, *The Taming of Chance* (New York: Cambridge University

book reconciles these two frameworks by recasting the 1860s to the 1920s as a period in which government bureaucracies, information networks, and professional forecasters came to accommodate the very uncertainties they had originally sought to conquer. Historians of science have documented the rise of probabilistic thinking by the turn of the twentieth century, but their accounts of the "taming of chance" emphasize the intellectual worlds of the scientific elite and the transmission of their theories.[25] By contrast, *Looking Forward* depicts the knowledge production of everyday life as a contested and mutually constitutive process in which myriad authoritative forecasters—from government scientists to commercial firms to farmers' associations to individual citizens—engaged in the making of knowledge about the future, which meant that popular and scientific forecasting practices sometimes shaped each other. Ultimately, this book offers a narrative of late nineteenth- and early twentieth-century America that incorporates both the organizational stability of knowledge infrastructures and the epistemological instability that emerged from new ideas about predictability and uncertainty.

*

The ubiquity of forecasts and our paradoxically skeptical dependence on them have become a standard feature of daily life, as forecasts of local, national, and global scope stream through the twenty-four-hour news cycle. Economists forecast rates of growth and employment, market analysts predict how stocks will perform for investors, pollsters predict results of elections months and years away, and climate scientists model the grim ecological future of our planet. Every day we consult forecasts and make decisions accordingly. From five-day weather forecasts to real-estate market outlooks to psychic readings, we seek and sometimes purchase predictions in whose accuracy we may have much or little faith. But a little over a century and a half ago, forecasting was more controversial than commonplace. In the decades after the Civil War, Americans battled over the accuracy and legitimacy of predictions and debated whether it was possible to look into the future with any degree of certainty.

Press, 1990); Alain Desrosières, *The Politics of Large Numbers: A History of Statistical Reasoning* (Cambridge, MA: Harvard University Press, 1998); Goodman, *Shifting the Blame*; Louis Menand, *The Metaphysical Club* (New York: Farrar, Straus & Giroux, 2001); Jackson Lears, *Something for Nothing: Luck in America* (New York: Viking Books, 2003); Puskar, *Accident Society*.

25. Hacking, *Taming of Chance*.

This book examines the origins of our reliance on forecasting and asks how we came to believe in the promise and accept the limitations of predicting the future.

Looking into the future was hardly a new phenomenon in the late nineteenth century. Traditions of religious prophecy, divination, and astrology date from antiquity, and almanacs and fortune-telling were not uncommon in colonial America. Recent research in behavioral economics and affective psychology posits that humans have always been forecasters, albeit not very good ones, as our decision-making processes are undermined by our inability to anticipate our future emotional responses to what we long for in the present, whether a new job or a new jacket. We may be wrong when we think we know what will make us happy, this research suggests, but we are by nature a future-oriented species.[26]

But this ostensibly timeless human propensity for prediction has its own history, which, as this book reveals, was transformed in the United States in the aftermath of the Civil War.[27] The scale and scope of forecasting expanded dramatically with the creation of the US Department of Agriculture's crop reporting network (1863) and the US Army Signal Service's weather observation network (1870), which set the stage for unprecedented competition with private local forecasters and the emergence of forecasting as a profession.[28] Forecasting became more systematized as government statisticians and meteorologists created standardized mechanisms for collecting data about the harvest and the weather and translating it into forecasts with economic value. Their attempt to impose a rational bureaucratic order on the countryside became a quest for predictive certainty as government officials came to believe, some-

26. See Daniel Gilbert, *Stumbling on Happiness* (New York: Alfred A. Knopf, 2006), esp. 3–25; Dan Ariely, *Predictably Irrational: The Hidden Forces That Shape Our Decisions* (New York: HarperCollins, 2008); Leonard Mlodinow, *The Drunkard's Walk: How Randomness Rules Our Lives* (New York: Pantheon Books, 2008); Nassim Nicholas Taleb, *The Black Swan: The Impact of the Highly Improbable* (New York: Random House, 2007).

27. For histories of literary futurism that identify a reconceptualization of *the future* in the late nineteenth century, see I. F. Clarke, *The Pattern of Expectation, 1644–2001* (New York: Basic Books, 1979); Karen Sue Kimball, "'The Curious Vagaries of Visionaries': A History of the American Future, 1880s–1910s" (PhD diss., Emory University, 1996).

28. Walter A. Friedman, *Fortune Tellers: The Story of America's First Economic Forecasters* (Princeton, NJ: Princeton University Press, 2014) locates the origin of professional forecasting in the early decades of the twentieth century, but, as this book reveals, professional agricultural statisticians and weather forecasters began their work in the 1860s and 1870s, respectively.

what paradoxically, that it was possible to produce crop estimates with exactitude and short-term weather forecasts with unfailing accuracy. As forecasters competed for scientific authority and professional credibility, they waged rhetorical battles that created a new vocabulary of forecasting. Nineteenth-century writers used the words *prediction*, *prophecy*, and *forecast* interchangeably, and *prophecy* and *forecast*—and their twentieth-century connotations of religion and science—were not yet separated by a wide intellectual or ideological divide. But the late nineteenth century signaled the divergence of the meanings of *prophecy* and *forecasting* as meteorologists and fortune-tellers alike fashioned themselves as modern scientific forecasters far removed from traditions of prophecy.

Yet this history of forecasting is not a standard narrative of professionalization in which local knowledge and informal practices were always replaced by professional scientific expertise and institutional authority. Rather, predictions produced by forecasters who had formal academic training and those who did not shaped each other to produce a distinct sensibility of the future that acknowledged the persistence of unpredictability. This book considers all forms of knowledge production as such and does not make retrospective judgments about which forecasts were scientific and which were not, for the simple reason that such rigid categorical distinctions did not yet exist.[29] Some readers might be surprised to find Coney Island fortune-tellers and utopian novelists sharing a book with agricultural statisticians, meteorologists, and market forecasters, but each, wrestling with the authority, value, and meaning of forecasting, helped to create the broader culture of prediction that characterized late nineteenth- and early twentieth-century America. A culture of prediction influenced the lives of farmers and speculators, underwriters and socialists, city-dwelling clerks and small-town consumers, often through the printed page. Anyone reading a newspaper in the late nineteenth and early twentieth centuries might well have encountered adjacent crop estimates, weather forecasts, market forecasts, and fortune-telling advertisements, all of which made claims to scientific expertise and authority. Of course, producers and consumers of these predictions often drew rhetorical boundaries between what they considered professional scientific forecasts and what they dismissed as amateurish, superstitious, or fraudulent prophecies, and indeed many parts

29. For a compelling statement of a historian's refusal to impose anachronistic distinctions between scientific and unscientific knowledge production, see Rebecca M. Herzig, *Plucked: A History of Hair Removal* (New York: NYU Press, 2015), 7–9.

of this book focus on efforts to construct and police boundaries between ostensibly rational and irrational ways of knowing the future.[30]

The irony underlying histories of forecasting is that more complex methods do not guarantee more accuracy. The history of prediction in the late nineteenth- and early twentieth-century United States is not a linear narrative of progress that inevitably yielded more precise forecasts. Whether a forecast was accurate and how to measure accuracy in the first place were not simple questions to answer. Amid a late nineteenth-century crisis of certainty, Americans found themselves both desirous and dismissive of forecasts whose value was not always clear.

CRISIS OF CERTAINTY

The infrastructures of Gilded Age capitalism seemed to promise predictability. Postbellum transportation and communication networks undergirded a national economy as four more transcontinental railroad lines and their branches, along with telegraph lines, spanned the country in the 1870s and 1880s, providing the material and management for a new industrial society.[31] New systematic approaches to business organization epitomized by Frederick Winslow Taylor's experiments with scientific management in the 1880s and 1890s sought to replace the inefficiencies of shop-floor culture controlled by workers with a centralized system that ensured predictable rates of production and profit.[32] And the construction of large-scale transportation, communication, and electrical systems created more predictable flows of people, products, information, and power. Along with passengers and freight, railroads carried an increas-

30. For the classic account of "boundary-work" in science, see Thomas F. Gieryn, "Boundary-Work and the Demarcation of Science from Non-Science: Strains and Interests in Professional Ideologies of Scientists," *American Sociological Review* 48, no. 6 (1983): 781–95; Thomas F. Gieryn, *Cultural Boundaries of Science: Credibility on the Line* (Chicago: University of Chicago Press, 1999).

31. Chandler, *Visible Hand*. For an influential revision of the long-standing characterization of railroad corporations as the nineteenth-century locus of system, order, and rationality, see Richard White, *Railroaded: The Transcontinentals and the Making of Modern America* (New York: W. W. Norton, 2011).

32. On Taylorism and the rise of scientific management, see Hugh G. J. Aitken, *Taylorism at Watertown Arsenal: Scientific Management in Action, 1908–1915* (Cambridge, MA: Harvard University Press, 1960); Robert Kanigel, *The One Best Way: Frederick Winslow Taylor and the Enigma of Efficiency* (New York: Viking, 1997).

ing time consciousness and, in 1883, divided the country into time zones that sought to, as standard railway time's foremost proponent William F. Allen put it, "reduce the present uncertainty to comparative if not absolute certainty."[33]

Yet the "specter of Uncertainty" kept reappearing—in the volatility of boom and bust, unpredictable supply and demand, falling commodity prices in global markets, uneven access to credit for farmers, and arbitrary wage cuts and irregular employment for industrial workers.[34] Business owners faced uncertain futures after the Panic of 1873, as thousands of enterprises failed each year, with some estimating a 95 percent rate of business failure in the 1870s and 1880s.[35] Gilded Age capitalism meant unfathomable wealth for a cohort of industrialists and financiers but crushing poverty for multitudes. Henry George's 1879 *Progress and Poverty* asserted that if eighteenth-century "visions of the future" had foreseen the astonishing material progress of the nineteenth century—the advent of the steamship, railroad, mechanized reapers and threshers, immense factories housing powerful steam hammers and precision machine tools—they would have predicted a "golden age" for all, not an "immense wedge" dividing the prosperous from the poor.[36] Workers facing the "modern uncertainty of employment" participated in nearly 37,000 strikes between 1881 and 1905; the Great Strike of 1877, Haymarket, Homestead, and Pullman were only the most violent manifestations of the uncertainties of the "labor question."[37]

An expanding population of industrial workers, one-third immigrants, experienced individual as well as collective economic uncertainty. In crowded tenement houses like the ones documented in Jacob Riis's 1890 *How the Other Half Lives*, the health and welfare of the working class were always fragile.

33. "Standard Railway Time," *Century* 26, no. 5 (1883): 797.

34. Levy, *Freaks of Fortune*; Alan Trachtenberg, *The Incorporation of America: Culture and Society in the Gilded Age* (New York: Hill and Wang, 1982), 39.

35. Popular anxiety over small business failure persisted throughout the nineteenth century, as evidenced by the staying power of the commonly cited (but never substantiated) statistic on a 97 percent failure rate. On the circulation of this dire statistic, see Sandage, *Born Losers*, 7–8.

36. Henry George, *Progress and Poverty: An Inquiry into the Cause of Industrial Depressions, and of Increase of Want with Increase of Wealth—The Remedy* (1879; repr., New York: D. Appleton, 1880), 7, 8, 11.

37. George L. Bolen, *Getting a Living* (New York: Macmillan, 1903), 453. On late nineteenth-century labor unionism and strikes, see Nell Irvin Painter, *Standing at Armageddon: A Grassroots History of the Progressive Era*, 2nd ed. (New York: W. W. Norton, 2008), chap. 2; Leon Fink, *The Long Gilded Age: American Capitalism and the Lessons of a New World Order* (Philadelphia: University of Pennsylvania Press, 2015), chap. 2.

Tenement reformer Lawrence Veiller noted "the uncertainty of the tenure of the home," lamenting that

> from the narrow space and miserable shelter of four bare walls, and a broken ceiling or leaky roof, a whole family may be expelled at a moment's notice upon the non-payment, at the precise time, of the weekly stipend of rent. Even the able-bodied, willing, and industrious husband and father may, from the loss of work occasioned by sickness or bad weather, see himself and family driven to the open street.[38]

Unemployment rates reached heights of an estimated 16 percent during the mid-1870s and mid-1880s, and 40 percent of industrial workers fell below the poverty line of five hundred dollars in annual income at the end of the 1880s.[39] Basil March, Howells's social conscience in *A Hazard of New Fortunes*, based his indictment of the "economic chance-world" on such uncertainties:

> It ought to be law as inflexible in human affairs as the order of day and night in the physical world, that if a man will work he shall both rest and eat, and shall not be harassed with any question as to how his repose and his provision shall come. . . . But in our state of things no one is secure of this. No one is sure of finding work; no one is sure of not losing it. I may have my work taken away from me at any moment by the caprice, the mood, the indigestion of a man who has not the qualification for knowing whether I do it well or ill.[40]

The "specter of Uncertainty" followed the industrial laborer to the factory, railroad, and mine, where workers faced a daily risk of commonplace yet unpredictable accidents in "an industrial-accident crisis of world-historical proportions."[41] Textile workers regularly lost fingers and limbs to fast-moving machinery, and coal miners and railroad men lost their lives at alarming rates—1,141 of every 100,000 brakemen and 300 of every 100,000 anthracite coal miners died on the job in 1890.[42]

38. Robert W. DeForest and Lawrence Veiller, eds., *The Tenement House Problem*, vol. 1 (New York: Macmillan, 1903), 74.

39. Trachtenberg, *Incorporation of America*, 90–91.

40. Howells, *Hazard of New Fortunes*, 396.

41. Witt, *Accidental Republic*, 22. On popular understandings of the frequency of industrial accidents, see Arwen Mohun, "On the Frontier of *The Empire of Chance*: Statistics, Accidents, and Risk in Industrializing America," *Science in Context* 18, no. 3 (2005): 337–57.

42. Witt, *Accidental Republic*, 27.

The American legal system and insurance industry also contended with, in aggregate, a nineteenth-century crisis of unpredictability in industrial accidents, steamboat explosions, and railroad collisions. In his narratives of New York tenement life, Riis dramatized the moments of mounting "dread and uncertainty" as concerned onlookers learned that their train is delayed because of a collision in the fog.[43] Accidents like this one also created confusion in the courtroom, as jurists struggled to determine culpability when the complexity of large-scale technological systems defied attempts to construct the "chains of causation" that traditionally indicated blame, contributing to "the proliferation of an epistemological uncertainty . . . that often failed not only to identify a given individual as blameworthy but to associate a given individual with a given act."[44] As accident law shifted to accommodate the uncertain causality of industrial accidents, the insurance industry expanded to deal with new forms and frequencies of industrial risk.[45] The postbellum insurance industry did not merely react to the catalogue of hazards that defined urban-industrial economic life but actively shaped a cultural reimagination of uncertainty by publishing trade periodicals with literary depictions of the unpredictable but ubiquitous accidents against which the reasonable American should be insured.[46]

The culmination of the insurance industry's literary turn came in 1896, when Bellamy's "specter of Uncertainty" won a starring role in a pamphlet published by executive Darwin P. Kingsley, who would become the president of New York Life in 1907.[47] Based on the *Looking Backward* scene in which protagonist Julian West returns in a nightmare from the utopian year 2000 to the inequalities of the Gilded Age, the pamphlet advertises life insurance as the only forward-looking and rational aspect of the late nineteenth century, "the one sane idea in a world full of madmen." But in Kingsley's version, the year 1887 is not a dream but a "monstrous" reality of "disorder and cruelty, humanity tearing at itself." Kingsley, Vermont farm boy turned social Darwinist, remained convinced (and perhaps relieved) that "the survival of the fittest is . . . not only universal as a law, but universal as a fact." And in the face of the "indescribable brutality" of economic life, Kingsley asserted, life insurance offered

43. Jacob A. Riis, *Out of Mulberry Street: Stories of Tenement Life in New York City* (New York: Century, 1898), 181.

44. Goodman, *Shifting the Blame*, 76, 66.

45. Lears, *Something for Nothing*, 200–201.

46. Jason Puskar, "William Dean Howells and the Insurance of the Real," *American Literary History* 18, no. 1 (2006): 29–58.

47. On Kingsley's life, see B. C. Forbes, *Men Who Are Making America* (New York: B. C. Forbes, 1916–17), 232–39.

the only remedy. A bridge between Bellamy's socialist future and Howells's "economic chance-world," life insurance served as a tool for risk management by allowing a man to "provid[e] some certainty not only for himself but for his family." Paralleling Bellamy's vision of a cooperative commonwealth based on principles of Christian solidarity, Kingsley described a community of risk that transcended racial and religious differences to collectively "bid defiance to that spectre called 'Uncertainty.'" Kingsley declared that policyholders would be "living after the fashion of Mr. Bellamy's dream" and pronounced "the question of how a man may, with some degree of certainty, carry his own personality into the next generation, and give his family some guarantee which will not lapse with his own life . . . answered."[48] We don't know how Bellamy, at the time ailing and writing *Equality*, his sequel to *Looking Backward*, reacted to Kingsley's pamphlet. He may have appreciated it given that he had previously described his utopia as "a universal insurance company," but considering his critique of the Gilded Age economy as a "wealth-making machine," he would likely have been dismayed by his ironic appearance in an advertisement elevating the life insurance industry to the pinnacle of corporate capitalism as "The First Business of the World."[49]

While the "specter of Uncertainty" represented risks to insure against, it also promised financial reward to those with the capital and nerve to engage in speculation, whether gambling and lotteries or commodity futures trading.[50] Although gambling aside from horse racing was prohibited in most states by the turn of the century and reformers managed to shut down lotteries in the Northeast, a thriving gambling culture persisted in the world of policy play, or "numbers," which amounted to small side bets placed on lotteries by crowds of players that included the poor and working classes, men and women, black and white.[51] The other end of the economic spectrum placed bets on the floors of commodity exchanges as a form of risk management during the ascen-

48. D. P. Kingsley, "The First Business of the World," pamphlet, 1896, folder 10, drawer 2, series 11, Edward Bellamy Memorial Archives, Edward Bellamy House, Chicopee, MA. Kingsley's pamphlet was subsequently published in *The Chronicle: A Weekly Journal Devoted to the Interests of Insurance, Manufacturers, and Real Estate* 60, no. 26 (1897): 322–23, and in D. P. Kingsley, *The First Business of the World and Other Addresses and Papers* (New York, 1903), 3–10.

49. Edward Bellamy, "Plutocracy or Nationalism—Which?," in *Edward Bellamy Speaks Again!* (Kansas City, MO: Peerage, 1937), 44; Bellamy, *Looking Backward*, 156.

50. On the increasing popularity of gambling and lotteries in the late nineteenth century, see Lears, *Something for Nothing*, 148, 173.

51. Fabian, *Card Sharps*, 112, 136–42.

dancy of futures trading, which dominated the exchanges by the 1890s. Bucket shops—close to one thousand betting parlors in cities and scattered across the countryside—"brought futures to the masses" by allowing small wagers on the speculative economy of futurity that had been institutionalized by the Chicago Board of Trade.[52]

Speculating on wholesalers' futures was just as profitable as speculating on wheat futures in the late nineteenth century. Assessing creditworthiness had become difficult because local, informal networks of risk assessment could not cover the nation's geographical span, increasingly mobile population, and expanding volume of trade. The credit reporting industry exploded, with R. G. Dun & Company reporting on over 20,000 businesses in 1859, 430,000 in 1870, and 1.2 million in 1900.[53] Dun and its rival, the Bradstreet Agency, faced so many new competitors in the last quarter of the century that, according to a Dun insider, credit reporting agencies were "getting as thick as blackberries."[54]

The late nineteenth century also witnessed an epistemic crisis in which deterministic explanatory frameworks no longer fit shifting philosophical, scientific, and religious conceptions of uncertainty.[55] In 1882, home economist and social reformer Helen Campbell noted unsettling "moments of perplexity and uncertainty" as

52. Fabian, *Card Sharps*, 156–57, 191; David Hochfelder, "'Where the Common People Could Speculate': The Ticker, Bucket Shops, and the Origins of Popular Participation in Financial Markets, 1880–1920," *Journal of American History* 93, no. 2 (2006): 335–58; quotation in Jonathan Ira Levy, "Contemplating Delivery: Futures Trading and the Problem of Commodity Exchange in the United States, 1875–1905," *American Historical Review* 111, no. 2 (2006): 316.

53. Rowena Olegario, "Credit Reporting Agencies: A Historical Perspective," in *Credit Rating Systems and the International Economy*, ed. Margaret J. Miller (Cambridge, MA: MIT Press, 2003), 122–23. On credit reporting and selfhood in the nineteenth-century United States, see Sandage, *Born Losers*, chaps. 4–6; Josh Lauer, "From Rumor to Written Record: Credit Reporting and the Invention of Financial Identity in Nineteenth-Century America," *Technology & Culture* 49, no. 2 (2008): 301–24.

54. John H. Smith to R. G. Dun & Co., [1877?], Dun Scrapbook, p. 88, Dun & Bradstreet Collection, quoted in James H. Madison, "The Evolution of Commercial Credit Reporting Agencies in Nineteenth-Century America," *Business History Review* 48, no. 2 (1974): 182.

55. On the nineteenth-century intellectual shift away from religious and scientific certainties, see Paul Jerome Croce, *Science and Religion in the Era of William James: Eclipse of Certainty, 1820–1880* (Chapel Hill: University of North Carolina Press, 1995). On the late nineteenth-century decline of certainty in popular understandings of psychology and sociology, see Samuel Jonathan Elworthy, "The Social Origins of Uncertainty: Popular Struggles over Science and Truth in America, 1870–1914" (PhD diss., Rutgers, The State University of New Jersey, 1999).

> venerable religious "truths," long established social "laws," well known historical and scientific "facts," under the remorseless analysis of our times, had seemed years ago to dissolve before my eyes. . . . One by one, old landmarks had vanished. Speculation came, with no answer to its questions. One *ism* after another presented itself, seeming at first to meet the demand for truth; then paling and fading away under the light of investigation.[56]

Darwinism loomed largest in such a parade, as evolution, probabilism, pragmatism, and social science all produced new formulations of uncertainty during the late nineteenth-century "erosion of determinism."[57] As Jackson Lears has observed, "Chance was becoming a legitimate category of scientific thought; certainty was becoming problematic."[58] The decline, by the turn of the twentieth century, of traditional models of causation in which the past determined the future signaled that a positivist universe had yielded to a probabilistic universe, as "[a] space was cleared for chance."[59]

Yet it was not easy to clear a space for chance in daily life, as William James acknowledged in an 1880 address to the Harvard Philosophical Club on "The

56. Helen Campbell, *The Problem of the Poor: A Record of Quiet Work in Unquiet Places* (New York: Fords, Howard & Hulbert, 1882), 7–8.

57. Hacking, *Taming of Chance*, 1. Here I invoke evolution as an example of a theory that destabilized existing knowledge frameworks by virtue of its reliance on chance. As Naomi Oreskes has pointed out, Darwin's account of natural selection did not make specific predictions and indeed depicted an evolutionary process "contingent on unpredictable environmental change." Naomi Oreskes, "Why Predict? Historical Perspectives on Prediction in Earth Science," in Daniel Sarewitz, Roger A. Pielke, Jr., and Radford Byerly, Jr., eds., *Prediction: Science, Decision Making, and the Future of Nature* (Washington, DC: Island Press, 2000), 27.

58. Lears, *Something for Nothing*, 212.

59. Ian Hacking, "Nineteenth-Century Cracks in the Concept of Determinism," *Journal of the History of Ideas* 44, no. 3 (1983): 455–75; quotation in Hacking, *Taming of Chance*, 1. On late nineteenth-century intellectual shifts toward a new acknowledgement of chance, see Stigler, *History of Statistics*; Porter, *Rise of Statistical Thinking*; Lorenz Krüger, Lorraine J. Daston, and Michael Heidelberger, eds., *The Probabilistic Revolution*, vol. 1 (Cambridge, MA: MIT Press, 1987); Lorenz Krüger, Gerd Gigerenzer, and Mary S. Morgan, eds., *The Probabilistic Revolution*, vol. 2 (Cambridge, MA: MIT Press, 1987); Haskell, *Emergence of Professional Social Science*; Dorothy Ross, *The Origins of American Social Science* (New York: Cambridge University Press, 1991); Menand, *Metaphysical Club*. On literary naturalism's new narratives of unpredictability that depicted characters reckoning, often unsuccessfully, with forces of chance they could neither predict nor control, see Puskar, *Accident Society*. For an analysis of antebellum literary formulations of chance, see Maurice S. Lee, *Uncertain Chances: Science, Skepticism, and Belief in Nineteenth-Century American Literature* (New York: Oxford University Press, 2012).

Sentiment of Rationality." We cannot escape a "permanent presence of the sense of futurity in the mind," James noted, and awaiting an uncertain future, whether it held joys or sorrows, was always "haunting."[60] At the Harvard Divinity School in 1884, James grappled again with the philosophical implications of future uncertainties in an address titled "The Dilemma of Determinism," his most influential reappraisal of chance. James noted at the outset that accepted definitions of determinism and indeterminism hinged on the human ability, or lack thereof, to anticipate:

> If we are determinists, we talk about the infallibility with which we can predict each other's conduct; while, if we are indeterminists, we lay great stress on the fact that it is just because we cannot foretell each other's conduct, either in war or statecraft or in any of the great and small intrigues and businesses of men, that life is so intensely anxious and hazardous a game.[61]

The future-oriented thought James described to his Cambridge audiences in the early 1880s was evident in contemporary writing, prophetic traditions, and new forms of predictive knowledge. "Tales of the future" proliferated in the late nineteenth century as readers encountered the proto-science-fiction fantasies of European writers Jules Verne and H. G. Wells in the pages of American periodicals and explored the futuristic visions of the nearly two hundred literary utopias that followed the publication of *Looking Backward* in 1888.[62] In 1891, an *Andover Review* editorial on "The Spirit of Expectancy"

60. William James, *The Sentiment of Rationality* (1882; repr., New York: Longmans, Green, 1905), 77, 78.

61. William James, "The Dilemma of Determinism," *Unitarian Review* 22, no. 3 (1884): 200. "The Dilemma of Determinism" was subsequently published in *The Will to Believe* in 1896. On its significance to the intellectual redefinition of chance in the late nineteenth century, see Lears, *Something for Nothing*, 221–23; Robert D. Richardson, *William James: In the Maelstrom of American Modernism* (2006; repr., Boston and New York: Houghton Mifflin, 2007), 246–47; Puskar, *Accident Society*, 161–62.

62. On "tales of the future," see Clarke, *Pattern of Expectation*, 141. On the role of periodicals in shaping turn-of-the-century discourse on the American future, see Joseph J. Corn and Brian Horrigan, *Yesterday's Tomorrows: Past Visions of the American Future* (1984; repr., Baltimore: Johns Hopkins University Press, 1996), 5–6. On the unprecedented number and significance of late nineteenth-century American literary utopias, see Kenneth Roemer, *The Obsolete Necessity: America in Utopian Writings, 1888–1900* (Kent, OH: Kent State University Press, 1976). For a comprehensive account of literary futurism in Europe and the United States, see Clarke, *Pattern of Expectation*.

observed that an affinity for anticipation had become so commonplace that "we seem to live in the thought of the future more than in the power of the present."[63] In his 1881 treatise titled *American Nervousness*, George Beard cited the "Habit of Forethought" in his cultural diagnosis of neurasthenia, arguing that a preoccupation with the future was both the cause and cost of progress. In his account of "this forecasting, this forethinking . . . this sacrifice of the present to the future," Beard, like countless other proponents of scientific racism, invoked a hierarchy from civilized to primitive societies to justify his racialized logic of futurity in which Anglo-American elites possessed the capacity for forethought and the nonwhite "barbarian" did not.[64] But looking into the future was in fact not solely the activity of the white, male intellectual elite, as Beard had claimed. For example, female fortune-tellers and spirit mediums of different races and ethnicities performed their predictive labors in private parlors and packed theaters, selling glimpses of the future to an upper-, middle-, and working-class consuming public. And religious futures were revealed by biblical prophecy writers, revivalists, and evangelists who circulated their premillennial predictions of the Second Coming through printed texts and speaking tours and found a growing audience of believers from all classes.[65]

Visions of the future were central to the ideologies of technoscientific progress that permeated late nineteenth-century American culture.[66] Technological futurism was on display at World's Fairs as visitors to the Philadelphia Centennial Exhibition and Chicago's White City toured monuments to discovery and innovation and celebrated the possibilities embodied in spectacular halls

63. "The Spirit of Expectancy," *Andover Review* 15, no. 88 (1891): 426.

64. George M. Beard, *American Nervousness, Its Causes and Consequences: A Supplement to Nervous Exhaustion (Neurasthenia)* (New York: G. P. Putnam's Sons, 1881), 128–29, 131.

65. Paul Boyer, *When Time Shall Be No More: Prophecy Belief in Modern American Culture* (1992; repr., Cambridge, MA: Belknap Press, 1999), 80–100.

66. Americans in the late nineteenth century commonly encountered the idea of the future framed by nationalist imperatives of technical innovation and geographic expansion. See, for example, James Parton et al., *Sketches of Men of Progress* (New York: Hartford Publishing Company, 1870–71); Charles L. Flint et al., *One Hundred Years Progress of the United States* (Hartford, CT: L. Stebbins, 1871); Theodore D. Woolsey et al., *The First Century of the Republic: A Review of American Progress* (New York: Harper & Brothers, 1876). On ideologies of progress, see Leo Marx and Bruce Mazlish, eds., *Progress: Fact or Illusion?* (Ann Arbor: University of Michigan Press, 1996). On technological futurism, see Joseph J. Corn, ed., *Imagining Tomorrow: History, Technology, and the American Future* (Cambridge, MA: MIT Press, 1986); Howard P. Segal, "The Technological Utopians," in Corn, *Imagining Tomorrow*, 119–36.

of machinery and electrified palaces.[67] Inventors, engineers, and writers made frequent and often utopian predictions about the transformative powers of technical innovations in the 1890s and well into the twentieth century.[68] As H. G. Wells argued in his 1901 *Anticipations*, it was science, not fiction, that should be the domain of "modern prophecy."[69] In 1912, *Good Housekeeping* magazine featured "the prophecy of Thomas A. Edison" on the power of electricity to transform "the housewives of the future from drudges into engineers," men and women into equals, and future generations of children into "mental, . . . physical and moral prodigies."[70]

To Robert H. Thurston, head of Cornell's school of engineering, the future of American progress could be precisely calculated.[71] In 1895, Thurston explained in the pages of the *North American Review* how charting data on metrics, including steam power, agricultural production, per capita wealth, and wages, "constitute[d] the best guide in predicting a probable future for our industrial and social system." In Thurston's fourteen charts, all trends pointed upward in a "curve of human progress" in which "the immediate future becomes knowable with a degree of accuracy and certainty." The charts indicated not only rising standards of American production, wealth, and happiness, but also the authority of scientific knowledge about the future. Characterizing his study as "a truly logical and scientific form of prophecy," Thurston concluded, "Science, and science only, often can, and frequently does, by a perfectly accurate and correct method, give us clairvoyant views of the immediate, if not often of the remote, future."[72]

67. On aesthetic experiences of electrification, see David E. Nye, *American Technological Sublime* (Cambridge, MA: MIT Press, 1994), chaps. 6 and 7. On the racial and imperial politics of world's fairs, see Robert W. Rydell, *All the World's a Fair: Visions of Empire at American International Expositions, 1876–1916* (Chicago: University of Chicago Press, 1984).

68. For a quantitative analysis of technological forecasting at the turn of the century, see George Wise, "Technological Prediction, 1890–1940" (PhD diss., Boston University, 1976); George Wise, "The Accuracy of Technological Forecasts, 1890–1940," *Futures* 8, no. 5 (1976): 411–19.

69. H. G. Wells, *Anticipations of the Reaction of Mechanical and Scientific Progress upon Human Life and Thought* (New York: Harper & Brothers, 1901), 4.

70. Thomas A. Edison, "The Woman of the Future," *Good Housekeeping Magazine*, October 1912, 440, 444.

71. On economic indicators as a new form of statistical knowledge in nineteenth-century America, see Eli Cook, *The Pricing of Progress: Economic Indicators and the Capitalization of American Life* (Cambridge, MA: Harvard University Press, 2017).

72. Robert H. Thurston, "The Trend of National Progress," *North American Review* 161 (September 1895): 311, 297, 312. As Naomi Oreskes has observed, science is not inherently

While statistics quantified the future of American progress, surveys mapped and promoted its westward direction. Following the publication of the thirteen-volume Pacific Railway Surveys between 1855 and 1860 came a variety of maps that made visual arguments for westward expansion, including John Disturnell's 1867 climatological and agricultural maps of North America as well as physician and Colorado booster Charles Denison's 1877 "Climatic Map of the Eastern Slope of the Rocky Mountains," which allowed homesteaders and health seekers to predict how arable or salutary a particular destination would be.[73] As former director of the US Geological Survey Clarence King observed in 1892, "We live in the future tense. Prediction is the hobby of the age."[74]

Westward migration also depended on predictive judgments about a region's agricultural potential, like the climatological dictum "the rain follows the plow," which famously assured nineteenth-century homesteaders that the economic development of the arid Plains—cultivating the land, planting trees, and building railroads and telegraphs—would increase rainfall. Popularizers of the theory included University of Nebraska professor of natural sciences Samuel Aughey, head of the US Geological Survey Ferdinand Hayden, and superintendent of the Department of Geology and Mineralogy in the Nebraska Academy of Sciences (and railroad advisor) Charles Dana Wilber, who coined the phrase in his 1881 book titled *The Great Valleys and Prairies of Nebraska and the Northwest*.[75] Marshaling experimental results on soil absorption and tables of rainfall statistics, Wilber outlined the ecological processes by which cultivation would transform the desert of the Plains into a lush garden, noting

---

focused on predicting the future: in the eighteenth and nineteenth centuries, physics and astronomy were considered predictive, but the earth sciences were understood to be explanatory until the twentieth-century rise of a "hypothetico-deductive" scientific method that privileged temporal prediction. Oreskes, "Why Predict?," 11, 24, 31, quotation on 31.

73. Susan Schulten, *Mapping the Nation: History and Cartography in Nineteenth-Century America* (Chicago: University of Chicago Press, 2012), 112–16.

74. Clarence King, "The Education of the Future," *Forum*, March 1892, 20, quoted in Kimball, "'The Curious Vagaries of Visionaries,'" 12.

75. Julie Courtwright, "On the Edge of the Possible: Artificial Rainmaking and the Extension of Hope on the Great Plains," *Agricultural History* 89, no. 4 (2015): 540; Gary D. Libecap and Zeynep Kocabiyik Hansen, "'Rain Follows the Plow' and Dryfarming Doctrine: The Climate Information Problem and Homestead Failure in the Upper Great Plains, 1890–1925," *Journal of Economic History* 62, no. 1 (2002): 93–94. On the role of Aughey and Wilber in developing the idea, first articulated by trader Josiah Gregg in the 1840s, that cultivation increased rainfall, see John F. Freeman, *High Plains Horticulture: A History* (Boulder: University Press of Colorado, 2008), 52.

that "in this miracle of progress, the plow was the avant courier—the unerring prophet—the procuring cause."[76] Although the theory was discredited after drought struck Kansas in 1893 and government officials denied any climatological evidence that rain did in fact follow the plow, statistics reporting higher precipitation in Kansas in the early 1880s had seemed initially to confirm the prophecy.[77]

With the western frontier closed by the superintendent of the Census in 1890 and more famously by Frederick Jackson Turner in 1893, the nation's expansionists anticipated a future of global markets and increasingly aggressive foreign policy. Before the opening of the Columbian Exposition in May 1893, the American Press Association asked prominent citizens to contribute to a "Chapter of Forecasts" by making predictions of American life one hundred years hence. Among the forecasts that were published in hundreds of newspapers were those of government officials and industrialists who predicted an expansion of economic and transportation networks that would secure the United States' rightful position as overseer of the Western hemisphere. Merchant W. R. Grace and William Eleroy Curtis, director of what would become the Pan-American Union, envisioned railway connections from New York or Chicago to Buenos Aires, and Treasury official Asa C. Matthews imagined an American republic spanning both continents as "the most perfect civilization . . . that the world ever knew."[78] And on the other side of the globe, American journalist George Kennan predicted in a series of influential articles and lectures in the late nineteenth century that the Russian people would

76. C. D. Wilber, *The Great Valleys and Prairies of Nebraska and the Northwest* (Omaha: Daily Republican, 1881), 70. For a literary analysis of Wilber's invocation of biblical imagery and Manifest Destiny in this passage, see Henry Nash Smith, "Rain Follows the Plow: The Notion of Increased Rainfall for the Great Plains, 1844–1880," *Huntington Library Quarterly* 10, no. 2 (1947): 188–89.

77. Libecap and Kocabiyik Hansen, "'Rain Follows the Plow,'" 94.

78. David Walter, introduction, in *Today Then: America's Best Minds Look 100 Years into the Future on the Occasion of the 1893 World's Columbian Exposition*, ed. David Walter (Helena, MT: American and World Geographic Publishers, 1992), 13–21; W. R. Grace, "A Great Era for South America"; Asa C. Matthews, "The United States of the Americas"; William Eleroy Curtis, "United States to Dominate the Hemisphere"; Erastus Wiman, "A New Field Opens—The Canadian North"; Matthew C. D. Borden, "American Cotton Will Control World Markets"; Warner Miller, "The Nicaraguan Canal Is Inevitable," all in Walter, *Today Then*, 29–31, 49–50, 87–88, 116–20, 146–47, 150–51, quotation on 50. On Curtis's role in the Pan-American movement, see Benjamin A. Coates, "The Pan-American Lobbyist: William Eleroy Curtis and U.S. Empire, 1884–1899," *Diplomatic History* 38, no. 1 (2014): 22–48.

eventually reject tsarist autocracy and embrace the principles of American democracy.[79]

The "habit of forethought" was evident in a late nineteenth-century culture of prediction that encompassed imagined literary futures, maps and charts indicating national progress, and aspirations of pan-Americanism. But anticipations of an electrified future or the predictive maxim that "the rain follows the plow," issued in response to singular events or processes, were different from the forecasts of everyday life that are the subject of this book. *Looking Forward* focuses on predictions that were issued regularly and systematically in response to the changing conditions of the weather, harvest, market, and daily life. These forecasts are historically significant not only because they were ubiquitous but also because they fueled debates over who could claim the professional authority and expert knowledge to predict the future in late nineteenth- and early twentieth-century America.

OVERVIEW OF THE BOOK

Looking Forward integrates histories of science, technology, environment, capitalism, and culture, and the following chapters draw on sources including the archival records of federal agencies and local organizations, government documents, legal decisions, personal papers and private correspondence, newspapers and periodicals, novels, short stories, and poems. Each of the book's main chapters, which are organized thematically but also chronologically, examines a different kind of forecasting.

The first chapter focuses on the crisis of certainty in cotton forecasting that began during the American Civil War, when the US Department of Agriculture first began to aggregate figures on crop condition and acreage in order to stabilize markets and insulate farmers from the power of speculators. The USDA's crop reporting system hinged on the premise that a disciplined cadre of statisticians and crop correspondents operating within an efficient bureaucracy would produce more accurate agricultural statistics. But in the volatile cotton economy of the 1890s, the competing forecasts of the USDA, a controversial British forecaster, and an association of Georgia cotton growers yielded more, not less, uncertainty in the market and eventually led government officials to redefine crop estimating as a statistically uncertain endeavor.

79. David S. Foglesong, *The American Mission and the "Evil Empire": The Crusade for a "Free Russia" since 1881* (New York: Cambridge University Press, 2007), 14–27.

The second chapter examines the politics of rural access to another late nineteenth-century information network, the first national weather service, which brought twenty-four-hour forecasts into everyday life. In response to farmers' demand, the Signal Service attempted to extend its short-term forecasts and storm warnings to rural areas beyond the reach of its telegraph network in the 1870s and 1880s, but many communities continued to face the daily uncertainties of when and indeed whether the "Probabilities" would arrive. The problem of forecasting in the countryside persisted until the advent of state weather agencies in the 1880s and early 1890s and then the introduction of Rural Free Delivery and rural telephone networks at the turn of the century, all of which revealed that the federal government's perennial problem of forecasting in "the distant parts of the country" found its solution in the countryside itself.[80]

The third chapter turns from short-term to long-range weather forecasting and traces conflicts between government forecasters and commercial "weather prophets" who sold predictions a month, a season, or a year in advance, threatening the professional authority of the new US Weather Bureau and its twenty-four- to forty-eight-hour forecasts. Willis L. Moore, chief of the Weather Bureau from 1895 to 1913, led a vigorous campaign to discredit "long-rangers" as unscientific frauds, after which the bureau began to produce its first weekly forecasts, which featured the very uncertainties it had formerly condemned in the predictions of the weather prophets. The Weather Bureau's entry into long-range forecasting in 1908 signaled an institutional acceptance of more uncertainty than the empiricism of late nineteenth-century forecasting had allowed.

Yet the acceptance of predictive uncertainty was not universal. The fourth chapter examines ideas about uncertainty and predictability in the work of three very different prophets—commodity price forecaster Samuel Benner, utopian novelist Edward Bellamy, and meteorologist Henry Helm Clayton—who sold short-term and long-term predictions of the economic future. The popularity of their forecasts into the 1930s and even beyond reveals the enduring appeal of predictive certainties in the realm of the market and political economy despite the increasing acceptance of probabilism in science and culture at the turn of the twentieth century.

The fifth chapter considers the policing and prosecution of fortune-telling, a booming business in American cities where fortune-tellers sold promises of

80. Quotation in *Annual Report of the Chief Signal-Officer to the Secretary of War for the Year 1877* (Washington, DC: Government Printing Office, 1877), 129.

love and money to clients of all classes while trying to avoid arrest under selectively enforced antidivination laws. In the 1890s and the first two decades of the twentieth century, state courts created a new framework for fortune-telling through a series of decisions that grappled with the Elizabethan legal concept of "pretending to tell fortunes" and ultimately arrived at a new acceptance of occult predictions that acknowledged their own indeterminacy.

*

This story of a crisis of certainty and the emergence of professional forecasting begins in the countryside, where rural producers and the US Department of Agriculture's Division of Statistics began a concerted effort during the Civil War to aggregate objective statistics on crop condition and acreage to counter speculators' false forecasts and market manipulations. *American Agriculturalist* editor Orange Judd, who enlisted his subscribers in a precursor to the federal government's crop reporting network, complained in 1862 that

> shrewd speculators, who have on hand a large stock of old grain, often circulate newspaper reports to the effect that owing to bad weather, insects, small breadth, etc., there will not be half a crop gathered. On the other hand, as the harvest begins, another class intending to become grain buyers, are interested in magnifying the yield for the purpose of depressing prices. . . . In short, there is such an entire absence of reliable statistics that all are in a state of doubt and uncertainty, and none more so than non-commercial producers.[81]

Judd's words rang especially true in the case of cotton, king of the nineteenth-century commodities, as USDA officials, a British "cotton prophet," and southern growers all competed to make the most accurate and influential forecasts of the season's yield. But mitigating the rampant uncertainty over millions of bales was not as easy as cotton forecasters predicted it would be.

81. *American Agriculturalist*, March 1862, 72, quoted in Walter H. Ebling, "Why the Government Entered the Field of Crop Reporting and Forecasting," *Journal of Farm Economics* 21, no. 4 (1939): 725.

CHAPTER 1

Cotton Guesses

In the fall of 1899, the crowd on the floor of the New York Cotton Exchange reportedly burst into laughter at a poem satirizing the yield forecasts of Henry M. Neill, the decade's most renowned and most vilified "cotton prophet."[1] "Strange," the poet observed, "that this city farmer, regardless rain or shine . . . makes crop that's always 'fine.' " When the Brazos River floods, Neill predicts that cotton pickers wearing "diving suits" will harvest "an extra million bales" in Texas. Then, when drought hits the rest of the South, Neill foretells, with a wink, a "monster crop" that will mean "the price must further drop." At the end of the poem, Neill insists—just as the real Neill had in August 1899—that the year's yield will surpass eleven million bales and possibly even twelve. The poem, written by Louise Prudden Hunt, wife of prominent Georgia grower Benjamin W. Hunt, concludes with this lament: "Oh, Mr. Neill, this cotton (so queer the south ne'er knew), / These phantom fleecy millions no planter picked but you."[2] Imaginary cotton bales were no poetic fancy but rather the center of a decade-long struggle among southern growers, commercial forecasters, and the federal government to produce the most accurate and authoritative cotton statistics. Indeed, the poem's swipe at Neill's perennially bearish

1. Quotation in "Neill Worked with Secret Circulars," *Atlanta Constitution*, November 7, 1899.

2. Quotations in "Cotton Crop of Mr. Neill," *Arkansas Democrat*, November 28, 1899. On Benjamin W. Hunt, see Georgia Department of Agriculture, *Georgia: Historical and Industrial* (Atlanta, 1901), 255. Thanks to Michelle Bechtell for sharing biographical information on Louise Prudden Hunt and Benjamin W. Hunt.

estimates encouraged, according to the *Arkansas Democrat*, "those who were fighting the south's battle against Mr. Neill," a battle for control of the market in which predictions determined prices.[3]

The cotton exchange poem also reveals the economic and epistemic crises of cotton forecasting in the 1890s, a decade in which Henry Neill was widely acknowledged as the primary market mover. A recognized cotton authority in Britain during the American Civil War, Neill achieved "cotton prophet" status on both sides of the Atlantic for his uncannily accurate predictions of the three largest yields in history in the 1890s. But in the context of plunging agricultural commodity prices and economic depression, Neill's forecasts came to symbolize the ruination of the cotton economy to southern growers, who organized a campaign of statistical resistance designed to break Neill's power over prices in what the *New-York Tribune* termed a "battle of crop estimates."[4]

The battle unfolded in a series of contests, which this chapter traces, over claims to statistical authority, objectivity, and predictive certainty in cotton forecasting. Yet despite such claims, the statistics produced by Neill, the US Department of Agriculture, and growers' associations did not combine to stabilize prices, curtail speculation, or rationalize markets, but rather had the opposite effect of perpetuating uncertainty and heightening concerns over market volatility.[5] The outcomes of these statistical battles led the USDA to reconceptualize crop forecasting as no longer a late nineteenth-century quest for statistical certainty that sought to eliminate judgment. Rather, by the early twentieth century, it was seen as a more flexible mode of statistical judgment that acknowledged the persistence of unpredictability, thereby rationalizing uncertainty into modern agricultural commodity markets. "Rationalizing uncertainty" has a dual meaning, referring to the late nineteenth-century attempt to impose statistical order and bureaucratic rationality on the uncertainty of

3. "Cotton Crop of Mr. Neill," *Arkansas Democrat*, November 28, 1899.

4. "Heavy Fall in Cotton," *New-York Tribune*, October 20, 1895.

5. For an excellent market-oriented analysis of cotton futures trading at the turn of the twentieth century, see Bruce E. Baker and Barbara Hahn, *Cotton Kings: Capitalism and Corruption in Turn-of-the-Century New York and New Orleans* (New York: Oxford University Press, 2015). Their account argues for the efficacy of federal regulation, culminating in the 1914 Cotton Futures Act, which signaled a victory for government reform and New Orleans bulls over the corrupt practices of New York bear-market speculators. This chapter, which focuses on the production and circulation of cotton forecasts rather than on the exchanges themselves, traces a longer history, beginning with the American Civil War, in which the federal government's cotton statistics did not make markets more rational or more efficient.

future cotton yields as well as to the early twentieth-century redefinition of uncertainty as an inescapable feature of crop forecasting. Of course, the implications of uncertainty depended on one's position in the "worldwide web of cotton," as incomplete figures on shipments and yield bedeviled brokers and spinners; inaccuracies in crop reports frustrated statisticians, lawmakers, and growers; and price fluctuations thwarted growers and rewarded speculators.[6]

Histories of nineteenth-century commodity exchange have depicted the savvy and unscrupulous speculator as a persistent source of market volatility and the object of moral critiques of gambling.[7] But these accounts overlook the work of the crop forecaster as a market-manipulating statistical middleman who, like the speculator, sometimes profited at the expense of rural producers.[8] Cotton forecasters clearly influenced markets—sometimes intentionally, sometimes not. And speculators wielded not objective, static economic data but rather dynamic and at times manipulated statistics that were produced by a combination of government statisticians, commercial interests, and farmers themselves.

Similarly, knowledge infrastructures are a vital but often invisible part of cotton's "global network of land, labor, transport, manufacture, and sale."[9] In the late nineteenth century, government and commercial crop estimates in cotton-producing nations circulated around the globe as the quadruplex telegraph, undersea cable, and stock ticker enabled nearly instantaneous transmission of

6. On the global history of cotton and capitalism, see Sven Beckert, "Emancipation and Empire: Reconstructing the Worldwide Web of Cotton Production in the Age of the American Civil War," *American Historical Review* 109, no. 5 (2004): 1405–38.

7. William Cronon, *Nature's Metropolis: Chicago and the Great West* (New York: W. W. Norton, 1991); Ann Fabian, *Card Sharps, Dream Books, and Bucket Shops: Gambling in Nineteenth-Century America* (Ithaca, NY: Cornell University Press, 1990); Steve Fraser, *Every Man a Speculator: A History of Wall Street in American Life* (New York: HarperCollins, 2005); David Hochfelder, "'Where the Common People Could Speculate': The Ticker, Bucket Shops, and the Origins of Popular Participation in Financial Markets, 1880–1920," *Journal of American History* 93, no. 2 (2006): 335–38.

8. On cotton forecasters' ability to manipulate prices, see Jamie L. Pietruska, "'Cotton Guessers': Crop Forecasters and the Rationalizing of Uncertainty in American Cotton Markets, 1890–1905," in *The Rise of Marketing and Market Research*, ed. Hartmut Berghoff and Uwe Spiekermann (New York: Palgrave Macmillan, 2012), 49–72; Baker and Hahn, *Cotton Kings*, chap. 2.

9. Sven Beckert, *Empire of Cotton: A Global History* (New York: Alfred A. Knopf, 2014), xx. On slavery, capitalism, and the global cotton economy, see also Walter Johnson, *River of Dark Dreams: Slavery and Empire in the Cotton Kingdom* (Cambridge, MA: Belknap Press, 2013).

market news and prices.[10] In national and global markets, knowledge of local crop conditions had minimal value, since the price a farmer could expect at harvest was determined not in relation to his neighbor's crop but rather to the crop's aggregate condition and yield.[11] In this context, agricultural statistics became increasingly significant as a mechanism that enabled the valuation of an individual farmer's crop relative to crops a county, a region, or half a world away.[12] As the USDA's chief statistician John Hyde explained at the turn of the century, agricultural statistics produced a new form of knowledge that was indispensable to commodity markets:

> The aggregation[s] of individual facts . . . are usually too large to be perceptible to the senses. Men know by sight a bushel of wheat or potatoes, but practically they can know millions of bushels only by representative figures. It would, therefore, appear that statistics are as necessary to our larger comprehension

10. Hochfelder, "'Where the Common People Could Speculate,'" 336–38.

11. Leon M. Estabrook, *Government Crop Reports: Their Value, Scope, and Preparation* (Washington, DC: Government Printing Office, 1915), 9.

12. The history of crop forecasting in the United States has drawn little scholarly attention. Internal histories published by the USDA have narrated the expansion and professionalization of government crop reporting as a steady march toward ever more accurate and economically valuable agricultural statistics. These institutional histories, along with early twentieth-century commentary on crop reporting and two recent scholarly analyses, have identified a 1905 "cotton leak" scandal that uncovered fraud within the USDA's crop estimating program as the origin of modern agricultural statistics, an unfortunate but necessary moment of bureaucratic reform that eradicated corruption and produced reliable, objective information. This chapter focuses instead on competing and often contradictory cotton statistics issued by the USDA's Division of Statistics, private commercial forecasters, and farmers' associations and traces contests over statistical objectivity, itself a constructed and historical category. For internal histories of crop reporting, see USDA, *The Crop and Livestock Reporting Service of the United States, Misc. Pub. No. 171* (Washington, DC: USDA, 1933); Emerson M. Brooks, "The Founding of the Crop Reporting Service: A Review of the Struggle to Establish an Agricultural Estimating Service in the U.S. Department of Agriculture, 1863–1903," manuscript, 1962, USDA National Agricultural Statistics Service Library; USDA Statistical Reporting Service, *The Story of U.S. Agricultural Estimates, Misc. Pub. No. 1088* (Washington, DC: USDA, 1969); Frederic A. Vogel, "The Evolution and Development of Agricultural Statistics at the United States Department of Agriculture," *Journal of Official Statistics* 11, no. 2 (1995): 161–80. For scholarly analyses of crop reporting, see Robert E. Snyder, "Federal Crop Forecasts and the Cotton Market, 1866–1929," *Journal of Southwest Georgia History* 7 (1989–1992): 40–58; Emmanuel Didier, "Do Statistics 'Perform' the Economy?," in *Do Economists Make Markets? On the Performativity of Economics*, ed. Donald MacKenzie, Fabian Muniesa, and Lucia Siu (Princeton, NJ: Princeton University Press, 2007), 276–310.

as are vision, touch, and other senses to our so-called personal knowledge of things.[13]

However, the "larger comprehension" of the objective aggregate was not, in fact, that far removed from the subjective judgment that yielded "personal knowledge." Both comprised the knowledge infrastructure that, in addition to relaying figures on millions of cotton bales and prices of cotton in cents per pound, actively constructed market information that could exacerbate uncertainty in the world of cotton exchange.

"THE UNCERTAINTY OF ANY FURTHER SUPPLIES FROM THE SOUTH"

The poetic satire on the floor of the New York Cotton Exchange in 1899 was not the first time that Henry Neill was accused of conjuring cotton bales. The outbreak of the American Civil War signaled, alongside the irreparable political crisis over slavery and its expansion, a crisis of predictability in cotton estimating in Britain. By the eve of the war, cotton textile production dominated global manufacturing, and the labor of enslaved men and women in the American South produced the overwhelming majority of cotton used by spinners throughout Europe. Neill Bros., the firm owned and operated by brothers Henry and William Neill, predicted no easy or quick resolution to the war and could not anticipate how much cotton the American South would produce, if it could be exported, and at what price.[14] A combination of Confederate export bans and Union blockades shut off the flow of cotton to Europe, causing the "cotton famine" that plagued Lancashire and other European textile production centers with mill shutdowns, soaring unemployment, and food riots.[15] Henry Neill himself was reportedly caught blockade running aboard the

13. John Hyde, *Of What Service Are Statistics to the Farmer?* (Washington, DC: Government Printing Office, 1898), 258–59.

14. The circulars issued by Neill Bros., excerpted in newspapers on both sides of the Atlantic, combined analysis of figures on prices and shipments with speculation regarding the outcome and duration of the war. See, for example, "The Cotton Circular of Messrs. Neill Brothers, of New York," *Glasgow Herald*, November 9, 1861; "The Cotton Supply," *Charleston Tri-Weekly Courier*, July 3, 1862; "Neill Brothers' Cotton Circular," *Liverpool Mercury*, October 4, 1862; "Messrs. Neill's Circular on the Cotton Supply," *Glasgow Herald*, May 4, 1863; "Messrs. Neill's Cotton Circular," *Glasgow Herald*, November 3, 1864; "Neill Brothers & Co.'s Cotton Circular," *Glasgow Herald*, February 4, 1865.

15. Beckert, "Emancipation and Empire," 1408–10.

Cumbria, a steamer headed from Liverpool to New Orleans that was captured off the coast of Charleston in 1862.[16] The cargo seized included "heavy cloth, saltpeter, [and] sixty cases of Enfield rifles," but Neill was likely more concerned with bringing cotton (or family members) back from the port of New Orleans than he was with delivering weapons to the Confederacy, considering that he was born into a family of abolitionists.[17] Neill was released and sailed back to Britain, where his firm continued to issue circulars addressing the shortages and rising prices caused by "the uncertainty of any further supplies from the South for some time to come."[18]

The war rendered estimates of American yield "mere guesses," as the Neills acknowledged in a December 1863 circular.[19] After the war ended, Neill Bros., whom Frederick Law Olmsted described as "the most pains-taking collectors of information about the cotton crop in the country," dispatched three reporters to survey the cotton states and then assembled their observations into a cautious estimate of 1.5 to 1.8 million bales for the September 1866 to September 1867 season.[20] The value of yield estimates, tentative and otherwise, soared in the aftermath of the war as a global network of cotton production and trade sprung up, linking growers and merchants from India, Egypt, and Brazil into the structures of global capitalism. As merchants and mill owners across the world sought "a secure and predictable supply of inexpensive cotton," private forecasters like the Neills competed with the USDA to make the supply of American cotton ever more predictable.[21]

16. "Local Intelligence," *Preston Guardian* (Lancashire), June 18, 1862.

17. *Sheffield and Rotterdam Independent*, June 17, 1862, as quoted in Jim Powell, "The Neills and the American Civil War," http://www.jim-powell.net/Documents/131011neills—-the-civil-war.aspx. I am grateful to Jim Powell, a descendant of William and Henry Neill, for investigating the *Cumbria* and for calling my attention to the abolitionist politics of the Neill family.

18. On Neill's return to Britain after the *Cumbria* incident, see Jim Powell, "The Neills and the American Civil War," http://www.jim-powell.net/Documents/131011neills—-the-civil-war.aspx; quotation in "General Markets," *New York Times* (hereafter *NYT*), August 1, 1861.

19. "The Cotton Circular of Messrs. Neill Brothers & Co.," *Glasgow Herald*, December 3, 1863.

20. Frederick Law Olmsted, *The Cotton Kingdom: A Traveller's Observations on Cotton and Slavery in the American Slave States*, vol. 1 (New York: Mason Brothers, 1862), 17; "Messrs. Neill's Cotton Report," *Glasgow Herald*, December 23, 1865; "Neill Brothers & Co.'s Cotton Report," *Glasgow Herald*, January 26, 1866.

21. Beckert, "Emancipation and Empire," 1405, quotation on 1418. On the postbellum southern cotton economy, see Gavin Wright, *The Political Economy of the Cotton South: Households, Markets, and Wealth in the Nineteenth Century* (New York: W. W. Norton, 1978); Gavin Wright, *Old South, New South: Revolutions in the Southern Economy Since the Civil War* (New York:

In 1866, Boston cotton manufacturer Edward Atkinson accused Neill Bros. of inflating their yield estimates for the South. Atkinson and the Neills feuded in the pages of the *Boston Daily Advertiser* and in private correspondence after Atkinson charged the Neills with issuing "diametrically opposite" instructions in October 1865. They advised the public, who was told to anticipate 2 to 3 million bales on hand (not the 1.5 million bales generally expected) to wait; at the same time, they instructed private subscribers, who were given a less optimistic estimate, to buy. Atkinson, convinced that depressed cotton prices would inhibit "the restoration of order in the South," wrote indignantly to the Neills that their public circular was "so written as to depress the market, the other to stimulate it." The *Liverpool Mercury* and the *Boston Daily Advertiser* printed extended comparisons of the Neills' private and public circulars from October 1865, which Atkinson pointed to as evidence for his "distrusting" of Neill's numbers.[22]

The Neills did not deny having issued two separate circulars and indeed justified their existence. "The value of our information to our friends," the Neills reportedly wrote to their clients, "will depend upon its being confined to a limited number, and we have therefore concluded not to print it in our circular, but to reserve it for those for whom we are receiving orders."[23] Moreover, the Neills made no pretenses of statistical objectivity. In a letter to the *New York Times*, William Neill reminded readers "that Messrs. Neill Brothers & Co. are chiefly agents for English cotton buyers. . . . It will therefore be well to consider our position very carefully."[24]

Basic Books, 1986); Gilbert C. Fite, *Cotton Fields No More: Southern Agriculture, 1865–1980* (Lexington: University Press of Kentucky, 1984); Pete Daniel, *Breaking the Land: The Transformation of Tobacco, Corn, and Rice Cultures Since 1880* (Urbana and Chicago: University of Illinois Press, 1985).

22. Edward Atkinson to William Neill, February 19, 1866, Letterbooks, 1853–1905, vol. 2, Edward Atkinson Papers, Massachusetts Historical Society; *Boston Daily Advertiser*, December 2, 1865; Atkinson to William Neill, October 29, 1866, Letterbooks, 1853–1905, vol. 2, Atkinson Papers. In his classic account of statistical objectivity, Theodore M. Porter observes that "reliance on numbers and quantitative manipulation minimizes the need for intimate knowledge and personal trust." Theodore M. Porter, *Trust in Numbers: The Pursuit of Objectivity in Science and Public Life* (Princeton, NJ: Princeton University Press, 1995), ix. The story of Henry Neill illustrates the extent to which subjective judgment and personal trust still figured significantly in the authority-at-a-distance Neill attained in both New Orleans and Liverpool.

23. Edward Atkinson, Letter to the editor, *Boston Daily Advertiser*, February 24, 1866.

24. Atkinson to the editor of the *New Orleans Times*, January 9, 1866, Letterbooks, 1853–1905, vol. 2, Atkinson Papers.

The feud between Atkinson and the Neills ended in little more than finger pointing. Both complained of being unfairly targeted, and the Neills insisted that their circulars had been selectively excerpted and their meaning altered in the press. Atkinson apologized to the Neills for basing his judgment on ostensibly edited letters published in the *Liverpool Mercury* but privately doubted that their content had been misrepresented.[25] But allegations of self-interested cotton statistics would continue throughout the late nineteenth century, and Atkinson's complaint that Neill's cotton figures received "undue prominence" from the *Boston Daily Advertiser* would be echoed by southern newspapers, especially the *Atlanta Constitution*, which lamented Neill's singular authority in the world of cotton forecasting.[26]

"UNCERTAINTY IS THE MOTHER OF SPECULATION"

Long before the Atkinson-Neill dispute over objectivity in cotton circulars, various organizations and individuals sought to collect disinterested agricultural statistics that could be used to make predictive judgments in commodity markets. In the first two decades of the nineteenth century, the Massachusetts Society for Promoting Agriculture and Virginia's Albermarle Agricultural Society, motivated by imperatives of scientific agriculture and the logic of improvement, undertook quantitative assessments of local farming practices. State-sponsored agricultural surveys, in New York in 1825 and in Massachusetts from 1837 to 1841, extended these efforts, and large-scale, systematic crop reporting emerged after the Panic of 1837, when the US Census, the US Patent Office, and the Maryland Agricultural Society launched experimental crop surveys between the late 1830s and the mid-1850s, which were complemented by occasional newspaper accounts from correspondents reporting on what they had seen and heard of the upcoming harvest. *American Agriculturalist* editor Orange Judd, who had published readers' descriptions of crop conditions in the mid-1850s, aggregated his readers' monthly reports of acreage and

25. Atkinson to Maurice Williams, February 21, 1866; Atkinson to William Neill, February 19, 1866; Atkinson to William Neill, October 29, 1866; Edward Atkinson to G. Nordhoff, March 12, 1866, all in Letterbooks, 1853–1905, vol. 2, Atkinson Papers.

26. Atkinson to William Neill, February 13, 1866, Letterbooks, 1853–1905, vol. 2, Atkinson Papers; "Neill Worked with Secret Circulars," *Atlanta Constitution*, November 7, 1898.

estimated yield from May to September 1862 in what amounted to the first published forecasts of the coming harvest.[27] Judd enlisted over 1,500 correspondents from over twenty states in a collective effort against the statistical uncertainties of agricultural commodity markets and speculators' intentionally misleading forecasts. But Judd, who had always believed that the federal government should oversee the collection of agricultural statistics, ended his project in September 1862 and yielded to the USDA, which had been established in May of that year.[28]

Immediately upon its founding, the USDA began to create a knowledge infrastructure to produce objective and authoritative agricultural statistics. The first commissioner of the USDA, Pennsylvania's Isaac Newton, recognized that data furnished by the decennial census, already two years old by the time it was published, rendered yield per acre, total acreage, and soil condition "questions of the highest magnitude but of which no one could speak with any certainty."[29] In 1863, the Division of Statistics was formed within the USDA and began issuing monthly reports on acreage under cultivation and crop conditions, and in 1864, it replaced "an indiscriminate and indefinite number" of crop reporters with dedicated county correspondents who disseminated circulars to their assistants and then compiled county-level reports to be sent to Washington, DC.[30] This network of county-level volunteers continued its work until the early 1880s, when it was augmented by a corps of state statistical agents who were paid employees of the Division of Statistics.[31]

27. USDA, *Story of U.S. Agricultural Estimates*, 5–14; "Prospects of the Crops in Illinois," *NYT*, July 7, 1855; Walter H. Ebling, "Why the Government Entered the Field of Crop Reporting and Forecasting," *Journal of Farm Economics* 21, no. 4 (1939): 724; Vogel, "The Evolution and Development of Agricultural Statistics," 163.

28. Ebling, "Why the Government Entered the Field of Crop Reporting and Forecasting," 725–26.

29. US Commissioner of Agriculture, *Report, 1862* (Washington, DC: USDA, 1863), 575, quoted in USDA, *Story of U.S. Agricultural Estimates*, 21.

30. USDA, *Story of U.S. Agricultural Estimates*, 25; Division of Crop and Livestock Estimates, *Crop and Livestock Reporting Service of the United States*, 4.

31. As it was expanding the reach of its national crop reporting network, the Division of Statistics also began looking abroad. In the early 1870s, the USDA's chief statistician visited agricultural statistics offices in London, Paris, Berlin, and Vienna, and in the early 1880s, the USDA created an agency in the Consul General's office in London to collect data for the USDA's European Crop Report. In June 1888, Congress mandated that all US Consular offices furnish to the State Department monthly reports of local crop conditions, which were then published in the Division of Statistics' monthly report as "Notes on Foreign Agriculture." Division of Crop and

This professionalization of state-level crop reporting occurred within a culture of certainty, a conviction that a disciplined cadre of statisticians and crop correspondents, operating within an increasingly efficient and expanding government bureaucracy, would yield more accurate agricultural statistics. The global depression set off by the Panic of 1873 sent cotton prices into freefall, and the price of grain relative to cotton fluctuated wildly, "introducing a new degree of uncertainty into rural producers' precarious lives."[32] With agricultural commodity prices in steep decline worldwide, USDA chief statistician Jacob Richards Dodge announced that "what is wanted in statistics is more of thought and less of flurry, more industry and less precipitancy, sounder judgment and less zeal without knowledge."[33] Administrative capacity and statistical precision went hand in hand, as Dodge noted in the early 1880s when he celebrated "improvement in our statistical machinery," a steady march toward ever more certainty, and heightened popular enthusiasm for statistics in general.[34] This quest for certainty in agricultural statistics was articulated by a USDA statistician writing at the turn of the century:

> It can not be too strongly emphasized that the farmer's profits are increased by everything which lessens doubt in regard to future values. If such values could be ascertained in advance with absolute certainty, harvest prices would not differ very materially from the prices later in the season. . . . Everything therefore, which tends towards certainty, as regards either supply or demand, is distinctly advantageous to the farmer.[35]

The production of these statistical certainties hinged on the ability of the USDA's crop reporting network to reach far into the countryside, where it extracted estimates of acreage, condition, and yield, then transmitted the raw data to the "center of calculation" in Washington, DC, where clerks and statisticians tabulated, aggregated, averaged, weighted, and verified the figures

---

Livestock Estimates, *Crop and Livestock Reporting Service of the United States*, 4; USDA, *Story of U.S. Agricultural Estimates*, 32, 35.

32. Beckert, "Emancipation and Empire," 1434.

33. Quoted in USDA, *Story of U.S. Agricultural Estimates*, 33.

34. US Commissioner of Agriculture, *Report, 1881 and 1882*, 666–67, quoted in Henry C. Taylor and Anne Dewees Taylor, *The Story of Agricultural Economics in the United States, 1840–1932: Men, Services, Ideas* (Ames: Iowa State College Press, 1952), 190.

35. "Government Crop Reports and Speculation in Farm Products," *Crop Reporter* 1, no. 4 (August 1899): 3.

into county, state, and national totals.[36] By the turn of the century, the USDA's network of crop reporters included 41 full-time, paid state statistical agents and their 7,500 assistants, 2,400 volunteer county correspondents and their 6,800 assistants, and 40,000 volunteer township or district correspondents.[37] But the vast majority of statistics on yield and condition came pouring in from individual volunteer correspondents who provided snapshots of production on their own farms. By the early twentieth century, almost a quarter of a million correspondents submitting close to 2.5 million reports comprised the USDA's crop reporting network, what the *New York Times* subsequently hailed as a "big army of correspondents engaged in the collection of facts that are of vital importance to the business welfare of the nation."[38]

The Division of Statistics sent out even more reports than it received. At the turn of the century, the USDA distributed by telegraph and mail over 1.5 million copies of its thirteen regular and six special crop reports. Exactly at a predetermined hour on designated days (noon for cotton and 4:00 p.m. for grain), the USDA released its summary crop report to Western Union and Postal Telegraph-Cable telegraphers and journalists, who relayed the numbers to brokerage firms, commodity exchanges, and newspapers across the country. Beginning in 1901, the USDA mailed postcard summaries within three hours of the report's official release to each of the nation's seventy-seven thousand post offices, chiefly for the benefit of rural communities outside the reach of telegraph networks. The USDA also published crop estimates in the approximately 1.5 million copies of the *Crop Reporter* it printed at the turn of the century.[39]

But it is important to note that the USDA's crop statistics in the late nineteenth and early twentieth centuries were technically not forecasts, although they functioned as such in the marketplace. USDA cotton correspondents reported on acreage and condition at the start of the growing season in May; on condition from June through September; and on average yield per acre, abandoned acreage, and cost of picking in November. (See figure 1.1.) Of

36. Bruno Latour, *Science in Action: How to Follow Scientists and Engineers through Society* (Cambridge, MA: Harvard University Press, 1987), chap. 6. Emmanuel Didier also invokes Latour's "center of calculation" in his account of the USDA's interwar agricultural statistics program. See Didier, "Do Statistics 'Perform' the Economy?," 279.

37. USDA, *Story of U.S. Agricultural Estimates*, 42; USDA, *Yearbook of the United States Department of Agriculture, 1897* (Washington, DC: Government Printing Office, 1898), 269.

38. "Government's Crop Report the Work of 130,000 People," *NYT*, June 1, 1913.

39. "Annual Report of the Statistician for the Fiscal Year 1901," *Crop Reporter* 3, no. 8 (December 1901): 1–2.

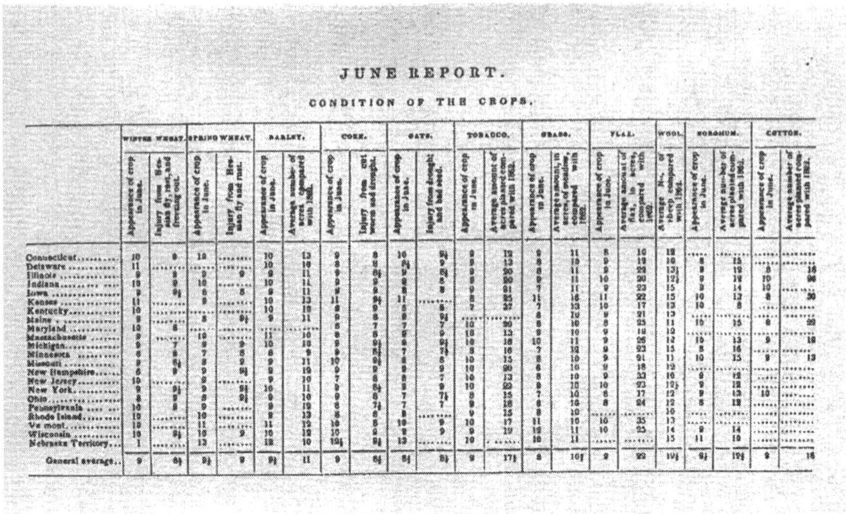

FIGURE 1.1. US Department of Agriculture, *Monthly Report of the Condition of the Crops*, July 10, 1863. From a pamphlet reproduced by the National Agricultural Statistics Service, July 1988.

course, it was fairly simple to calculate probable yield by combining acreage and condition figures with historical yield data, which brokers, speculators, and newspapers regularly did. The USDA's cotton statistics did not quantify numbers of bales until early December, when the USDA published estimated end-of-season yield by state that, after 1902, could then be checked against the yield totals tabulated from ginners' records by the federal census.[40]

The purpose of data collection, tabulation, and distribution on such a grand scale, according to USDA officials, was "to throw light on future conditions and do away, as far as possible, with uncertainties as to supply and demand" in order to reduce speculators' control over prices.[41] USDA statisticians re-

40. Charles William Burkett and Clarence Hamilton Poe, *Cotton: Its Cultivation, Marketing, Manufacture, and the Problems of the Cotton World* (New York: Doubleday, Page, 1906), 252–53, 258; Margo J. Anderson, *The American Census: A Social History* (New Haven, CT: Yale University Press, 1988), 114.

41. "Government Crop Reports and Speculation in Farm Products," *Crop Reporter* 1, no. 4 (August 1899): 3. Crop estimates were used in myriad ways throughout the late nineteenth century by those who were neither producers nor consumers of agricultural commodities. Manufacturing firms, agricultural tool makers, and hardware companies used crop reports to strategically market their products to farmers who harvested bountiful crops. Railroad corporations relied on crop estimates to determine how many cars would be required to ship commodities to market, and banks commonly used crop reports to anticipate seasonal demand for farm credit.

marked that the speculator's "advantage lies in greater certainty" and pointed to the importance of agricultural statistics in reducing that advantage. As head statistician John Hyde explained in 1897,

> The more closely the production of any given crop is ascertained—in other words, the less the uncertainty existing as to the amount available for consumption and export—the smaller will be the risk attending the operations of the merchant and trader, the less will be the inducement to speculation, and the more stable will be the value of the commodity. In every sphere of human thought uncertainty is the mother of speculation.[42]

Yet despite, and indeed because of, the standardization and efficiency of its crop reporting bureaucracy, the USDA faced significant challenges in translating the raw statistical data reported by hundreds of thousands of individual farmers; state, county, and township crop correspondents; and specially appointed traveling field agents into information that would "reveal true conditions."[43] Farmers fearing higher tax assessments allegedly underreported their acreage to census representatives and USDA statisticians.[44] USDA officials also had to contend with uncertainties resulting from inaccurate assessment of a crop's appearance in the field, figures provided from a locality that was not typical of the reporting area as a whole, skewed averages because of improper weighting, extreme local variation in weather and crop condition, and reports submitted incomplete, late, or not at all.[45] Public criticism—from the rural press and, in the 1880s and 1890s, the National Board of Trade and the Chicago Board of Trade—dogged the Division of Statistics, which found itself defending the relevance and accuracy of its work against charges that its monthly crop estimates proved more useful to foreign buyers and "served no useful purpose" for domestic commercial and agricultural interests.[46]

But the principal sources of uncertainty in the USDA's crop estimates were not uncooperative farmers or careless statisticians but rather the Division of Statistics' methods of calculation. Crop reporters estimated acreage

---

Estabrook, *Government Crop Reports*, 9–10; USDA, *Crop and Livestock Reporting Service of the United States*, 2.

42. USDA, *Yearbook of the United States Department of Agriculture, 1897* (Washington, DC: Government Printing Office, 1898), 264–65.

43. *Yearbook of the USDA, 1897*, 265.

44. USDA, *Story of U.S. Agricultural Estimates*, 34–36.

45. "Difficulties Encountered by the Crop Statisticians," *Crop Reporter* 1, no. 3 (July 1899): 2.

46. USDA, *Story of U.S. Agricultural Estimates*, 38.

and monthly condition based not on the fields they saw stretching out before them but rather compared to historical acreage data and to an ambiguous category of "normal" crop condition.[47] Individual crop reporters used decennial census acreage data as a baseline for their acreage estimates, so that, as one USDA statistician remarked, "every return of area has in it all the uncertainty of the census determination."[48] Every year until the next census, crop reporters estimated acreage as a percentage of their previous year's estimate, so inaccuracies for a single year would result in a "cumulative error," which could be significant. For example, with baseline census data of ten million acres planted, if crop reporters underestimated acreage by 2 percent each year, in ten years the census figures and the Bureau of Statistics figures would differ by nearly two million acres, and the yield per acre and total yield estimates would be similarly skewed. And simply calculating acreage posed a challenge in the case of irregularly shaped fields, which the USDA addressed by providing geometry lessons in the *Crop Reporter*, a monthly handbook for crop correspondents first published in 1899.[49] (See figure 1.2.)

A second source of uncertainty emerged from condition estimates, in which farmers assessed the current state of their crop not compared to the previous year's but based on an abstract "normal" crop, which the USDA defined as 100 percent crop yield—neither bumper crop nor crop failure—without any significant adverse effects from weather, disease, or insects.[50] Farmers quantified their crop conditions relative to a "normal" base ten (e.g., a condition figure of eleven indicated a crop 10 percent better than the previous year's, while a nine indicated 10 percent worse). But "normal" did not mean "average," as many crop correspondents mistakenly assumed. The *Crop Reporter* repeatedly reminded its readers that "normal" was neither "average" nor "perfect," "neither

47. Henry Parker Willis, "The Adjustment of Crop Statistics," *Journal of Political Economy* 11, no. 1 (1902): 6. On the "normal" and condition categories in interwar agricultural surveys, see Didier, "Do Statistics 'Perform' the Economy?," 286–88.

48. Henry Farquhar, "Crop Conditions and Prospects," *North American Review* 161, no. 466 (1895): 320.

49. Estabrook, *Government Crop Reports*, 18; "Methods for Estimating Areas of Land," *Crop Reporter* 2, no. 2 (June 1900): 4. Initially the *Crop Reporter* was issued to crop correspondents only, but it was made available to the general public beginning in May 1900 in response to popular demand. Over 1.4 million copies were published during the 1901 fiscal year. Taylor and Taylor, *Story of Agricultural Economics*, 209–11; "Report of the Statistician for the Fiscal Year 1900," *Crop Reporter* 2, no. 10 (February 1901): 2; "Annual Report of the Statistician for the Fiscal Year 1901," *Crop Reporter* 3, no. 8 (December 1901): 1.

50. Didier, "Do Statistics 'Perform' the Economy?," 288.

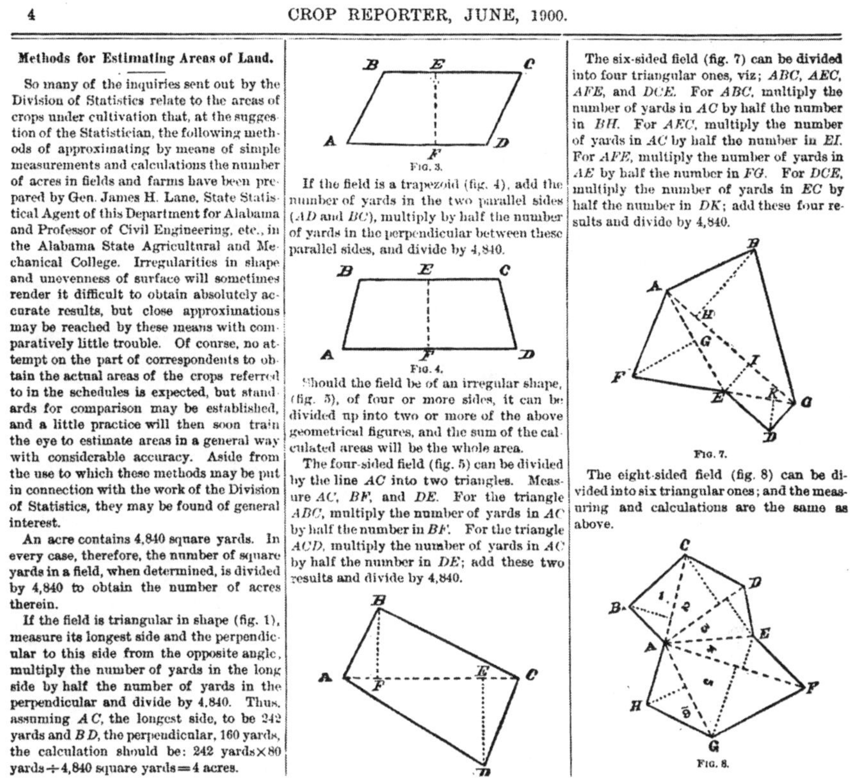

4 CROP REPORTER, JUNE, 1900.

Methods for Estimating Areas of Land.

So many of the inquiries sent out by the Division of Statistics relate to the areas of crops under cultivation that, at the suggestion of the Statistician, the following methods of approximating by means of simple measurements and calculations the number of acres in fields and farms have been prepared by Gen. James H. Lane, State Statistical Agent of this Department for Alabama and Professor of Civil Engineering, etc., in the Alabama State Agricultural and Mechanical College. Irregularities in shape and unevenness of surface will sometimes render it difficult to obtain absolutely accurate results, but close approximations may be reached by these means with comparatively little trouble. Of course, no attempt on the part of correspondents to obtain the actual areas of the crops referred to in the schedules is expected, but standards for comparison may be established, and a little practice will then soon train the eye to estimate areas in a general way with considerable accuracy. Aside from the use to which these methods may be put in connection with the work of the Division of Statistics, they may be found of general interest.

An acre contains 4,840 square yards. In every case, therefore, the number of square yards in a field, when determined, is divided by 4,840 to obtain the number of acres therein.

If the field is triangular in shape (fig. 1), measure its longest side and the perpendicular to this side from the opposite angle, multiply the number of yards in the long side by half the number of yards in the perpendicular and divide by 4,840. Thus, assuming *A C*, the longest side, to be 242 yards and *B D*, the perpendicular, 160 yards, the calculation should be: 242 yards×80 yards÷4,840 square yards=4 acres.

Fig. 3.

If the field is a trapezoid (fig. 4), add the number of yards in the two parallel sides (*AD* and *BC*), multiply by half the number of yards in the perpendicular between these parallel sides, and divide by 4,840.

Fig. 4.

Should the field be of an irregular shape, (fig. 5), of four or more sides, it can be divided up into two or more of the above geometrical figures, and the sum of the calculated areas will be the whole area.

The four-sided field (fig. 5) can be divided by the line *AC* into two triangles. Measure *AC*, *BF*, and *DE*. For the triangle *ABC*, multiply the number of yards in *AC* by half the number in *BF*. For the triangle *ACD*, multiply the number of yards in *AC* by half the number in *DE*; add these two results and divide by 4,840.

The six-sided field (fig. 7) can be divided into four triangular ones, viz; *ABC*, *AEC*, *AFE*, and *DCE*. For *ABC*, multiply the number of yards in *AC* by half the number in *BH*. For *AEC*, multiply the number of yards in *AC* by half the number in *EI*. For *AFE*, multiply the number of yards in *AE* by half the number in *FG*. For *DCE*, multiply the number of yards in *EC* by half the number in *DK*; add these four results and divide by 4,840.

Fig. 7.

The eight-sided field (fig. 8) can be divided into six triangular ones; and the measuring and calculations are the same as above.

Fig. 8.

FIGURE 1.2. "Methods for Estimating Areas of Land," *Crop Reporter*, June 1900, 4.

deficient on the one hand nor extraordinarily heavy on the other," and "not being everywhere the same."[51] Bureau of Statistics chief Leon Estabrook eventually conceded the "apparent vagueness" of the "normal" category as well as the fact that normal crop yields often varied by state. USDA statisticians also maintained that farmers' judgments of the elusive "normal" were bound to be impressionistic in the absence of accurate historical statistics.[52]

Despite perennial confusion over the meaning of a "normal" crop, the USDA retained its condition category into the twentieth century based on its assumption that farmers were disinclined to objectively quantify the condition of their crops. As Estabrook explained, the "normal" category was effective

51. USDA, *Crop Reporter* 1, no. 1 (May 1899): 3. The official definition of "normal" was reprinted in the *Crop Reporter* in August 1899 and in July 1900.

52. Estabrook, *Government Crop Reports*, 20; Farquhar, "Crop Conditions and Prospects," 316.

because it was "psychological" in that "the average farmer thinks of his crop as 'crops' and not in mathematical terms of percentages or averages."[53] But the subjective judgment of "normal" was a liability according to critics like cotton broker and *Commerce and Finance* editor Theodore H. Price, who argued that the "imaginary normal" was fertile ground for self-interested, bullish underestimation. The "normal" category and crop reports more generally were, Price insisted, unscientific and unreliable, based on the work of crop reporters charged with the hopeless task of "describ[ing] the present and its promise in figures that are supposed to be a percentage of an impossible perfection."[54]

Along with the statistical uncertainties produced by cumulative error in acreage estimates and the ambiguity of the "normal" category for crop condition came the economic and environmental uncertainties of cotton itself. Cotton was the undisputed king of the postbellum southern economy, as tenant farmers and sharecroppers became increasingly enmeshed in global markets and constrained by local credit systems. Despite perennial cries of "overproduction" and repeated calls for diversification, sharecroppers and tenant farmers continued to cultivate a cash crop that was particularly susceptible to environmental damage from unpredictable rainfall, killing frosts, crop disease, and insects, especially the boll weevil, whose well-documented march across the Rio Grande in 1894 resulted in decades of cotton crop devastation.[55] Cotton was deemed "capricious" and the least predictable of the speculative crops, and, as an Alabama newspaper remarked, cotton yield was notoriously difficult to estimate since it "is subject to vicissitudes up to the hour of picking."[56]

The USDA's crop estimates competed with numerous private assessments of current crop conditions and yield forecasts. In the early 1870s, the New

53. Estabrook, *Government Crop Reports*, 20.

54. Theodore H. Price, "The Value and Defects of Government Crop Reports," *Commerce and Finance*, August 16, 1916, 915.

55. Beckert, "Emancipation and Empire," 1427; Fite, *Cotton Fields No More*, 48–49; Daniel, *Breaking the Land*, 6. On the physical threat and cultural meanings of the boll weevil, see James C. Giesen, *Boll Weevil Blues: Cotton, Myth, and Power in the American South* (Chicago: University of Chicago Press, 2011).

56. Burkett and Poe, *Cotton*, 256; "Mr. Henry Neill's Downfall," *Birmingham Age Herald*, March 12, 1899. In theory, cotton's attributes should have made its yield easier to calculate—growers did not consume significant amounts of their crops before marketing; cotton had a clearly defined shipping and marketing infrastructure; cotton bales did not decompose like other crops; and cotton could be quantified by river and rail shippers and by mills—but this was not the case. Willis, "The Adjustment of Crop Statistics," 7.

Orleans Cotton Exchange's superintendent, Henry G. Hester, established a local network of correspondents who provided information on acreage, weather, and crop conditions for the monthly reports that won him wide acclaim as "a wizard in cotton statistics" and a "prophet [who] has honor in his own country."[57] Credit reporting agency R. G. Dun had its own system for soliciting crop forecasts, as did the rival Bradstreet Agency. Bradstreet, lacking the federal government's army of crop reporters, mailed circulars directly to cotton growers, factors, and buyers that asked for estimates of the season's yield.[58] In 1893, Atlanta's Samuel M. Inman, head of one of the world's largest cotton brokerage firms, pronounced crop estimating "only a consensus of guesses. . . . Sometimes they hit it and sometimes they miss it. That's about the way it goes."[59]

"THE LEADING STATISTICIAN OF THE SOUTH"

By 1890, this array of forecasts had an even greater influence on the abstractions of finance capitalism than on the physical sale of crops because of the rise of futures trading, which transformed commodity exchange in the late nineteenth century. By 1885, Chicago's ratio of grain futures to actual grain traded was an estimated fifteen or twenty to one, and by 1900, New York and Chicago exchanges traded a volume of agricultural futures seven times greater than the volume of the harvest.[60] The dominance of futures trading posed an epistemic "crisis" for late nineteenth-century boards of trade, commodity

57. L. Tuffly Ellis, "The New Orleans Cotton Exchange: The Formative Years, 1871–1880," *Journal of Southern History* 39, no. 4 (1973): 548–50; quotations in Meigs O. Frost, *"Hester Says—" An Intimate Personal Sketch of the World's Greatest Cotton Authority* (New Orleans: Theo. H. Harvey Press, 1926), Jones Hall Louisiana Research Collection, Tulane University Libraries.

58. "Crops in Western New-York," *NYT*, July 13, 1891; "Don't Believe It," *Atlanta Constitution*, December 23, 1893.

59. Tammy Harden Galloway, *The Inman Family: An Atlanta Family from Reconstruction to World War I* (Macon, GA: Mercer University Press, 2002), 26; quotation in "Don't Believe It," *Atlanta Constitution*, December 23, 1893.

60. Some futures trading was noted as early as 1853 in Chicago, but futures trading did not become institutionalized into the Chicago Board of Trade (est. 1848) until after the Civil War, which brought a dramatically increased demand for oats and pork for the Union Army and thus a higher volume of futures trading, which in turn led to standardized futures contracts and

exchanges, bucket shops (establishments for betting, but not actually trading, on exchanges), and the legal system, all of which struggled to adapt to what Congress termed "fictitious dealings" in which the physical exchange of capital for goods had been supplanted by an abstract cascade of transactions based not on current commodity prices but on predictions of future price differentials.[61] While the Chicago Board of Trade waged its ultimately successful war on the bucket shops, antifutures sentiment infused the rhetoric of agrarian radicals who railed against rampant speculation in the "wind wheat" that had become irrevocably divorced from the fields where it was grown.[62]

Standing amid this whirlwind of commodity speculation and government and private crop statistics was Henry Neill, who, despite criticism of his allegedly fictitious cotton forecasts, achieved credibility on both sides of the Atlantic by maintaining what he called a "close correspondence" by cable and mail with his brother in London.[63] In 1847, the eighteen-year-old Neill left Belfast and the Great Famine behind for the United States, where he would live in New York, New Orleans, and Mobile, Alabama (his future wife's residence).[64] In 1857, Henry Neill and his brother William founded the firm of Neill Bros., cotton merchants with offices in New Orleans, London, and Liverpool. Neill, a cotton trader as well as a forecaster, was a member of the New York and New Orleans Cotton Exchanges, from where he exerted considerable influence on both spot and futures markets in New Orleans, New York, and Liverpool. Newspapers hailed Neill as "the greatest cotton crop estimate expert in the world," "the leading statistician of the South," "the accepted authority on cotton," and, less effusively, "the cotton guesser." Neill counted over one thousand subscribers—almost all of them in England—to his cotton letters, which he

formal trading regulations. Cronon, *Nature's Metropolis*, 124, 126; Fraser, *Every Man a Speculator*, 252.

61. Jonathan Ira Levy, "Contemplating Delivery: Futures Trading and the Problem of Commodity Exchange in the United States, 1875–1905," *American Historical Review* 111, no. 2 (2006): 307–10, quotation on 310. On congressional hearings titled *Fictitious Dealings in Agricultural Products*, see Jonathan Levy, *Freaks of Fortune: The Emerging World of Capitalism and Risk in America* (Cambridge, MA: Harvard University Press, 2012), 245–46.

62. On commodity speculation and the futures market, see Cronon, *Nature's Metropolis*; Levy, "Contemplating Delivery"; Hochfelder, "'Where the Common People Could Speculate'"; Fabian, *Card Sharps, Dream Books, and Bucket Shops*; Fraser, *Every Man a Speculator*.

63. "Cotton Crop Forecast," *Washington Post* (hereafter *WP*), September 9, 1895; "Heavy Fall in Cotton," *New-York Tribune*, October 20, 1895.

64. Jim Powell, "Family History," http://www.jim-powell.net/Family-History.aspx.

sent out four times each year, and his eagerly anticipated forecasts reached a wider audience as newspaper excerpts.[65]

In the 1890s, Neill's forecasts influenced a cotton economy in which farmers stood little chance to profit as they watched cotton and wheat prices continue to fall, cotton from twenty-three cents a pound in 1870 to seven cents a pound in 1894, and wheat from $1.37 a bushel in 1870 to fifty cents in 1894.[66] The price of cotton was largely determined on the New York and Liverpool Cotton Exchanges, not in the small local markets that dotted the South, where prices were roughly consistent. The New York Cotton Exchange and the New Orleans Cotton Exchange, both founded in 1871, were linked to each other by telegraph and by transcontinental cable to the Liverpool Cotton Exchange.[67]

The 1890s brought record cotton yields—8.6 million bales in 1890–1891, 9 million bales in 1891–1892, 9.9 million bales in 1894–1895, 11.2 million bales in 1897–1898, and 11.27 million bales in 1898–1899—and an "almost catastrophic" drop in price.[68] New York prices reached near-record lows in 1894–1895 (5 9/16 cents), 1897–1898 (5 3/4 cents), and 1898–1899 (5 5/16 cents), and after shipping costs, some growers bemoaned a price of less than five cents per pound. "What are the causes of this depression in the price of cotton?" asked the *New Orleans Daily Picayune* in October 1897. "It is . . . not lack of demand, actual or prospective, which has caused the depression. The real trouble has been the estimate of Mr. Henry Neill forecasting the yield at 10,300,000 bales,

65. "Henry M. Neill Killed," *NYT*, September 13, 1906; "Cotton Crop Estimate This Year," *NYT*, November 4, 1892; "The Feeling About Neill Is Mixed," *Atlanta Constitution*, November 14, 1899; "Henry Neill's Power Broken," *Atlanta Constitution*, March 8, 1903.

66. Fraser, *Every Man a Speculator*, 200.

67. Fite, *Cotton Fields No More*, 49; Kenneth J. Lipartito, "The New York Cotton Exchange and the Development of the Cotton Futures Market," *Business History Review* 57, no. 1 (1983): 52; Ellis, "The New Orleans Cotton Exchange," 547. For an excellent account of the competition between the New York and New Orleans Cotton Exchanges and the advent of federal regulation of cotton markets, see Baker and Hahn, *Cotton Kings*.

68. Throughout the 1880s, average prices in New York had remained steady, hovering between 10 and 11.5 cents per pound (for middling upland cotton) in all but two years, but prices declined sharply in the 1890s, with middling upland cotton averaging between six and just over eight cents a pound in all but one year of the decade. James L. Watkins, *King Cotton: A Historical and Statistical Review* (New York: James L. Watkins & Sons, 1908), 23, 30. Quotation in Theodore Saloutos, "The Southern Cotton Association, 1905–1908," *Journal of Southern History* 13, no. 4 (1947): 494.

and even more under favorable circumstances."[69] The *Atlanta Constitution*, Neill's chief but by no means only critic, complained that his predictions exerted a disproportionate influence on the market:

> Whenever he issued a bulletin the effect was startling. One blast from his horn has been known to knock the price of cotton down 50 points, which meant an apparent reduction of nearly $20,000,000 in the value of the crop. The spinners had the most absolute faith in his predictions, and his influence so dominated the market that, regardless of what conditions might seem to indicate, the price of the south's great staple was practically dominated by this one man's opinion.[70]

An Alabama newspaper expressed similar sentiments, pronouncing Neill "the dictator of the country's chief export crop."[71] And Neill could dictate prices without even publishing a word, as rumors of his forthcoming estimate or revision commonly moved markets in Liverpool, New York, and New Orleans.[72]

Neill was widely regarded as a wizard of the cotton exchange, an immensely powerful figure whose cotton letters—presented in the language of economic rationality—had seemingly magical effects. Neill, whom newspapers commonly referred to (both approvingly and pejoratively) as a "prophet," was depicted by his detractors at the *Atlanta Constitution* as something of a sorcerer "whose word was magic" during "a time when the name of Neill was something to conjure with, and when the farmers of the land stood in awe of him. . . . When Neill waved his hand prices subsided . . . and the lights went out."[73] Neill's contemporaries were well aware of forecasts' power over markets, as

69. Only twice before, in 1841–1842 and 1844–1845, had New York prices plunged to five cents per pound, a result of extremely large yields and surplus. Watkins, *King Cotton*, 17, 23–24, 30–31; "The Situation in Cotton," *New Orleans Daily Picayune*, October 25, 1897.

70. "Henry Neill's Power Broken," *Atlanta Constitution*, March 8, 1903.

71. "Mr. Henry Neill's Downfall," *Birmingham Age Herald*, March 12, 1899.

72. "Commercial and Markets," *Liverpool Mercury*, November 1, 1892; "The Cotton Market," *Galveston Daily News*, October 10, 1895, October 8, 1896, November 12, 1896.

73. "Henry Neill's Power Broken," *Atlanta Constitution*, March 8, 1903; "Will Keep the Graves Green," *Atlanta Constitution*, February 14, 1900. In a study of the USDA's agricultural statistics in the 1920s and 1930s, Emmanuel Didier critiques theories of the performativity of markets for their implication that statistics wholly create the conditions of the economy, rather like "the waving of some magic wand." Didier, "Do Statistics 'Perform' the Economy?," 304–5, quotation on 305. But as contemporary newspaper accounts reveal, Neill's forecasts were understood in precisely those terms.

evidenced by a Hubbard, Price & Co. circular in 1891 that noted Neill's pessimistic appraisal of a cotton crop in need of rain and then observed that "the future course of prices remains dependent upon the currency which bad crop accounts attain."[74]

Likewise, good crop accounts attained a currency of their own during the record-setting seasons of 1894–1895, 1897–1898, and 1898–1899, in all of which Neill had predicted larger-than-average yields. As a Birmingham newspaper observed at the decade's end, "Two or three lucky crop-guesses makes a prophet." Although Neill was widely known and sometimes derided for his cotton "guesses," early in the season he acknowledged the indeterminacy of yield forecasting. In an August 1894 circular, Neill predicted at least 8.9 million bales, "qualif[ying] this belief by saying that it is impossible at this early date to forecast the growing crop with certainty, and that . . . this suggestion of the crop is given as a conservative calculation of probabilities."[75] The 1896–1897 season yielded about 1 million bales over the prevailing estimates and 300,000 to 400,000 bales less than Neill's estimate, which was thought "absurdly large" at the time.[76] In spring 1897, Neill Bros. noted in a Manchester trade publication that Neill's overestimating the 1896–1897 crop had saved "millions" for European buyers, a comment that drew a sharp rebuke from the *Atlanta Constitution*. Neill also reportedly declared, "Yes, I missed the crop 2,000,000 bales but I saved the cotton manufacturers of New England $5,000,000."[77] With one "lucky crop-guess" already to his name by April 1897, Neill retorted, "Who can charge me with willfully falsifying if I come within 250,000 of the crop nearly a year in advance?"[78]

Not surprisingly, British investors and spinners cheered Neill's consistently optimistic forecasts. In March 1894, a letter to the editor of the *Manchester Courier* offered this profession of faith in the New Orleans cotton prophet: " 'Do you believe in Henry Neill?' a man asked me the other day. 'He is not only the best crop authority,' I replied, 'but there is absolutely no other. Lancashire is indebted—has been again and again highly indebted—to Mr. Neill.' "[79]

74. "The Cotton Markets," *Atlanta Constitution*, May 27, 1891.

75. "Largest Cotton Crop Known," *NYT*, August 15, 1899; "Mr. Henry Neill, Cotton Expert," *Birmingham Age Herald*, November 11, 1899; "An Enormous Crop of Cotton," *Macon (GA) Telegraph*, July 31, 1894.

76. "Rome's Cotton Receipts," *Atlanta Constitution*, April 6, 1897; "Enormous Cotton Crop of 1898," *Chicago Daily Tribune*, September 10, 1898. Quotation in "Rome's Cotton Receipts."

77. "Small Faith in Neill," *Raleigh News and Observer*, August 24, 1899.

78. "Mr. Neill's Crop Estimate," *Atlanta Constitution*, April 1, 1897.

79. "Atwood Violett & Co's. Cotton Letter," *Atlanta Constitution*, March 8, 1894.

But not everyone believed in Henry Neill, particularly the cotton growers and brokers in the United States who grew increasingly critical of his influence throughout the 1890s.[80]

"A RECKLESS GUESSER"

Neill's detractors included 150 delegates who walked through four inches of snow in January 1895 to attend the Southern Cotton Growers convention in Jackson, Mississippi, where they established the Cotton Growers' Protective Association, with Hector D. Lane, Alabama's commissioner of agriculture, as president. Delegates proposed a concerted acreage reduction of 25 to 50 percent from the previous year that had yielded a record-setting 9.9 million bales, and the association passed resolutions staunchly opposing futures trading and supporting the Hatch Bill, which sought (unsuccessfully) to end futures trading through prohibitive taxation.[81]

The Cotton Growers' Protective Association articulated a statistical founding mission, seeking to "combine the American cotton growers for their mutual benefit and protection, and to gather and disseminate statistics and information useful to its members." Each member, in addition to paying a fee of two cents per acre cultivated in 1894 and signing an agreement to reduce his acreage by 25 percent in 1895, was required to submit an acreage and condition report to his county chapter in mid-June.[82] But the association's plan would be thwarted by Henry Neill.

After the Neill Bros. circular of July 1895 ranged upward of 8 million bales to bear the market, a skeptical Hector Lane, president of the newly formed Cotton Growers' Protective Association, issued a more conservative forecast of 6.5 to 7 million bales. In September, Lane blamed a variety of harmful factors, including drought, rust, worms, heavy rain, and shedding, for the below-average crop conditions before declaring, "I don't pose as a statistician, and no one but a reckless guesser can make any accurate forecast of the size of the crop before frost." The reckless guessers, Lane charged, were Britain's "resident

80. "Cotton Again Black-Eyed," *Dallas Morning News*, August 18, 1898; *Columbus (GA) Enquirer-Sun*, December 15, 1899, 4.

81. "King Cotton's Future," *WP*, January 10, 1895; Watkins, *King Cotton*, 31. The Hatch Bill passed the House and the Senate but failed to become law when Congress adjourned before a House vote on Senate amendments. On the regional support for and opposition to the bill, see Cedric Cowing, *Populists, Plungers, and Progressives: A Social History of Stock and Commodity Speculation, 1890–1936* (Princeton, NJ: Princeton University Press, 1965), 18–23.

82. "To Cut the Crop," *Atlanta Constitution*, January 22, 1895.

alien professional crop-estimating representative, Mr. Neill, and his able bear coadjutors," who convinced spinners across Europe to distrust the prevailing estimates of poor crop conditions and lower-than-average acreage.[83] And the acreage that had been planted in early 1895 was indeed, as Lane had urged, much lower: every major cotton-producing state did reduce its acreage that year, for an overall reduction of 3.5 million acres from the 23.7 million acres planted in 1894.[84]

In January 1896, Hector Lane denounced the bearish rumors that growers would cultivate substantially more acreage using twice as much fertilizer in the coming season, which would yield 10 to 12 million bales (including 4 million bales in Texas alone). To counter the "reckless despots and wholesale despoilers of [southern] prosperity," Lane rallied the association to a meeting in Memphis to join what he described without a hint of irony as "the struggle to emancipate our people from a bondage that is as oppressive and ignominious as was ever inflicted by the czar upon a serfdom."[85]

Lane addressed somewhere between six hundred and one thousand delegates in Memphis, where he railed against the five-cent cotton economy responsible for "a vortex of poverty and destitution." The problem, Lane declared, was that market forces of supply and demand were overpowered by "the commercial fakir and flim-flam artist" and his "false prophecy, cunning manipulation and wilfull misrepresentation." Lane exhorted his cheering audience to pledge a collective acreage reduction to defeat "cotton 'bears'" like Neill. A Mississippi grower in attendance declared agricultural statistics "indispensable" and urged southern state legislatures "to provide bureaus of information and statistics" for cotton, corn, oats, and other crops.[86] But such statistical initiatives and Lane's futile call for acreage reduction—acreage actually increased by 3 million acres in 1896 to 23.3 million—could not vanquish the bears, and market watchers continued to decry Neill's power over prices.[87]

83. Quotations in "Cotton Crop Forecast," *WP*, September 9, 1895. The USDA's Division of Statistics recorded a 70.8 percent condition average for September 1895 and 65.1 percent for October 1895, the lowest figures for that two-month period on record. Watkins, *King Cotton*, 289–92.

84. Watkins, *King Cotton*, 288.

85. "To Regulate Cotton Planting," *Atlanta Constitution*, January 4, 1896; quotations in "Cotton Growers," *Atlanta Constitution*, January 4, 1896.

86. "Advice to Cotton Planters," *WP*, January 22, 1896; quotations in American Cotton Growers' Protective Association, *Proceedings of the American Cotton Growers' Protective Association* (Memphis, TN: Tracy, 1896), 2, 4, 21.

87. Watkins, *King Cotton*, 288.

In April 1897, the *Atlanta Constitution* complained that the timing of Neill's forecasts—after the crop had been planted in the late winter months—left growers unable to adjust their acreage.[88] In September that year, Neill predicted the largest cotton crop in history: over 4 million bales from Texas and Indian Territory and a total of 11.5 to 12 million bales across the South. In October 1897, a South Carolina newspaper blamed anticipation of an enormous crop for lowering prices, insisting that "the estimate of its quantity now having the greatest influence upon the trade is that of Mr. Henry Neill." Shortly thereafter, Neill published a forecast in Liverpool of 10.5 million bales, which dropped prices further on the Liverpool and New York exchanges. The *Atlanta Constitution* cited skeptics within the cotton industry and issued more conservative forecasts, despite professing that it "prefer[red] to comment on facts rather than make predictions." But the facts of the 1897–1898 season confirmed Neill's predictions: 11.2 million bales, the largest yield ever, easily eclipsed the previous high of 9.9 million bales in 1894–1895. New York prices dipped to 5.75 cents a pound and averaged 6.22 cents, a full 1.5 cents lower than the 1896–1897 average.[89]

THE "ALLEGED STATISTICIAN" OF LIVERPOOL

The final two cotton seasons of the nineteenth century marked the dramatic conclusion of Henry Neill's reign, the end of what the *Atlanta Constitution* had lamented as "the domination of the English autocrat of New Orleans over the cotton market."[90] In 1898, growers planted just under 25 million acres, the most ever and nearly 650,000 acres more than the previous year. In August, Neill's circular "assured" a crop of at least 10.5 million bales with the possibil-

88. "Mr. Neill's Crop Estimate," *Atlanta Constitution*, April 1, 1897. The cotton cultivation cycle remained the same throughout the nineteenth century: planters broke the land in the late winter and planted cotton rows, then enslaved people or farm laborers thinned and weeded the crop once it had sprouted, until July, when field labor stopped. Cotton bolls yielded their lint in September and October, when cotton picking began. After the seeds were separated from the lint by the cotton gin, the lint was baled for market, and the seeds were kept for future planting and sold to cottonseed oil companies. Before the 1880s, leftover cottonseeds were disposed of. Daniel, *Breaking the Land*, 6.

89. "Enormous Cotton Crop of 1898," *Chicago Daily Tribune*, September 10, 1898; "Cotton Should Go Higher," *Columbia (SC) State*, October 22, 1897; "The Price of Cotton," *Atlanta Constitution*, November 12, 1897; Watkins, *King Cotton*, 31.

90. "The Feeling about Neill Is Mixed," *Atlanta Constitution*, November 14, 1899.

ity of between 1 and 1.5 million more.[91] Neill's November estimates indicated between 11,750,000 and 12,000,000 bales, then climbed to 12,150,000 (assuming a mild winter), and peaked at 12,250,000. As Neill's figures rose, prices plummeted, sinking to near-record lows in December and forcing some growers to part with the bulk of their crop for less than five cents a pound.[92]

British buyers could name their price in 1898, until a harsh winter set in, "probably the worst the south has ever known," according to the *Atlanta Constitution*. The South witnessed record temperature lows (zero degrees in Alabama) during a series of historic cold waves in February 1899, and although the US Weather Bureau issued cold-wave warnings that saved hundreds of thousands of dollars in crops and livestock, damage ran into the millions. The cotton crop sustained heavy losses.[93]

But, according to some observers, the winter weather could not mask the underlying explanation for the fall-off in cotton receipts across the South: the forecasts of Henry Neill. In March 1899, the *New York Sun* proclaimed that

> the truth can no longer be concealed that the diminished movement of cotton is due to exhaustion, and that the cotton crop of 1898–99 has been grossly overestimated; that the prophecy of Mr. Neill of a 12,000,000 bale crop, which was the ground for this expectation, was simply a reckless guess; and that the fright caused by this great exaggeration led the American cotton planter to sell his cotton for millions of dollars less than it was worth.[94]

The *Sun*'s complaint was confirmed by railroad and shipping receipts: the crop for 1898–1899, though not Neill's 12 million bales, was still a record 11.2 million. It surpassed the previous year's yield by just under 75,000 bales, but its market value was over $14 million less. Growers in 1897–1898 averaged 5.71 cents

91. James L. Watkins, *The Cotton Crop of 1898–99* (Washington, DC: Government Printing Office, 1900), 25, 32; quotation in "Cotton Crop of 1898–'99," *Columbus (GA) Enquirer-Sun*, March 22, 1899.

92. "Mr. Neill and the Cotton Crop," *Atlanta Constitution*, March 11, 1899; "Mr. Henry Neill's Downfall," *Birmingham Age Herald*, March 12, 1899.

93. "Mr. Henry Neill's Downfall," *Birmingham Age Herald*, March 12, 1899; "Mr. Neill and the Cotton Crop," *Atlanta Constitution*, March 11, 1899; US Weather Bureau, *Report of the Chief of the Weather Bureau, 1898–99* (Washington, DC: Government Printing Office, 1900), 3–7.

94. "Mr. Neill and the Cotton Crop," *Atlanta Constitution*, March 11, 1899. Newspaper estimates of how much Neill's bear-market prophecies cost Southern growers who rushed their crop to market at depressed prices in fall 1898 ranged from millions of dollars to fifty million to hundreds of millions. "Mr. Henry Neill's Downfall," *Birmingham Age Herald*, March 12, 1899.

per pound for their 11-million-bale cotton, but only 5.27 cents the following year.[95]

Producers were not Neill's only victims, the *Atlanta Constitution* noted, citing fertilizer companies, sellers of livestock, farm implement manufacturers, and Liverpool speculators who lost heavily on New York futures contracts for spring delivery. Neill's soaring estimates of 12 million bales and beyond for the 1898–1899 season cost him some credibility. "The cotton pope has been shoved off his throne," an Alabama newspaper announced.[96]

But Neill had not been dethroned entirely. Despite proclamations that "one failure exposes the prophet" and that "Neillism had been disproved" in the 1898–1899 season, Neill's forecasts continued to command attention.[97] The 1899–1900 season marked the first time that the United States eclipsed Britain in the consumption of raw cotton (a result of surging domestic textile demand for department store and mail-order sales), but, as a Division of Statistics official later recalled, "the most extraordinary feature of the year was the wide range of prices from the beginning to the close of the season." Opening spot prices in September 1899 hovered around 6 cents per pound and around 9.5 cents per pound by the end of August 1900.[98] This variability was due in no small part to the dueling crop estimates of the USDA's Division of Statistics and Henry Neill. As the New Orleans Cotton Exchange's Henry Hester observed in September 1900, the wide range of cotton estimates that year—from 8.5 million to over 11 million bales—was also extraordinary.[99]

To open the season, the Division of Statistics issued acreage and condition figures that market watchers converted into a prospective yield of 9 to 9.5 million bales. Neill disagreed, predicting in an August 1899 circular a record 12 million bales, citing optimal temperature, rainfall, and growth that would "make a great crop certain."[100] A North Carolina newspaper reported "little confidence" in Neill's forecast, and the president of the New York Cotton Exchange protested, "I don't see how Mr. Neill or any other man can pretend to say this early in the year what the crop will be. . . . The possibilities of the crop, when

95. Watkins, *King Cotton*, 31; James L. Watkins, *The Cotton Crop of 1899–1900* (Washington, DC: Government Printing Office, 1901), 41; Watkins, *Cotton Crop of 1898–99*, 25.

96. "Mr. Neill and the Cotton Crop," *Atlanta Constitution*, March 11, 1899; "Mr. Henry Neill's Downfall," *Birmingham Age Herald*, March 12, 1899.

97. "Mr. Neill and the Cotton Crop," *Atlanta Constitution*, March 11, 1899; "Mr. Henry Neill's Downfall," *Birmingham Age Herald*, March 12, 1899.

98. Watkins, *Cotton Crop of 1899–1900*, 5.

99. "Cotton Report of Mr. H. G. Hester," *Crop Reporter* 2, no. 5 (September 1900): 6.

100. "Largest Cotton Crop Known," *NYT*, August 15, 1899.

all conditions are favorable, are almost infinite, but account must be taken of possible droughts, scorching temperature and other things likely to hurt the crop."[101] And according to some, unpredictable weather was far less injurious than Neill's forecasts. As an Alabama newspaper fumed in September 1899, "Henry Neill has damaged the cotton farmers more than all the boll worms, rust, floods and drouths. By the time Neill dies with old age the cotton men will learn that he was always influenced by private gain. Neill is a professional bear."[102] Although the American press considered Neill exclusively a bear, he occasionally published bullish figures.[103] But Neill professed statistical objectivity in an 1895 interview, declaring, "I do not issue estimates as a bull or a bear, but as an estimator anxious above all to maintain my reputation for accuracy in my forecasts."[104]

On September 1, 1899, the Division of Statistics reported that the crop condition was the second-worst ever on that date (68.5 percent, the lowest but for the 64.2 percent of September 1896). Neill insisted, two days later, that his 12-million-bale prediction was correct, arguing that the previous year's weather and yield were indeed comparable to the present year's. Neill defiantly blamed New York firms for issuing conservative reports that compelled southern growers to hold their cotton off the market. According to Neill, the weather remained ideal, picking extensive, and the cotton grade high.[105]

The Division of Statistics' condition report worsened in October 1899, but Neill held firm. He issued a circular that again compared the current season to the previous year, when a drought did not preclude a large crop.[106] Division of Statistics cotton expert and former planter James L. Watkins would later report that Neill's insistence of "at least 11,000,000 bales certain" as a "minimum estimate" was intended "as if to break the force of such [government] announcement."[107] Indeed, that year the *New-York Tribune* lamented "divided"

101. "Small Faith in Neill," *Raleigh News and Observer*, August 24, 1899; "Neill Overshoots Himself," *Atlanta Constitution*, August 18, 1899.

102. "Henry Neill Is a Bear," *Birmingham Age Herald*, September 5, 1899.

103. "Cotton Market Higher," *Galveston Daily News*, November 24, 1891; "Commercial and Markets," *Liverpool Mercury*, April 25, 1895.

104. "Heavy Fall in Cotton," *New-York Tribune*, October 20, 1895.

105. Watkins, *King Cotton*, 289–93; "Bears Attempt to Break Spot," *Atlanta Constitution*, October 17, 1899.

106. Watkins, *King Cotton*, 293; "This Year's Cotton Crop," *NYT*, October 11, 1899.

107. Watkins, *Cotton Crop of 1899–1900*, 29; Watkins, *King Cotton*, 3. For the same complaint about the timing of Neill's circular, see "Another Neill Estimate," *Dallas Morning News*, October 12, 1899.

cotton markets as a result of competing commercial and government statistics. Of course, Neill was not the sole influence—declining Liverpool prices and the impending Boer War also depressed the market—but the *Dallas Morning News* identified the "real" cause of declining prices as "the promulgation of a circular by Neill, in which he says a crop of 11,000,000 bales is assured."[108] The *Houston Post* counted this forecast among Neill's "outrageous and inexcusable prophecies," citing a consensus of 9 to 10 million bales at most.[109] New York cotton firms shared these frustrations, with Price, McCormick pointing to "the evident unreasonableness" of Neill's estimate, and Atwood, Violett castigating Neill as the worst "conspirator against the South."[110]

What the *Atlanta Constitution* had previously denounced as Neill's "fabric of guess work" unraveled in late October 1899 when the USDA put Neill's numbers under scrutiny.[111] Division of Statistics head John Hyde evaluated Neill's use of US Weather Bureau rainfall data as evidence that a drought in August and September 1898 had not had an adverse effect on the 1897–1898 crop that was nearly 2.5 million bales larger than the previous year's and 1.3 million bales larger than the previous high of 1894–1895. The 1899–1900 crop, so Neill's logic went, would be even larger.[112] But using historical weather data to support his claims about future cotton yield proved to be Neill's undoing.

Upon comparing the Weather Bureau's figures with Neill's, Hyde concluded that Neill had manipulated the bureau's tabulations to suit his forecast. Neill had selectively assembled rainfall statistics for the Atlantic and Gulf states from July, August, and September 1899 to convince nervous buyers that the present year's rainfall was in fact significantly greater than in 1897 and that drought was not a present danger. The opposite was true, as Hyde noted: all but three cotton states had experienced *less* rain in August and September 1899 than they had two years earlier.[113]

108. "The Cotton Prospect," *New-York Tribune*, August 16, 1899; "Another Neill Estimate," *Dallas Morning News*, October 12, 1899.

109. "Neill and the Cotton Crop," *Raleigh News and Observer*, October 15, 1899.

110. "Cotton Growers' Convention—Neill Answered," *Raleigh News and Observer*, October 14, 1899; "Bears Attempt to Break Spot," *Atlanta Constitution*, October 17, 1899.

111. "Mr. Neill and the Cotton Crop," *Atlanta Constitution*, March 11, 1899.

112. "This Year's Cotton Production," *NYT*, October 25, 1899. For Neill's rainfall figures, see "He Still Sticks to It," *Charleston Weekly News and Courier*, October 14, 1899.

113. "How Neill Makes Estimates," *Atlanta Constitution*, November 2, 1899; "Mr. H. M. Neill's Cotton Views Criticized," *New Orleans Daily Picayune*, October 26, 1899.

Hyde tried to re-create Neill's calculations but failed since Neill had not specified which Weather Bureau stations had reported the rainfall statistics he cited or which states he had assigned to which region. Hyde did determine, however, that Neill had included rainfall from the last week of July in his tabulations for August and September, a bit of arithmetical sleight-of-hand that boosted the 1899 totals for the Atlantic and Gulf Coast states higher than those of 1897. A storm of anti-Neill sentiment subsequently burst forth in the pages of the *Atlanta Constitution*, which crowed, "Mr. Neill must have invented his rainfall to suit his own purposes. His figures were grossly incorrect, and his rainfalls were mythical." But Neill defended his comparison to the 1897–1898 crop as "perfectly logical and proper," insisting that late July and October rainfall of the current season had minimized the crop damage that was, overall, "less serious than in 1897."[114]

Neill's rebuttal did not dissuade the producerist *Atlanta Constitution* from denouncing him as a "commercial Judas," "an enemy of the public welfare," a "prophet [who] is an agent of evil," and "a demon of destruction." Amid the barrage of the *Constitution*'s epithets was an attack on Neill for wreaking economic and epistemic havoc on the cotton market. The *Constitution* condemned him for making fraudulent predictions, for "cloth[ing] Falsehood in the garb of Fact," pronouncing him "the alleged statistician" who relied not upon "ascertained facts" but on "empty probabilities" that compelled growers to sell their crop below true market value.[115]

Undeterred by allegations of statistical impropriety, Neill stood firm. On November 10, the Division of Statistics released its first end-of-season yield estimate of a "maximum of 9,500,000 bales," and four days later, Neill Bros. issued a statement that Neill had "not reduced his former minimum estimate of 11,000,000 by a single bale."[116] At the end of November, Neill explained away lower-than-expected receipts by blaming growers for holding cotton off the market for higher prices.[117] And planters, like those in Louisiana's Red River parish, were "up in arms" over Neill's 11-million-bale estimate, pronouncing

114. "This Year's Cotton Production," *NYT*, October 25, 1899; "Mr. Neill's Fakes," *Atlanta Constitution*, November 1, 1899; "Cotton Statistician Neill Replies to Mr. Hyde," *New Orleans Daily Picayune*, November 2, 1899.

115. "The Fall of a Prophet of Evil," *Atlanta Constitution*, November 7, 1899; "How Neill Makes Estimates," *Atlanta Constitution*, November 2, 1899.

116. Watkins, *Cotton Crop of 1899–1900*, 29.

117. "Mr. Neill on Cotton," *Wall Street Journal*, November 29, 1899.

him "a public enemy of our great cotton-growing industry" and organizing to defeat him.[118]

In December 1899, the Division of Statistics indicated that the year's yield would not exceed 8,900,000 bales, but Neill held at 11,000,000. By January 1900, after most of the crop had gone to market and Neill's figure clearly appeared to be off the mark, Neill claimed that planters were still withholding cotton and adjusted his figure to a minimum of 10 million bales. Neill's insistence, from August 1899 through the turn of the century, on an improbably high yield disadvantaged southern growers, who sold most of their cotton in the first half of the September-to-September commercial year. USDA statistician James Watkins lamented that government statistics were no match for Neill's when it came to affecting the market and thereby cost growers "two to three cents per pound."[119]

Neill's false cotton prophecies in 1899 cost some growers their profits and cost him his reputation as "the great cotton crop authority of New Orleans."[120] Neill's 11-million-bale estimate was more than 1.5 million bales over the 1899–1900 total of 9,436,416 bales. Neill's analogy between the current crop and the previous year's had been disproven twice, first when the chief statistician Hyde uncovered the falsified rainfall totals and then when the Division of Statistics recorded the current year's crop as 2,046,367 bales less than the 1898–1899 crop. After Neill's exposure, prices rose dramatically: the value of the 1899–1900 crop was nearly thirty million dollars more than the previous year's 11.3-million-bale crop, with New York prices averaging 8.69 cents per pound in 1899–1900, as compared to 6 cents the previous year.[121]

Although Neill continued to publish forecasts in the early twentieth century, a skeptical public put less stock in his predictions. But when Neill fell silent the year after his exposure and had issued no cotton figures by September 1900, some still sought a prophecy from the "false prophet." As a Fort Worth newspaper remarked, "An estimate from him would be received with some interest. Not that it would command respect, but people would just like to know what he has to say."[122]

118. *Arkansas Democrat*, November 20, 1899, 2.

119. Watkins, *Cotton Crop of 1899–1900*, 29.

120. "Rome's Cotton Receipts," *Atlanta Constitution*, April 6, 1897.

121. Watkins, *King Cotton*, 31; Watkins, *Cotton Crop of 1899–1900*, 7, 30.

122. "The Fall of a Prophet of Evil," *Atlanta Constitution*, November 7, 1899; *Fort Worth Morning Register*, September 4, 1900, 4.

Yet Henry Neill was not the only one whose cotton statistics influenced commodity exchanges. The USDA's crop estimates, as statistician George K. Holmes explained to an audience of Delaware Grangers in September 1900, occasionally drove prices higher and lower. Yet Holmes contrasted the government's "more trustworthy" figures with those of private forecasters, announcing that the Division of Statistics was "a steadier of prices" and its reports "a corrective" to a market easily swayed by inaccurate commercial forecasts.[123]

Holmes's words soon came back to haunt him when the USDA issued corrective cotton figures that unsteadied the market. The Division of Statistics' December 1900 estimate of 10.1 million bales, 500,000 bales higher than expected, threw the New York Cotton Exchange into a frenzied sell-off in which prices fluctuated so quickly they defied recording. Perhaps chastened by his exposure as "the great fakir of cotton estimates," Henry Neill came in at 9.6 million bales, lower than the rest of the commercial estimates, which averaged 9.75 million bales. When the Division of Statistics released its figures at noon on December 10, brokers raced to unload their options for January delivery, and prices plummeted. In the ensuing chaos, cotton lost two dollars a bale, and across the trading floor "half a dozen prices were quoted for the same month at the same instant." As the *Dallas Morning News* headline blared, the "Cotton Pit Was Wild."[124]

STATISTICAL RESISTANCE

Henry Neill's market manipulations in the 1899–1900 cotton season spurred southern growers to renewed collective action. Georgia county chapters and the Alabama state chapter of the Cotton Growers' Protective Association were established in summer 1900, when, as the *Macon Telegraph* announced, a "great movement . . . spread like wildfire over the cotton planting regions."[125] Orchestrating this movement was Harvie Jordan, president of the Georgia

123. "Practical Results of the Work of the Department of Agriculture: Address Delivered at the Annual Convention of the State Grange of the Patrons of Husbandry at Brandywine Springs, Del., September 6, 1900, by George K. Holmes, Of the Division of Statistics," Crop Report and Historical Materials, 1914–1918, box 1, Records of the National Agricultural Statistics Service (RG 355), National Archives College Park, College Park, MD.

124. "Mr. Neill's Fakes," *Atlanta Constitution*, November 1, 1899; "Cotton Pit Was Wild," *Dallas Morning News*, December 11, 1900.

125. "Bibb Is to Be Organized First," *Macon (GA) Telegraph*, June 3, 1900; "Alabama Growers Organize," *Atlanta Constitution*, July 19, 1900.

Cotton Growers' Protective Association and the architect of an agenda of statistical resistance against Henry Neill, king of the cotton bears.

Jordan planned to establish county suborganizations that would function as local versions of the USDA's national crop reporting network and quantify physical bales of cotton to counter Neill's "phantom" bales. His primary goal was "to obtain correct, statistical information in regard to the cotton crop," which would be collected weekly from cotton ginners and secretaries of the suborganizations in order to provide accurate figures of weekly sales as well as current yield forecasts. Jordan declared that "this information will cover facts, and will check the annual false estimates sent out by such men as Neil [*sic*], who last season, caused the farmers of the South to lose over a hundred million dollars in the sale of their cotton at low prices."[126] Jordan envisioned for growers a greater hand in determining the price of their cotton and thus greater economic independence from American cotton exchanges and British buyers and speculators. The movement, Jordan announced, was "the grandest . . . the South has ever undertaken," but he did not believe that statistics alone would be enough to "break the shackles . . . to Wall Street and Liverpool gamblers."[127] He also secured the support of the Georgia Bankers Association in creating a local subtreasury to provide credit at 75 percent of cotton's market value, allowing growers to store their crop off the market in warehouses and wait for prices to rise.[128] The Alabama Cotton Growers' Protective Association organized an identical statistics and subtreasury initiative in July 1900, and letters poured in from across the South in support of Jordan's vision. Jordan traveled widely, urging growers to establish local branches of the association and market their cotton gradually to achieve the elusive ten cents per pound. By late August 1900, the growers' movement had spread throughout Georgia, with a majority of the state's counties reporting their own branches.[129] (See figure 1.3.)

126. "County Sub-Organizations," broadside, 1900, Dudley M. Hughes Papers, Hargrett Library Broadside Collection, 1900–1919, University of Georgia Libraries, http://www.libs.uga.edu/hargrett/rarebook/broadside/.

127. "Georgia Cotton Growers!" broadside, 190[?], Dudley M. Hughes Papers, Hargrett Library Broadside Collection, 1900–1919, University of Georgia Libraries, http://www.libs.uga.edu/hargrett/rarebook/broadside/.

128. "Georgia Bankers Will Aid Georgia Cotton Growers," broadside, 1900, Dudley M. Hughes Papers, Hargrett Library Broadside Collection, 1900–1919, University of Georgia Libraries, http://www.libs.uga.edu/hargrett/rarebook/broadside/; "Cotton Growers Are Organizing," *Atlanta Constitution*, August 26, 1900.

129. "Alabama Growers Organize," *Atlanta Constitution*, July 19, 1900; "Organization Everywhere," *Macon (GA) Telegraph*, July 8, 1900; "Upson County Cotton Men," *Atlanta Constitution*,

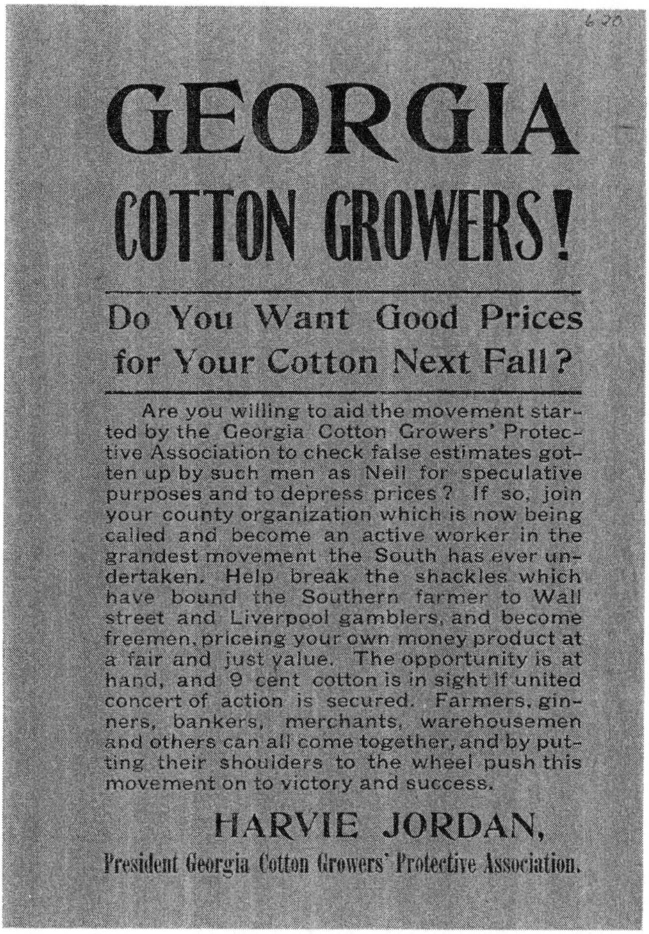

GEORGIA

COTTON GROWERS!

Do You Want Good Prices for Your Cotton Next Fall?

Are you willing to aid the movement started by the Georgia Cotton Growers' Protective Association to check false estimates gotten up by such men as Neil for speculative purposes and to depress prices? If so, join your county organization which is now being called and become an active worker in the grandest movement the South has ever undertaken. Help break the shackles which have bound the Southern farmer to Wall street and Liverpool gamblers, and become freemen, priceing your own money product at a fair and just value. The opportunity is at hand, and 9 cent cotton is in sight if united concert of action is secured. Farmers, ginners, bankers, merchants, warehousemen and others can all come together, and by putting their shoulders to the wheel push this movement on to victory and success.

HARVIE JORDAN,

President Georgia Cotton Growers' Protective Association.

FIGURE 1.3. In 1900, the Georgia Cotton Growers' Protective Association printed broadsides calling cotton growers to join a collective effort of statistical resistance against the bear-market forecasts of Henry Neill. "Georgia Cotton Growers!" broadside, 190[?], Dudley M. Hughes Papers, Hargrett Library Broadside Collection, 1900–1919, University of Georgia Libraries, http://www.libs.uga.edu/hargrett/rarebook/broadside/. Courtesy of Hargrett Rare Book and Manuscript Library/University of Georgia Libraries.

By October 1900, with the Cotton Growers' Protective Association strong in Georgia, Alabama, and South Carolina, Jordan urged the creation of a centralized bureaucracy to coordinate the collection and distribution of thousands

August 23, 1900; "Why Republicans Should Have It," *Macon (GA) Telegraph*, August 9, 1900; "Cotton Growers Are Organizing," *Atlanta Constitution*, August 26, 1900.

of crop reports. In February 1901, the Interstate Southern Cotton Growers' Protective Association was established by representatives from Georgia, North Carolina, South Carolina, and Mississippi, who elected Jordan president and charged him with the task of organizing state conventions throughout the South in order to "gather and distribute all information possible, and especially statistical information" on the cotton crop. The *Atlanta Constitution* had high hopes for the burgeoning association, anticipating that it would "develop into the greatest bureau of cotton statistics the world has ever known."[130]

But the *Constitution*'s prediction was not accurate. The Cotton Growers' Protective Association, never a powerful agricultural statistics bureau, was supplanted by the Southern Cotton Association, itself in existence only from 1905 to 1908.[131] But the newspaper record of Harvie Jordan's activities reveals that the Cotton Growers' Protective Association indeed began to build a local reporting network that functioned for at least a year or two. More importantly, the association's avowed statistical resistance to an unpredictable market—and especially Henry Neill's predictably high forecasts—set off a chain of events that would lead the USDA to redefine crop reporting as a process of predictive estimation that acknowledged the uncertainties of individual judgment.

"A LITTLE COTTON DEAL"

An uncommonly volatile cotton market in the first decade of the twentieth century revealed new controversies over crop forecasting, which emerged from a resurgence of cotton growers' action, a "cotton leak" scandal within the Bureau of Statistics, and investigations of the federal government's crop estimating work. Members of Congress, USDA officials, cotton growers' and manufacturers' associations, and the press all sought to measure the value of crop reports in a cotton market that was becoming increasingly difficult to predict. When the *New York Times* wondered in a July 1905 headline, "Are crop reports worthless?" (and concluded otherwise), it acknowledged the uncertainties of crop reporting, pronouncing it "folly to expect anything like uniformity or

130. "They Met and Had Talk on Cotton," *Atlanta Constitution*, October 6, 1900; "Organization of Cotton Growers," *Columbus (GA) Enquirer-Sun*, February 15, 1901; "Cotton Men to Protect Crops," *Atlanta Constitution*, February 15, 1901.

131. Theodore Saloutos, "The Southern Cotton Association, 1905–1908," *Journal of Southern History* 13, no. 4 (1947): 492.

accuracy from thousands of scattered observers describing things as they see them over millions of acres."[132]

Yet reports often proved worthless in a cotton market characterized by volatility, large yields, and increasing acreage. Prices fluctuated wildly between 1902 and 1905, especially in the 1903–1904 and 1904–1905 seasons. In October 1903, middling spot cotton brought 9 1/2 cents a pound in New York and 17 1/4 cents four months later, and New Orleans prices jumped from 9 1/8 to 16 7/16 cents during the same period. Yields surpassed 10 million bales each season from 1900 to 1904, driving average New York prices above ten cents per pound in 1902–1903 and 1903–1904, as well as from 1905 through 1908, with the range of prices in those seasons much wider than it had been in the previous decade. As the southern cotton economy rebounded, with prices reaching seventeen cents a pound in early 1904, growers cultivated ever more acreage, and thirty million acres planted in 1904 yielded a record 13.6 million bales, with prices dropping to six to seven cents a pound by the end of the year.[133]

With cotton prices fluctuating and Henry Neill discredited, a new cotton statistics expert, Katherine Giles, rose to prominence in the early years of the twentieth century. Newspaper accounts vary regarding how Giles got her start—as a former assistant to a New York cotton broker, a clerk for a grain statistician, or a USDA statistician herself—but she burst on the scene in 1902 and by 1907 was deemed "the recognized cotton authority of New York."[134] Credited with driving the value of cotton up five million dollars when her figures diverged from the government's in 1902 and then driving futures down sixty points in June 1904 with low condition estimates, Giles was remembered upon her retirement in 1917 as the "woman whose opinion ruled [the] market."[135] Giles's explanation of her success, which reportedly earned her ten thousand dollars a year, rested on her extensive network of over one thousand trusted planters who submitted monthly reports to her, and, like Henry Hester and

132. "Are Crop Reports Worthless?" *NYT*, July 10, 1905.

133. Watkins, *King Cotton*, 26–28, 31, 288.

134. Quotation in "Women Who Counted in Wall Street When Bottom Seemed to Be Dropping," *Atlanta Constitution*, November 24, 1907. For varying newspaper accounts of Giles's early career, see "Miss Giles is Cotton Queen," *Boston Daily Globe*, November 1, 1903; "Views of Woman Who Sways Cotton Market," *NYT*, July 25, 1904; "Miss Katherine Giles, Cotton Queen," *Literary Digest*, May 16, 1908, 737–38; "Wall Street Loses Giles, Cotton Expert," *Chicago Daily Tribune*, June 24, 1917.

135. "Woman's Report Hit Cotton Market Hard," *NYT*, June 2, 1904; quotation in "Wall Street Loses Giles, Cotton Expert," *Chicago Daily Tribune*, June 24, 1917.

Henry Neill before her, she steadfastly refused to engage in speculation.[136] Recognized by *Ladies' Home Journal* as one of eighteen women "holding men's jobs," Giles gained entry into the traditionally male professions of agricultural statistics and commodity futures trading by concealing her gender in market reports she signed simply, "K. M. Giles."[137] Newspapers celebrated her status as the country's lone female cotton statistician while at the same time emphasizing her "exceedingly feminine" physical appearance ("young and extremely good looking"), singing voice ("clear soprano tones"), and weekly performances in her church choir, thereby domesticating her presence in the masculine world of finance capitalism.[138] Katherine Giles, "cotton queen," was defined by turn-of-the-century gender norms but also transcended them.[139]

As Neill's statistical authority yielded to Giles's, Harvie Jordan resumed his impassioned arguments for acreage reduction in January 1905 when he spoke at a growers' meeting in New Orleans that led to the founding of the Southern Cotton Association. Jordan marched through a familiar agenda that included diversification and gradual marketing, but his clarion call was for a 25 to 40 percent acreage reduction in the 1905–1906 season. Anyone who did not comply, Jordan thundered, was "an enemy to himself and a traitor to his country." Other speakers echoed Jordan's argument, if not his fiery rhetoric, and similar endorsements came from New York and New Orleans Cotton Exchange officials as well as the Southern Bankers' Association. Although Jordan didn't get his 25 percent reduction, growers did plant nearly 14 percent less cotton in the 1905–1906 season (26.1 million acres) than in the previous year (30.1 million acres), bringing average prices up over two cents per pound in 1905.[140]

In 1905, the Southern Cotton Growers' Association had a hand in not only reducing acreage and raising prices but also exposing a sensational story of

136. "Wall Street Loses Giles, Cotton Expert," *Chicago Daily Tribune*, June 24, 1917; "Views of Woman Who Sways Cotton Market," *NYT*, July 25, 1904.

137. "Where They Are Holding Men's Jobs," *Ladies' Home Journal*, April 1916, 33; "Women Who Counted in Wall Street When Bottom Seemed to Be Dropping," *Atlanta Constitution*, November 24, 1907.

138. "Wall Street Loses Giles, Cotton Expert," *Chicago Daily Tribune*, June 24, 1917; "Women Who Counted in Wall Street When Bottom Seemed to Be Dropping," *Atlanta Constitution*, November 24, 1907.

139. "Cotton Queen Retires," *WP*, June 24, 1917.

140. *Report of the Commissioner of Corporations on Cotton Exchanges*, Part 5: *Influence of Producers' Organizations on Prices of Cotton* (Washington, DC: Government Printing Office, 1909), 325, 329–34, quotation on 330; Watkins, *King Cotton*, 288, 31.

cotton-report corruption. At the turn of the century, cotton traders alleged that speculators saw crop estimates before their release, and worse, that Census or USDA officials in cahoots with speculators were falsifying statistics.[141] The association discovered that its condition reports were significantly lower than the Bureau of Statistics' monthly estimates, by 7 percent in May 1905. The association then learned from New York cotton broker Louis C. Van Riper the details of a "cotton leak" within the Bureau of Statistics, and the association secretary informed Secretary of Agriculture James Wilson that he had uncovered a conspiracy.[142]

At the center of the intrigue stood USDA assistant statistician Edwin S. Holmes, who, over several months in late 1904, leaked reports a day early to Van Riper and a partner, who then turned a profit, often by selling short in anticipation of bearish figures. Holmes also apparently falsified June 1905 estimates to bear the market. Holmes, the two brokers, and a go-between agreed to split the profits, which, it was revealed during Holmes's trial, totaled at least two hundred thousand dollars (and some speculated that Holmes alone pocketed five hundred thousand dollars). Holmes's five-week trial laid bare the workings of the conspiracy, including Holmes's secret signal to communicate with his confederates on crop reporting day: from the Department of Agriculture building, he would adjust a window shade upward if something had prevented him from manipulating the figures and downward if the scheme would be even more profitable than expected.[143] Van Riper, the prosecution's

141. H. Parker Willis, "Cotton and Crop Reporting," *Journal of Political Economy* 13, no. 4 (1905): 522. For allegations of leaked USDA cotton statistics in 1902, see "Scandal in Cotton Market," *Chicago Daily Tribune*, December 5, 1902; "Evidence That It 'Leaked' Out," *Atlanta Constitution*, December 5, 1902; "Cotton Leak Denied," *WP*, December 6, 1902; "Did the Bureau Leak?" *Los Angeles Times*, December 6, 1902; "Secretary Wilson Discredits Charges," *New-York Tribune*, December 6, 1902. On other accusations of crop reporting corruption, see Snyder, "Federal Crop Forecasts and the Cotton Market," 44.

142. Willard L. Hoing, "James Wilson as Secretary of Agriculture, 1897–1913" (PhD diss., University of Wisconsin, 1964), 168; "Peckham Arrested," *WP*, August 24, 1905. For accounts of the "cotton leak," see Hoing, chap. 7; Snyder, "Federal Crop Forecasts and the Cotton Market," 44–47; Pietruska, "Cotton Guessers," 62–63; Baker and Hahn, *Cotton Kings*, 82–87.

143. "Peckham Arrested," *WP*, August 24, 1905; "Holmes Misses Court; Denies Cotton Charge," *NYT*, August 27, 1905; "Indicted," *Boston Daily Globe*, August 26, 1905; "Holmes, Cotton Report Juggler, Is Dismissed," *NYT*, July 9, 1905; "Cotton Leak Profits," *WP*, October 14, 1905; "Holmes Leaked and Grew Rich," *Atlanta Constitution*, May 12, 1907; "Stream of Gold Came from 'Leak,'" *Chicago Daily Tribune*, June 18, 1907; "Holmes' Cotton Code," *WP*, June 21, 1907.

star witness, testified that he didn't see anything wrong with the window shade signal because it was "legitimate to get information any way you can."[144]

What Van Riper's partner called "a little cotton deal" turned out to be a big deal during an era when progressive reformers and muckraking journalists regularly targeted corruption in business and government. Details of the conspiracy and the Holmes trial (which one writer likened to a Sherlock Holmes story) were splashed throughout daily newspapers, and President Theodore Roosevelt urged vigorous prosecution of Holmes, whom he pronounced "a far greater scoundrel than if he had stolen money from the government, as he used the government to deceive outsiders and to make money for himself and others."[145] And Roosevelt put the federal crop reporting system under scrutiny when he approved the creation of the Committee on Department Methods, known as the Keep Commission after its head, Assistant Treasury Secretary James H. Keep, which investigated every department, bureau, and division of the federal government from 1905 to 1909 in order to modernize government business practices and, according to Roosevelt, "put the country's housekeeping in order."[146] Although the resulting nineteen published volumes of findings on the day-to-day administration of government did not yield sweeping change in federal bureaucracy because the Keep Commission had no legislative powers, the commission played a key role in the USDA's redefinition of its crop reporting work.[147]

As the Keep Commission began its inquiry, Secretary of Agriculture Wilson mandated reforms to prevent the falsifying or leaking of crop reports. New protections included removing county names from tabulation sheets, locking up county and township figures from Texas and Georgia along with reports on all speculative crops, and disconnecting Bureau of Statistics telephones on the

144. "Name of Price Was Suppressed," *Atlanta Constitution*, June 20, 1907.

145. "Roosevelt Calls Man Scoundrel," *Chicago Daily Tribune*, July 16, 1905.

146. Harold T. Pinkett, "The Keep Commission, 1905–1909: A Rooseveltian Effort for Administrative Reform," *Journal of American History* 52, no. 2 (1965): 300–301, quotation on 303, from *Washington Star*, March 23, 1906.

147. The Keep Commission's reports, which circulated widely in government and business, modernized federal record-keeping practices, particularly in the shift toward vertical filing, card records, and more efficient correspondence filing. On the legacy of the Keep Commission, see Harold T. Pinkett, "Investigations of Federal Record-keeping, 1887–1906," *American Archivist* 21 (1958): 163–92; Oscar Kraines, "The President Versus Congress: The Keep Commission, 1905–1909," *Western Political Quarterly* 23, no. 1 (1970): 5–54.

morning of the crop report's release. But more significant was the formation of a crop-reporting board, a rotating committee of four from the Bureau of Statistics that would assemble in Washington under the watchful eye of the secretary or assistant secretary of agriculture, who would look on as each member calculated his own crop estimates individually before the committee agreed upon the final figures. No longer would a statistician be able to single-handedly manipulate crop reports before their release.[148] In 1909, an act of Congress would declare it unlawful for any government employee or associate to leak privileged crop information or to use such information to speculate on agricultural commodity markets. Violators faced a maximum fine of ten thousand dollars or a prison term of up to ten years, and any government officer or employee who made a false report would face a fine of five thousand dollars or a prison term of up to five years.[149]

After the introduction of the crop-reporting board, the American Cotton Manufacturers' Association, acting on behalf of growers who desired lower acreage estimates, called for a recalculation of the Bureau of Statistics' June 1905 acreage figures.[150] At noon on July 26, 1905, the crop-reporting board issued revised statistics that revealed that chief statistician John Hyde, under the influence of his corrupt assistant Holmes, had unwittingly overestimated acreage: the current acreage represented a 14.9 percent decrease from the previous year, not the 11.4 percent decrease initially reported (a revision on the order of 500,000 bales).[151]

The correction made an instant impact on the New York Cotton Exchange, sending futures contracts for October delivery soaring, but the revised figures didn't match other estimates: the *New York Journal of Commerce and Commercial Bulletin* indicated an acreage decrease of 12.3 percent, while growers pointed to an 18 percent decrease. According to economist Henry Parker

148. USDA, *Yearbook of the United States Department of Agriculture, 1905* (Washington, DC: Government Printing Office, 1906), 98–99; USDA, "Orders Governing the Preparation of Monthly Crop Reports of the Bureau of Statistics," July 8, 1905, reprinted in Willis, "Cotton and Crop Reporting," 523–27.

149. USDA, *Crop and Livestock Reporting Service of the United States*, 12–13.

150. For a thorough account of the acreage revision, see Taylor and Taylor, *Story of Agricultural Economics*, 217–20.

151. "Revised Report on Cotton Acreage, July 26, 1906," *Crop Reporter* 7, no. 4 (1905): 34, reprinted in Taylor and Taylor, *Story of Agricultural Economics*, 218; "The Cotton Question," *NYT*, November 26, 1905.

Willis, the correction did not inspire public confidence in the beleaguered Bureau of Statistics but was rather a "discrediting influence."[152]

Willis's assessment proved accurate. In September 1905, the New England Cotton Manufacturers' Association implored Roosevelt to investigate the USDA on the grounds that "the unreliability of Government crop reports and the leaks in the Bureau of Statistics have tended to increase speculation in cotton futures."[153] In late November, a *New York Times* headline announced "a season of confusion" due to "wide differences in government and expert 'mathematical crops'—estimates and facts." The article cited three different yield figures: 9,588,000 bales from the Southern Cotton Association, 10,900,000 bales from well-respected broker Theodore H. Price, and a typically high estimate of 11,500,000 bales from Henry Neill. Such a range was, the *Times* concluded, a fitting end to the season of dramatic and unpredictable events: the "cotton leak" scandal, the revised acreage estimate, and continued discrepancies between Census and Bureau of Statistics crop reports.[154] The publication of contradictory condition figures from the Bureau of Statistics and ginning estimates from the Bureau of the Census throughout October and November 1905 sent prices bouncing wildly up and down the New York Cotton Exchange. As the *New York Evening Post* wryly observed, "no faro-table ever presented such alluring uncertainties as this season's movement of cotton values."[155]

"NOTHING MORE THAN A CONSENSUS OF JUDGMENT"

These cotton market uncertainties drew the attention of lawmakers in December 1905, when the House Committee on Agriculture held hearings on a resolution proposed by Massachusetts congressman (and cotton manufacturer) William C. Lovering, who sought a revision of the USDA's December yield estimate of 10,167,818 bales (since that figure, much lower than the previous year's production, was based on July's revised acreage estimate).[156] Lovering denounced Bureau of Statistics figures on cultivated, abandoned, and net acre-

152. Taylor and Taylor, *Story of Agricultural Economics*, 219; H. Parker Willis, "Cotton and Crop Reporting," 528–29, quotation on 529.

153. Committee on Department Methods, *Report to the President by the Committee on Department Methods, Government Crop Conditions* (Washington, DC: Government Printing Office, 1906), 5.

154. "The Cotton Question," *NYT*, November 26, 1905.

155. "More Fascinating Than Poker," *WP*, November 18, 1905.

156. Taylor and Taylor, *Story of Agricultural Economics*, 219.

age as mere "guesses, or an average of guesses, possessing all the elements of uncertainty and subject to the whims and environments of the agents located in many thousand places." Yet, Alabama congressman Henry D. Clayton argued on behalf of growers that the bureau's cotton guesses, however uncertain, were better than none at all:

> If the Government does not make these estimates, or guesses, as he [Lovering] calls them, and refrains from saying anything at all, and abolishes this Bureau of the Agricultural Department, everybody knows that the great commercial interests of the country will engage in guessing for themselves. That is, the different people, like . . . Neal [*sic*] and others, will go on guessing, and it is better that the Government should do the guessing than that each individual, responsible to nobody, should do it.[157]

Lovering's resolution sought a seemingly straightforward bureaucratic solution to a statistical problem—the recalculation of the USDA's inaccurate December yield figures—but the hearings raised far more complex questions about the limits of estimation in the pursuit of predictive certainty.

Lawmakers were not alone in grappling with the epistemic problems of agricultural commodity markets in 1905. That year, the Chicago Board of Trade's crusade against what it deemed illegitimate bucket-shop gambling culminated in the Supreme Court decision in *Board of Trade v. Christie* that confirmed the legality of organized commodity futures trading but disallowed speculation of the kind practiced by the Christie Grain and Stock Co. of Kansas City, one of the nation's largest bucket shops, where patrons placed small bets on the movement of prices on commodity exchanges. "Speculation of [the former] kind by competent men," wrote Justice Oliver Wendell Holmes Jr. in his ruling in favor of the Board of Trade, "is the self-adjustment of society to the probable." The capitalist system, according to Holmes, was characterized by economic uncertainty: "Of course, in a modern market, contracts are not confined to sales for immediate delivery. People will endeavor to forecast the future, and to make agreements according to their prophecy."[158]

157. House Committee on Agriculture, *Hearing on House Joint Resolution 45, Directing the Secretary of Agriculture to Furnish Certain Information Concerning the Cotton Crop of the Season of 1905 and 1906*, 59th Cong., 1st sess. (December 19, 1905), 2, 10.

158. Board of Trade v. Christie Grain & Stock Co., 198 U.S. 236, 247 (1905). On the *Christie* decision's transformation of futures trading, see Levy, "Contemplating Delivery," 327–35; Lipartito, "The New York Cotton Exchange and the Development of the Futures Market," 67–68.

The Bureau of Statistics was at the same time beginning its own self-adjustment to the probable, a shift first articulated by the Keep Commission in its 1906 findings on government crop reporting. The commission concluded that the bureau's cotton estimates generally did not warrant harsh criticism but that the rest of the bureau's crop estimates needed to be improved or discontinued. Either way, the Keep Commission concluded that uncertainty was an inescapable part of crop reporting and the Bureau of Statistics. Predicting a crop's future was always an imprecise endeavor, the commission noted: "the use of judgment and the making of a more or less uncertain estimate of future yield is the best that can be done." The commission also recommended that when publishing monthly condition figures—a percentage reported to the nearest tenth (e.g., 74.9)—the bureau should abandon the decimal places that implied "mathematical exactness by a mathematical process."[159] Most significantly, the commission urged the bureau to issue cotton condition figures in total prospective yield in bales rather than letting commercial forecasters and speculators convert condition reports into often conflicting forecasts. Despite these recommendations, the commission realized that cotton yield forecasting would remain an uncertain endeavor and urged the Bureau of Statistics to abandon the mantle of "statistics" for two reasons: the Department of Commerce and Labor already had its own Bureau of Statistics, and, more importantly, knowledge about crop conditions was by definition not statistical in nature. Crop condition reports "are not statistics, but estimates," the commission declared.[160]

Bureau of Statistics officials came to agree, and in 1914, the Bureau of Statistics was renamed the Bureau of Crop Estimates to, as Chief Leon M. Estabrook subsequently put it, "indicate to the public more clearly the real nature of its work." The real nature of the bureau's work was not compiling immutable statistical certainties but rather assembling a predictive aggregation of approximations. Indeed, Secretary of Agriculture David F. Houston had initially wanted to rename the bureau the "Bureau of Agricultural Forecasts." The USDA's cotton statistics had always functioned as de facto forecasts in the hands of the commodity exchanges, commercial forecasters, and speculators who converted the government's monthly condition reports from base ten into a probable seasonal yield in millions of bales. However, it was not until 1911

159. Committee on Department Methods, *Report to the President . . . Government Crop Conditions*, 20–24, quotations on 10, 17.

160. Committee on Department Methods, *Report to the President . . . Government Crop Conditions*, 16, quotation on 24.

that the bureau decided to issue its own quantitative yield forecasts for an array of major crops.[161]

But cotton was not among them. When the USDA's crop-reporting board made its first forays into quantitative yield forecasting, it steered clear of cotton. "Try it out on grain before you touch cotton; because cotton is dynamite," Secretary of Agriculture Wilson reportedly warned, and the crop-reporting board started with wheat, oats, corn, tobacco, and rice before issuing its first official cotton forecast in 1914 in the newly established "Agricultural Outlook."[162] The USDA's quantitative yield forecasting marked a decisive epistemic shift toward the rationalizing of uncertainty into the federal government's production of predictive agricultural statistics.

In 1912, Bureau of Statistics chief Victor Olmsted acknowledged the inescapable uncertainty of not just cotton forecasting but crop reporting more generally:

> The reports issued from month to month do not purport to be other than estimates. . . . Every quantitative estimate of the bureau, whether relating to acreage and production of crops or numbers of live stock, is nothing more than a consensus of judgment of many thousands of correspondents and a limited number of agents.[163]

With this statement, the head of the Bureau of Statistics formally signaled the end of the USDA's pursuit of statistical certainty and cleared a space for uncertainty in government crop reporting.

Exactly seven years and one month after Henry M. Neill infamously predicted a cotton crop of an unprecedented 12 million bales—a forecast that led to his condemnation as a false prophet in 1899—he failed to anticipate the danger posed by an oncoming streetcar. Neill, standing too close while trying to hail the car in New Orleans on the morning of September 12, 1906, was struck by the front step and thrown to the ground. Neill succumbed to his

161. Taylor and Taylor, *Story of Agricultural Economics*, 229–30; quotation in Leon M. Estabrook, "Opening Address," in US Bureau of Crop Estimates, *Meeting of the Field Agents, Crop Specialists and Administrative Officials of the Bureau of Crop Estimates at the National Museum, Washington, D.C., January 22–27, 1917*, 1-B.

162. Nat C. Murray, "A Close-Up View of the Development of Agricultural Statistics from 1900 to 1920," *Journal of Farm Economics* 21, no. 4 (1939): 715–16; Snyder, "Federal Crop Forecasts and the Cotton Market," 52; Taylor and Taylor, *Story of Agricultural Economics*, 229n175.

163. USDA, *Yearbook of the United States Department of Agriculture, 1912* (Washington, DC: Government Printing Office, 1913), 91.

injuries later that evening, providing his critics at the *Atlanta Constitution* with this headline: "Well Known Cotton Statistician Meets Death by Accident."[164] Although the streetcar would end Neill's life, his reign as the preeminent "cotton prophet" had already been ended by a run-in with the US Weather Bureau, the federal agency responsible for compiling the rainfall data that Neill had manipulated in 1899 to justify his improbable 11-million-bale forecast. The US Weather Bureau, like its predecessor agency in the US Army Signal Service, relied on a national information network similar to that of the USDA's Bureau of Statistics to collect daily observations of temperature, precipitation, and barometric pressure that became the basis for short-term forecasts and storm warnings. But much like cotton statistics, the value and meaning of the daily "Probabilities," as the first government weather forecasts were called, were often uncertain.

164. "Henry M. Neill Killed by Car," *Atlanta Constitution*, September 13, 1906; "Henry M. Neill Killed," *NYT*, September 13, 1906.

CHAPTER 2

The Daily "Probabilities"

In 1872, the *Boston Evening Transcript* published a poem titled "Old Probabilities," a testimonial to the forecasting acumen of the newly established government weather service and its head, US Army chief signal officer Albert J. Myer, both widely known as "Old Probabilities." The poem began,

Who warns us of the coming storm,
And hints of currents cold or warm,
Which may affect the human form?
Old Probabilities.

Who tells the farmer when to sow,
To plough, to plant, to reap or mow,
That plenty may her gifts bestow?
Old Probabilities.

When men go on the sea in ships,
Who telleth with prophetic lips
The time to start upon their trips?
Old Probabilities.[1]

1. John J. Glover, "'Old Probabilities,'" *Boston Evening Transcript*, August 21, 1872, Signal Corps Newspaper Meteorological Reports, Sept. 1871–Feb. 1873 (hereafter Signal Service clipping book), box 1, Records of the Weather Bureau (RG 27), National Archives at College Park, College Park, MD (hereafter NACP).

But "Old Probabilities" was actually not old at all. Writer Mary Clemmer Ames observed in 1873 that "strange to say, he is a young man . . . [who] looks soldierly and trig," and a newspaper account reminded readers that " 'Old Prob.' alias General Myer" was hardly "a grizzled old wizard whose eyes are bleared by the lightnings" but rather "a fresh and vivacious youth of thirty" who had married well and lived well in the nation's capital.[2]

The "Synopsis and Probabilities," the official term for the federal government's twenty-four-hour summary and forecast, was not old either. As the poet mused, "Why took our grandsires, as it came, / Weather and wind of every name? / Because then quite unknown to fame / Were Probabilities."[3] Beginning in 1870, the US Army Signal Service used a network of trained observers and telegraph lines to systematically record and transmit local weather observations to its Washington headquarters, which then disseminated storm warnings and short-term forecasts across the nation for the first time, marking the introduction of the federal government's weather "Probabilities" into daily life.[4]

Contrary to the poet's rendition, the Signal Service did not "tell the farmer when to sow" but rather predicted only the next twenty-four hours' weather, resisting demand throughout the late nineteenth century for long-range monthly or seasonal forecasts. But in some rural areas, farmers heard nothing at all from "Old Probabilities." As the Signal Service expanded in the 1870s to serve agricultural as well as commercial regions, it struggled with the problem of distributing forecasts to the "distant parts of the country," where rural residents wondered whether forecasts would arrive regularly and on time, late, occasionally, or not at all.[5] The history of government weather forecasting in the countryside reveals that the daily "Probabilities" brought both predict-

2. Mary Clemmer Ames, *Ten Years in Washington: Life and Scenes in the National Capital, as a Woman Sees Them* (A. D. Worthington, 1873), 503; undated newspaper clipping, Signal Service clipping book. Cleveland Abbe, the nation's most prominent meteorologist, was the first "Old Probabilities," but the nickname became more commonly associated with Myer. C. Fitzhugh Talman, "Old Probabilities," "Why the Weather? A Science Service Feature," April 25, 1927, http://docs.lib.noaa.gov/rescue/whytheweather/1927/19270502.pdf.

3. Glover, " 'Old Probabilities,' " Signal Service clipping book.

4. The Signal Service used the term "probabilities" until December 1876, when it was changed to "indications." The "indications" were renamed "forecasts" in April 1899. National Oceanic and Atmospheric Administration, "A Brief History of the Boston Weather Bureau," http://www.erh.noaa.gov/box/NWShistory.html. But "probabilities" would remain in popular usage as a term for government weather forecasting throughout the late nineteenth century.

5. Quotation in *Annual Report of the Chief Signal-Officer to the Secretary of War for the Year 1877* (Washington, DC: Government Printing Office, 1877), 129.

ability and uncertainty. Institutional histories of the national weather service have chronicled a narrative of increasing technoscientific progress and organizational stability that produced forecasts of obvious economic and social value. This chapter recasts the history of the construction of the United States' first national meteorological infrastructure as a far more contested, inefficient, and disorderly process by examining, through the accounts of farmers and rural postmasters, how government weather forecasting functioned and was perceived in the countryside. The Signal Service's weather forecasting network represents not only the bureaucratic rationality of late nineteenth-century system building and the search for order but also irregularity and controversy surrounding the circulation of knowledge about the future.[6]

The politics of access to short-term weather forecasts reveals an unacknowledged source of agrarian discontent in the late nineteenth century, when rural producers sought not only capital, credit, and equitable shipping rates but also information that would mitigate weather-related risk of crop damage.[7] Signal Service forecasts arguably represented the most regular daily interaction between farmers and the state, linking rural producers to the federal government through their access, however uneven, to weather forecasting as a public good.[8] In this regard, the advent of government weather forecasting is an important episode in the history of "infrastructural" state power in nineteenth-century America.[9] In the 1870s and 1880s, the Signal Service sought to expand its weather reporting network as farmers petitioned for its extension to places it couldn't reach. At the turn of the century, the introduction of Rural Free

6. On the Signal Service's weather reporting network as an early manifestation of Robert Wiebe's "search for order," see James Rodger Fleming, "Storms, Strikes, and Surveillance: The U.S. Army Signal Office, 1861–1891," *Historical Studies in the Physical and Biological Sciences* 30, no. 2 (2000): 316.

7. On agrarian politics, see Charles Postel, *The Populist Vision* (New York: Oxford University Press, 2007); Elizabeth Sanders, *Roots of Reform: Farmers, Workers, and the American State, 1877–1917* (Chicago: University of Chicago Press, 1999); Robert C. McMath, Jr., *American Populism: A Social History, 1877–1898* (New York: Hill and Wang, 1993); Lawrence Goodwyn, *Democratic Promise: The Populist Moment in America* (New York: Oxford University Press, 1976).

8. On weather forecasting as a public good in late nineteenth-century America, see Erik D. Craft, "Private Weather Organizations and the Founding of the United States Weather Bureau," *Journal of Economic History* 59, no. 4 (1999): 1063–71.

9. William J. Novak, "The Myth of the 'Weak' American State," *American Historical Review* 113, no. 3 (2008): 763. On the centrality of postal communication networks to the early nineteenth-century American state, see Richard R. John, *Spreading the News: The American Postal System from Franklin to Morse* (Cambridge, MA: Harvard University Press, 1995).

Delivery and rural telephone lines largely solved the problem of distributing short-term forecasts to the countryside, where Americans became increasingly certain that they would indeed receive the daily "Probabilities."

As telegraphy and the mail carried government weather forecasting into rural areas, it met with mixed reactions, from enthusiastic to ambivalent to hostile. But farmers most commonly called for more timely distribution of short-term forecasts, even if, as the next chapter reveals, they were skeptical about forecasts' meaning and value. This chapter tells the story of the controversies over access to the first government weather service in the United States and the uncertainties surrounding the arrival of "Old Probabilities" in the countryside.

"THE CLERK OF THE WEATHER"

The less common nickname of "Old Probabilities" was "the clerk of the weather," a name that was perhaps more accurate, since late nineteenth-century weather forecasting was actually not probabilistic and relied more on bureaucratic and technical innovations than on theoretical advances in meteorology.[10] The national weather service could not have worked without the web of poles and wires that connected "island communities" into the postbellum telegraphic communications network or the fact that telegraphic messages traveled faster than the weather.[11] As Harvard paleontologist Nathaniel Shaler wrote, Signal Service forecasting hinged on

> the simple principle that weather always has a history; that it means conditions that pass from one region to another by certain laws of movement and at a certain rate. This general fact was long ago recognized by meteorologists, but it was not until the telegraph enabled knowledge to outstrip the storms that it was possible to make any use of it in foretelling the weather.[12]

Western Union, which had a near monopoly on the telegraph industry after its 1866 merger with the United States Telegraph Company and the American

10. "The Clerk of the Weather," *Daily Enquirer*, March 28, 1871; "The Weather Clerks," *Nebraska Daily Press and Herald*, December 5, 187[?]; "Clerk of the Weather," *Chicago Evening Journal*, February 10, 1872; "The 'Clerk of the Weather,'" *Buffalo Courier*, February 16, 1872; "The Weather Clerk," *Leavenworth (KS) Daily Commercial*, October 20, 1872, all in Signal Service clipping book.

11. Robert H. Wiebe, *The Search for Order, 1877–1920* (New York: Hill and Wang, 1967), xiii.

12. N. S. Shaler, "The Future of Weather Foretelling," *Atlantic Monthly*, November 1880, 645.

Telegraph Company, provided the majority of the commercial lines for government weather telegraphy, while the Signal Service built its own lines in the early 1870s in coastal and mountain areas as well as in Indian Territory.[13]

In an article titled "The Clerk of the Weather," a Richmond newspaper pronounced Myer a "compound man," with assistants for eyes, meteorological instruments for hands, and telegraph wires for nerves.[14] Indeed, Myer sat "at the center of an electric intelligence network spanning the nation," as James Rodger Fleming has described it, which enabled not only forecasting but also "domestic surveillance and social control."[15] (See figure 2.1.)

Throughout the 1870s, Signal Service weather observers telegraphed to Washington reports on temperature, precipitation, and risks of impending storms, frosts, and droughts, but occasionally included firsthand observations of labor strikes in cities and clashes between whites and Native Americans on the frontier.[16] The public also participated in such surveillance by reporting, for example, a gang of counterfeiters in Indiana and destructive swarms of locusts in Colorado.[17]

Myer's information network was a tightly organized bureaucracy consisting of military-trained weather observers, standardized meteorological instruments, and a system for conducting synchronous daily observations.[18]

13. For economic and business histories of telegraphy in the United States, see David Hochfelder, *The Telegraph in America, 1832–1920* (Baltimore: Johns Hopkins University Press, 2012); Richard R. John, *Network Nation: Inventing American Telecommunications* (Cambridge, MA: Belknap Press, 2010); Alexander James Field, "The Magnetic Telegraph, Price and Quantity Data, and the New Management of Capital," *Journal of Economic History* 52, no. 2 (1992): 401–13. On weather telegraphy, see James Rodger Fleming, *Meteorology in America, 1800–1870* (Baltimore: Johns Hopkins University Press, 1990), chap. 7; Fleming, "Storms, Strikes, and Surveillance," 319–20.

14. "The Clerk of the Weather," *Daily Enquirer*, March 28, 1871, Signal Service clipping book.

15. Fleming, "Storms, Strikes, and Surveillance," 320, 316.

16. Fleming, "Storms, Strikes, and Surveillance," 320.

17. H. Ames, Postmaster, Vinespring, Indiana, to Chief of Signal Service, March 1875; P. M. Hinman, Secretary, State Grange of Colorado, to the president of the United States, March 10, 1875, both in box 779, Meteorological Correspondence of the Signal Office, 1870–93, Agricultural Series, Letters Received, 1873–77, RG 27, NACP (hereafter Signal Office Agricultural Correspondence).

18. On "the structure of bureaucratic meteorology" and the centrality of field networks to meteorological science in the early twentieth century, see Jeremy Vetter, "Lay Observers, Telegraph Lines, and Kansas Weather: The Field Network as a Mode of Knowledge Production," *Science in Context* 24, no. 2 (2011): 259–80, quotation on 262. On the importance of the conceptual

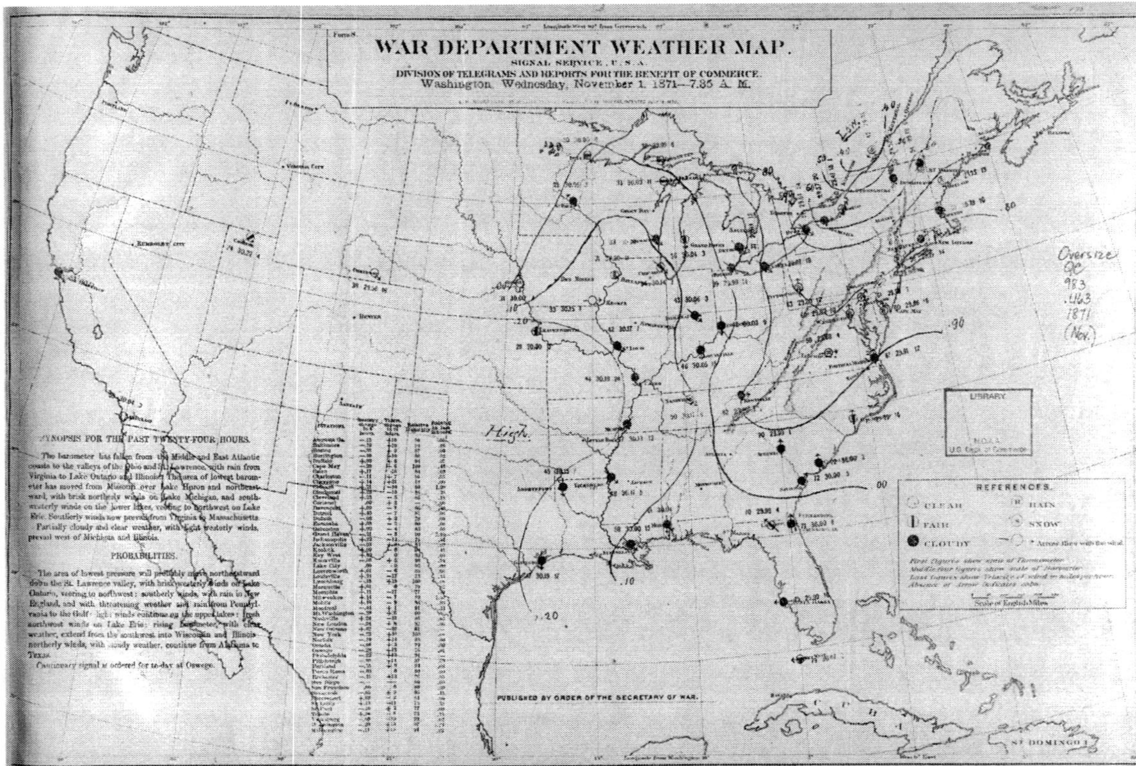

FIGURE 2.1. The US Army Signal Service's daily weather map in the 1870s, which included the synopsis of the previous twenty-four hours' weather and the probabilities for the next twenty-four hours' weather, indicated temperature, barometric pressure, wind velocity, and general conditions for each node in the new government weather telegraph network. "War Department Weather Map, Washington, Wednesday, November 1, 1871—7:35 a.m.," NOAA Central Library Data Imaging Project, http://www.lib.noaa.gov/collections/imgdocmaps/daily_weather_maps.html.

The weather service grew quickly, with over five hundred trained "observer-sergeants" on staff by the mid-1870s, and, by 1890, a total of 500 stations, which included 26 first-order stations (using self-recording instruments to make continuous observations), 118 second-order stations (making two observations daily), 34 third-order stations (making one nightly observation), 73 special display stations, and over 200 special stations in river and cotton regions.[19] Observers used thermometers, barometers, anemometers (to measure wind speed and direction), hygrometers (to measure humidity), rain gauges, and clocks to record measurements at 7:35 a.m., 4:35 p.m., and 11:35 p.m. (Washington time), then translated their data into a cipher that enabled a cheaper and ideally more accurate telegraphic transmission.[20] For instance, "Francisco. Emily. Alone. Barge. Churning." meant that San Francisco one evening reported a 29.4 barometric reading, 61°F temperature, and 18 percent humidity.[21] Under an agreement with Western Union, government messages received priority, ensuring the swiftest possible transmission to Washington, where clerks received a flurry of telegrams, decoded them, and charted the data on weather maps.[22] From these "timely miniature snapshots of surface weather," forecasters constructed the "Synopsis and Probabilities" for transmission back to Signal Service stations, newspapers, Western Union offices, scientific institutions, and commodity exchanges at home and abroad.[23] (See figure 2.2.) The "Probabilities," a twenty-four-hour projection of observed surface conditions

---

framework of infrastructure to the history of meteorology, see Paul N. Edwards, "Meteorology as Infrastructural Globalism," *Osiris* 21 (2006): 229–50.

19. Fleming, "Storms, Strikes, and Surveillance," 319; *Annual Report of the Chief Signal Officer of the Army to the Secretary of War* (Washington, DC: Government Printing Office, 1890), 203. For the number of different types of stations in operation each year between 1870 and 1890, see Erik D. Craft, "An Economic History of Weather Forecasting," *EH.net*, https://eh.net/encyclopedia/an-economic-history-of-weather-forecasting/.

20. "The Weather Clerk," *Leavenworth (KS) Daily Commercial*, October 20, 1872, Signal Service clipping book; Office of the Chief Signal Officer, *War Department Circular: The Practical Use of Meteorological Reports and Weather-Maps* (Washington, DC: Government Printing Office, 1871), 5–6.

21. Signal Service USA, *Telegraph Cipher* (1875), 2, box 1, Orders, Circulars, and Instructions; Office of the Chief Signal-Officer: Administrative Orders, Circulars, Memorandums, and Instructions, 1871–1891, RG 27, NACP.

22. On disputes between the Signal Service and Western Union over the transmission of government weather telegrams in the early 1870s, see Hochfelder, *Telegraph in America*, 58–62.

23. Quotation in Mark Monmonier, *Air Apparent: How Meteorologists Learned to Map, Predict, and Dramatize the Weather* (Chicago: University of Chicago Press, 1999), xi. This overview of weather telegraphy is drawn from Monmonier, *Air Apparent*, chap. 1; Edmund P. Willis and

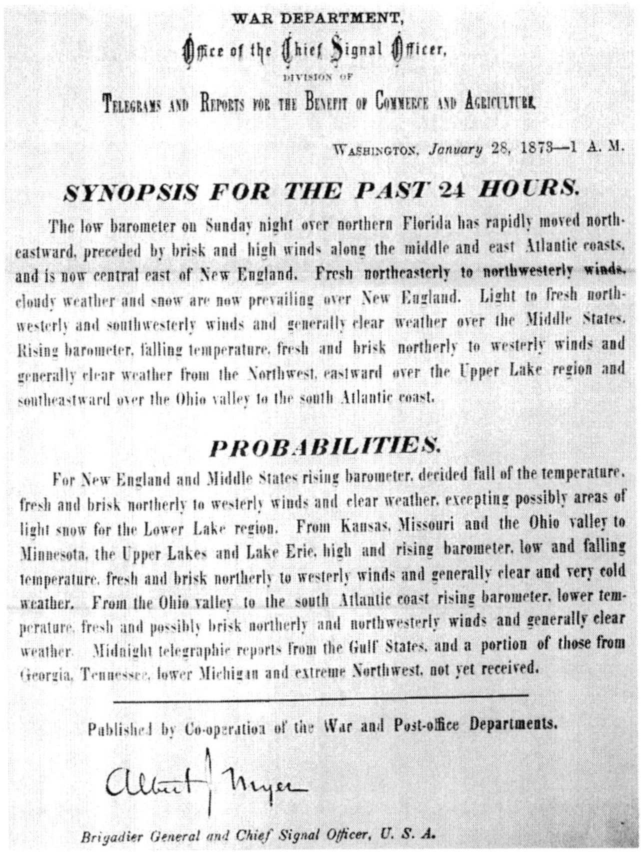

WAR DEPARTMENT,

Office of the Chief Signal Officer,

DIVISION OF

TELEGRAMS AND REPORTS FOR THE BENEFIT OF COMMERCE AND AGRICULTURE.

WASHINGTON, *January* 28, 1873—1 A. M.

SYNOPSIS FOR THE PAST 24 HOURS.

The low barometer on Sunday night over northern Florida has rapidly moved north-eastward, preceded by brisk and high winds along the middle and east Atlantic coasts, and is now central east of New England. Fresh northeasterly to northwesterly winds, cloudy weather and snow are now prevailing over New England. Light to fresh north-westerly and southwesterly winds and generally clear weather over the Middle States. Rising barometer, falling temperature, fresh and brisk northerly to westerly winds and generally clear weather from the Northwest, eastward over the Upper Lake region and southeastward over the Ohio valley to the south Atlantic coast.

PROBABILITIES.

For New England and Middle States rising barometer, decided fall of the temperature, fresh and brisk northerly to westerly winds and clear weather, excepting possibly areas of light snow for the Lower Lake region. From Kansas, Missouri and the Ohio valley to Minnesota, the Upper Lakes and Lake Erie, high and rising barometer, low and falling temperature, fresh and brisk northerly to westerly winds and generally clear and very cold weather. From the Ohio valley to the south Atlantic coast rising barometer, lower temperature, fresh and possibly brisk northerly and northwesterly winds and generally clear weather. Midnight telegraphic reports from the Gulf States, and a portion of those from Georgia, Tennessee, lower Michigan and extreme Northwest, not yet received.

Published by Co-operation of the War and Post-office Departments.

Albert J. Myer

Brigadier General and Chief Signal Officer, U. S. A.

FIGURE 2.2. The official "Synopsis and Probabilities" issued at 1:00 a.m. on January 28, 1873. Meteorological Correspondence of the Signal Office, Letters Received, 1873–1877, Agricultural Series, box 774, Records of the United States Weather Bureau, National Archives College Park.

forward in time and eastward in space, were defined by the chief signal officer as "announcements of the changes, considered from the study of the charts . . . as probably to happen within the twenty-four hours then next ensuing."[24]

The stations that transmitted data and received the daily "Synopsis and Probabilities" represented the authority of both the centralized government

William H. Hooke, "Cleveland Abbe and American Meteorology, 1871–1901," *Bulletin of the American Meteorological Society* 87, no. 3 (2006): 317.

24. Monmonier, *Air Apparent*, 81; *Annual Report of the Chief Signal-Officer to the Secretary of War for the Year 1874* (Washington, DC: Government Printing Office, 1874), 90.

reporting network and its local observers. The Signal Service garnered praise for its "unerring . . . prophecies . . . [that] have astounded even the most sanguine advocates of the system," and the *Detroit Free Press* exulted that "the signal men have hardly ever been wrong in their reckonings."[25] As the *Buffalo Courier* observed of "the weather clerk" in 1872, "this heretofore mythical personage has, in our time, and only quite recently, become a reality, and a very reliable authority."[26] One letter to the editor in March 1872 recounted a visit to "the courteous 'weather clerk' Mr. Buell," who warned that a snowstorm turning to rain would arrive in ten hours:

> I must say that this prognostication was really wonderful. The accuracy with which the veering of wind, the time of beginning and ending of the storm, strikes me at once that the signal service is performing a great and beneficial work for all mankind, and that the probabilities are not guessed at, but are drawn from laws which govern atmospheric disturbances. Heretofore I have had little faith in this service, for the reason that I knew nothing of the laws of and the working of the system; but with suggestions from the weather clerk I am now able to use the published reports to advantage, and can see clearly that this service is only in its infancy; that as the masses become educated and understand its value, it will be the most popular and useful bureau in the country.[27]

Buell's enthusiastic visitor easily overcame his initial skepticism, but Myer had had to work hard to make himself the clerk of the weather in the first place.

The need for a lake storm warning system was the impetus for the creation of the national weather service in 1870, when the danger posed to commercial shippers captured the attention of Congress, thanks to the efforts of eminent meteorologist Cleveland Abbe, meteorologist and civil engineer Increase Lapham, and Congressman Halbert Paine, a former student of renowned storm researcher Elias Loomis at Western Reserve College. In December 1869, Paine proposed a bill asking for federal oversight of a weather service for the Great Lakes and the Gulf and Atlantic coasts.[28] The US Army Signal Service was

25. "Local Items. The Signal Corps," *Washington Capital*, November 26, 1871; "Hit It Again," *Detroit Free Press*, November 29, 1871, both in Signal Service clipping book.

26. "The 'Clerk of the Weather,'" *Buffalo Courier*, February 16, 1872, Signal Service clipping book.

27. "Notes from the People. A Plea for the Signal Service," newspaper clipping, March 26, 1872, Signal Service clipping book.

28. Abbe had attempted to start a Midwestern regional weather observation and forecasting network in 1869, but, lacking necessary support from the Chicago Board of Trade, appealed

assigned to the task due to the effective campaigning of Myer, who sought institutional stability and funding after his embarrassing dismissal as head of the beleaguered corps during the Civil War.[29] Support for military administration of the new weather service was not unanimous, but proponents argued that the "*military precision and promptness*" of the Signal Service would produce the most accurate and reliable observations.[30]

The work of the first national weather service officially began on November 1, 1870, when twenty-four Signal Service observers made synchronous observations at 7:35 a.m. (Washington time). In October of the following year, the Signal Service's system of storm-warning flags was implemented in eight ports on the Great Lakes and sixteen ports on the Atlantic coast. This new system paid off immediately for Great Lakes shipping companies, who faced a significantly reduced risk of losing vessels in unforeseen storms.[31] Newspapers reprinted Signal Service tabulations of the number of vessels destroyed or damaged in winter lake storms, which decreased from 294 in November and December of 1870 to 173 the following year, and the number of lives lost on the Great Lakes dropped from 97 to 23 during the same period.[32]

---

to his friend Lapham, who had previously tried to start a Wisconsin state weather service, and Lapham petitioned Paine, enumerating the 1,914 ships that had crashed in stormy lake weather that year alone, costing 209 lives and $4.1 million. Charles C. Bates and John F. Fuller, *America's Weather Warriors, 1814–1985* (College Station: Texas A&M University Press, 1986), 9–10; Craft, "Private Weather Organizations," 1067.

29. Myer was dismissed as chief signal officer in 1863 after the failure of his tactical telegraphy systems and his clashes with secretary of war Edwin Stanton, and the Signal Corps's telegraphic equipment was temporarily absorbed by the US Military Telegraph. Reinstated after Stanton's tenure, Myer was committed to restoring the Signal Corps, the institution he had created. Paul J. Scheips, "'Old Probabilities': A. J. Myer and the Signal Corps Weather Service," *Arlington Historical Magazine* 5, no. 2 (1974): 29–43; Donald R. Whitnah, *A History of the United States Weather Bureau* (Urbana: University of Illinois Press, 1961), 18; Joseph M. Hawes, "The Signal Corps and Its Weather Service, 1870–1890," *Military Affairs* 30, no. 2 (1966): 68–76.

30. Quotation in Thompson B. Maury, "The Telegraph and the Storm: The United States Signal Service," *Harper's New Monthly Magazine*, August 1871, 405 (emphasis in original). On Cleveland Abbe's vehement opposition to the military administration of the weather service, see Hawes, "The Signal Corps and Its Weather Service," 70.

31. Craft, "Private Weather Organizations," 1068; Erik D. Craft, "The Value of Weather Information Services for Nineteenth-Century Great Lakes Shipping," *American Economic Review* 88, no. 5 (1998): 1059–76.

32. "Practical Value of the Signal Service," *Cleveland Leader*, n.d., Signal Service clipping book. See also "Value of the Signal Service," *Knoxville (TN) Daily Chronicle*, February 16, 1872, Signal Service clipping book.

In addition to collecting synchronous daily weather observations and statistics on vessels lost, the Signal Service also carefully monitored its public image. Clerks gathered newspaper clippings on the Signal Service that had been mailed in by observers across the country and pasted them into an oversized book that served as an archive of public opinion on the promise as well as the limitations of government weather forecasting.[33] The Signal Service earned much acclaim, and in the early 1870s newspapers pronounced "'Old Probabilities' . . . by long odds the most popular national institution," "a household word" and "a comfort of domestic life," and "such a necessity of American life, that . . . [i]f the service should cease suddenly, it would be a National calamity, something like a general stoppage of the mails." Of course, not everyone celebrated the precision of what one writer called "the weather guessing bureau."[34] The *New York Evening Post*, lamenting inaccurate forecasts in August 1872, declared that

> residents of the city have every morning had their hopes raised only to be lowered, and have added to their sufferings the disappointment which is felt at seeing what had come to be a trusted authority prove delusive. The old almanacs, which run down in their tables an entire month with such a sentence, if it was August, as "about this time look out for flies"; or, "overcoats will not be needed"—we say the old almanacs were hardly more ridiculous than the 'probabilities' recently telegraphed all over the country by the War Department. . . . What is more important, we have all left our umbrellas up-town, by reason of our faith in the prognostications of the weather prophet.[35]

And as Americans came to rely on daily weather forecasts, they also began to consider the nature of forecasting itself.

The efficient machinery of "the Washington weather-factory"—its military precision, far-reaching and lightning-fast telegraph lines, and army of

33. On the cultural practices of newspaper clipping, see Ellen Gruber Garvey, *Writing with Scissors: American Scrapbooks from the Civil War to the Harlem Renaissance* (New York: Oxford University Press, 2012).

34. *Indianapolis Daily Sentinel*, December 7, 187[?]; "The Weather Reports," *New York Times* (hereafter *NYT*), December 8, 1872; "Old Probabilities," *Detroit Post*, December 11, 1872; "'Old Probabilities.' The Machinery of the Weather Guessing Bureau," newspaper clipping, April 28, 1872, all in Signal Service clipping book.

35. "The Weather Makers," *New York Evening Post*, August 16, 1872, Signal Service clipping book.

observers and weather clerks—cultivated the expectation of predictive certainty.[36] Author Mary Clemmer Ames marveled at the Signal Service's ability to track precisely the path of a severe storm from San Francisco on its journey from Utah to the Midwest in February 1871: "The President of the Pacific Railroad has not more perfectly under his eye and control the train that left San Francisco, today, than General Meyer [*sic*] had the storm just described."[37] The implication that the unpredictability of weather could be tamed was echoed by commentary suggesting that Signal Service weather forecasts promised more than mere "probabilities." The *Washington Capital* reported in November 1871 that "the chief of the corps is confident that soon the heading 'Probabilities' in his reports will be changed to 'Certainties,' and a fellow going out of an evening will positively know whether to wear a cane or an umbrella," and *Coleman's Rural World* remarked that "no one can for a moment fail to discern the immense value these reports, with their accompanying 'probabilities' (which might more properly be called 'certainties,' so accurate have they become), will become to our farmers when they can be laid before them in sufficient time to take advantage of their warnings."[38] Believing that a combination of telegraphy and meteorology would conquer chance, contemporary observers exclaimed that "the 'probabilities' of the weather . . . are not founded upon any known or unknown law of chance or guess-work," "meteorology is now an exact science," and "the scientist becomes a prophet."[39]

"FOR THE BENEFIT OF COMMERCE AND AGRICULTURE"

The official name of the new government weather service was "The Division of Telegrams and Reports for the Benefit of Commerce," and in June 1872, Congress officially extended the reach of weather forecasts and storm warnings

36. "Local Matters. A Word for the Ladies," *Pittsburgh Evening Leader*, March 20, 187[?], Signal Service clipping book.

37. Clemmer, *Ten Years in Washington*, 502.

38. "Local Items. The Signal Corps," *Washington Capital*, November 26, 1871; "The Importance of the Science of Meteorology," *Coleman's Rural World* (St. Louis), November 23, 1872, both in Signal Service clipping book.

39. " 'Old Probabilities.' Something about Storm Science," *Boston Times*, January 21, 1872; "Clerk of the Weather," *Chicago Evening Journal*, February 10, 1872, both in Signal Service clipping book.

"for the benefit of commerce and agriculture."[40] Signal Service clerks penciled in "and agriculture" on publications that would need to be reprinted, but expanding a national meteorological infrastructure accordingly would not be so simple.[41] The name change did not resolve what some perceived as a fundamental incompatibility between military oversight of the weather service and the interests of rural producers, while other skeptics, including an indignant reader of the *Washington Capital*, questioned the adaptation of naval meteorology to land, the former being "applicable only to commerce, being totally inapplicable to agriculture and floods."[42]

Previous approaches to weather reporting were not necessarily inapplicable to agriculture, but they had never been applied, given that the origins of the national weather service were not in the countryside. The US Army first systematized weather observation and reporting in 1814 through surgeons' diaries of climate and disease, and by 1853, the surgeon general supervised ninety-seven camps in a weather-reporting network.[43] Navy lieutenant Matthew Fontaine Maury's pioneering analysis of ship logs and shipmasters' weather observations in the 1840s and 1850s produced wind, whaling, and storm charts that were widely acclaimed for their value in mitigating risk for commercial shippers, who saved time and money with more efficient international routes, and Maury himself "urg[ed] the extension to the land—for the benefit of farmers, the shipping in our ports, and the industrial pursuits of the country generally."[44] Just as state agricultural societies had experimented with crop-reporting systems to meet the growing demand for agricultural

40. Patrick Hughes, *A Century of Weather Service: A History of the Birth and Growth of the National Weather Service, 1870–1970* (New York: Gordon and Breach, 1970), 8, 21.

41. Annotated copy of War Department, *The Practical Use of Meteorological Reports and Weather-Maps* (Washington, DC: Government Printing Office, 1871), box 1, Orders, Circulars, and Instructions; Office of the Chief Signal Officer: Administrative Orders, Circulars, Memorandums, and Instructions, 1871–1891, RG 27, NACP.

42. A. Watson, letter to the editor, *Washington Capital*, December 14, 1871, folder "Committee on Appropriations, Dept. of War (4) Telegraphic Weather Service (SC)," HR 42A-F3.24, House Committee on Agriculture, Records of the United States House of Representatives, RG 233, National Archives Building, Washington, DC (hereafter NAB).

43. Bates and Fuller, *America's Weather Warriors*, 12.

44. Hughes, *Century of Weather Service*, 19; Bates and Fuller, *America's Weather Warriors*, 6–7. Quotation in "The Progress of Marine Geography," *Journal of the American Geographical and Statistical Society*, part 1 (New York, 1860), 6, quoted in Maury, "The Telegraph and the Storm," 401.

statistics before the Civil War, state and county agricultural societies petitioned the House Committee on Agriculture in the mid-1850s to adopt Maury's system in farming regions, envisioning, in the words of the New York State Agricultural Society, an "immediate benefit."[45]

Yet some opposed the adaptation of Maury's system, especially the Smithsonian's first secretary, Joseph Henry, who argued that it was too simple to use inland and that Maury lacked the expertise to oversee such a project. From 1849 to 1874, the Smithsonian presided over a weather-reporting network of unprecedented scope, collecting observations by mail and telegraph from as many as six hundred volunteers located primarily east of the Mississippi, as well as in Canada and Latin America, and published climatological data and, beginning in 1856, a daily weather map.[46] Henry's network seemed a promising model for an agricultural weather service by 1860, when the superintendent of the Census noted that "the want of meteorological knowledge, and consequent want of adaptation of our industry to the laws of climate, both general and local, is a frequent source of loss to the farmer."[47] But Henry's project suffered major setbacks when the Civil War cut off postal and telegraphic communication with southern observers and the Smithsonian was devastated by a fire in January 1865. Henry's pioneering weather-reporting network was never fully restored and was ultimately absorbed by the Signal Service's Division of Telegrams and Reports for the Benefit of Commerce in 1874.[48]

From its inception, the national weather service relied heavily on civilian expertise and personnel, which became a liability in later debates over military oversight of what many perceived to be essentially a civilian organization.

45. Jefferson County Agricultural Society, "Petition for the adoption of Lieutenant M. F. Maury's system of meteorological observation," February 1856; New York State Agricultural Society, "Petition for the adoption of Lieutenant M. F. Maury's system of meteorological observations," February 1856, both in folder "Adoption of Lieutenant M. F. Maury's System of Meteorological . . . Feb. 14 to Feb. 28, 1856," HR 34A-G1.1, House Committee on Agriculture, RG 233, NAB.

46. Fleming, *Meteorology in America*, 75–76, 143. The numbers of Smithsonian observers fluctuated between 1849 and 1874, from 150 initially to 500 in 1860 and 1870, with fewer than 300 during the Civil War.

47. Joseph C. G. Kennedy, Preliminary Report on the Eighth Census, 1860, H.R. Rep. No. 116-37, at 101 (1862).

48. On the effects of the war and the Smithsonian fire on Henry's meteorological work, see Fleming, *Meteorology in America*, 146–47. On Henry's administration of the Smithsonian weather-reporting network, see Fleming, *Meteorology in America*, chap. 4.

Meteorology was not a separate academic discipline in the late nineteenth century, when meteorologists were either self-taught, educated in various university programs indirectly related to meteorology, or trained by the US Army Signal Corps. University courses offered on weather and climate were usually basic introductory courses in geology, geography, physics, or agriculture departments. No consensus existed on what kind of science meteorology was—applied or theoretical—and where its disciplinary home should be—natural history or physical science. Government scientists tended to define it as a theoretical endeavor in the physical sciences, whereas university programs linked it with one of two natural history traditions—zoology/botany and geology/geography. Most university training occurred in the context of voluntary student meteorological societies or college observatories, or in ad hoc training programs in geography and geology departments at Harvard University and Johns Hopkins University, as well as a graduate program run by Weather Bureau officials at what later became George Washington University. The most organized program of meteorological education was within the Signal Service, which opened a Signal Corps school at Fort Myer, Virginia, in 1881, a school considered to be the country's "first postgraduate program in meteorology."[49]

One of the first and most difficult problems Myer faced was how to satisfy the steadily increasing demand for rural weather forecasts. Although "frequent . . . letters and communications from societies and associations" requested more observation stations in the countryside, the Signal Service's funding limited its efforts to serve rural populations to publishing circulars on the prospects of an agricultural weather service, circulating weather charts and bulletins to the USDA, and telegraphing weather reports to newspapers.[50] For those who lived outside the reach of the government's information network, the Signal Service envisioned a solution combining local observation, instrumentation, and individual skill. In 1871, the Signal Service published a circular

49. Bates and Fuller, *America's Weather Warriors*, 10–11; James Rodger Fleming, introduction; David D. Houghton, "Meteorology Education in the United States after 1945"; quotation in William A. Koelsch, "From Geo- to Physical Science: Meteorology and the American University, 1919–1945," 512, all in *Historical Essays on Meteorology, 1919–1995*, ed. James Rodger Fleming (Boston: American Meteorological Society, 1996).

50. US Department of War, *Annual Report of the Secretary of War* (Washington, DC: Government Printing Office, 1871), 272. For the annual budget for the Signal Service's weather reporting network, which grew from just over $32,000 in 1870 to over $1 million by 1879, see Craft, "An Economic History of Weather Forecasting," *EH.net*, https://eh.net/encyclopedia/an-economic-history-of-weather-forecasting/.

suggesting that "the navigator, the agriculturalist, or the student, can supplement . . . by the readings of his own instruments and his local knowledge, the reports and information furnished by the United States, and is fitted to arrive at intelligent conclusions as to the data before him." To this end, the Signal Service sold blank weather maps, for $2.75 per hundred, on which farmers could combine the daily "Probabilities" with their own data on temperature, barometric pressure, wind direction, and precipitation, thereby superimposing local observations on the aggregation of hundreds of individual synchronous observations.[51]

This vision of "every man his own weather prophet," however, did not mean that the Signal Service would accept unofficial observations, like those volunteered by the Pike County (Georgia) Agricultural Society's weather committee in 1871, which the Signal Service explained "would not be of use for the immediate purposes of this division."[52] Myer stressed the need for complete military control of weather information, ordering observers to "confine themselves strictly to the instructions issued from this Office, and . . . not, under any circumstances, publish or cause to be published forecasts or predictions of the weather not originated by this Office."[53]

The range of responses to official Signal Service forecasts revealed enthusiasm, skepticism, and ambivalence on the part of those who consulted the daily "Probabilities" out of curiosity but did not trust in them. Postmaster Solomon Stinemetz of Clear Spring, Maryland, echoed many other postmasters when he wrote in 1874 that "the reports are highly appreciated by the public," and postmasters who reported regular morning delivery of the Synopsis and Probabilities could often say that "very regularly it is consulted by crowds as soon as it is put up."[54] The Signal Service received many letters attesting to farmers' particular interest in daily forecasts, including the one from a Green Spring,

51. US Signal Service, *Practical Use of Meteorological Reports and Weather-Maps*, 8–9, quotation on 8.

52. "Every Man his own Weather Prophet," *Hearth and Home*, October 21, 1871, Signal Service clipping book; *Annual Report of the Chief Signal Officer to the Secretary of War for the Fiscal Year Ended June 30, 1871* (Washington, DC: Government Printing Office, 1871), 143.

53. *Annual Report of the Chief Signal-Officer to the Secretary of War for the Year 1872* (Washington, DC: Government Printing Office, 1873), 142.

54. Solomon Stinemetz to Chief Signal Officer, May 9, 1874, box 777; Ferd. Stoesser to Garrick Mallery, April 17, 1874, box 777; A. B. Smith to Capt. H. W. Howgate, March 11, 1873, box 774, all in Signal Office Agricultural Correspondence.

Pennsylvania, postmaster who asked, "Please send to this Office the Weather Probabilities quite a lot of the Farmers desire to be posted."[55]

In 1873, Pennsylvanian Lewis Bollman penned a lengthy letter reflecting on his twelve years' experience as a farmer, during which time he read books on meteorology and made daily local observations "to search out some rules by which to direct my acts as a farmer." Bollman relied to some extent on his own short-term forecasts, based on experiential generalizations like an east wind and starry night meant rain before dawn, but he concluded that "all else resolved itself into a confirmation of the farmer's rule—*never trust the weather*." Bollman did, however, no doubt due to his stint as a statistician in the US Department of Agriculture, trust in government weather forecasting. "The first thing I usually do in looking over my daily paper," he wrote, "is to read the 'Weather Probabilities,' to see, especially, whether the prevailing winds in any one of your divisions of our country, so influenced the portion I live in that I could determine what the weather would be on the succeeding day."[56] Bollman's meditative letter to the Signal Service was unusual, but his search for a rational, anticipatory form of knowledge to guide daily work on the farm was commonplace. The daily "Probabilities" led farmers to renounce time-honored traditions of weather knowledge, as a Craigville, New York, postmaster reported: "At first the Farmers here turned up their noses & went through their almanacs. I have hung this sign with Report in it (by my PO) and you should all see them at 1.20 P.M. go for this sign one old weather prognosticator says cuss me if it dont hit every time what's coming next from Washington."[57] A Tennessee postmaster offered the strongest statement of the epistemological authority of the Probabilities when he wrote, "Incredulity on the part of the people has been conquered completely; and *now* the daily reports are consulted with as much simple faith as our grandfathers used to look into 'Millers Almanac' for prognostication of the weather;—with this

55. On farmers' demand for daily forecasts, see Wm. G. Ferris to the Office of the Chief Signal Officer, March 11, 1873, box 774; C. W. Baird to Chief Officer Signal Service, April 25, 1874, box 777; F. H. Hasmary to Chief Signal Officer, April 21, 1874, box 777, all in Signal Office Agricultural Correspondence. Quotation in G. W. Swigert to Gen. A. J. Myer, April 25, 1874, box 777, Signal Office Agricultural Correspondence.

56. Lewis Bollman to Bureau of Meteorology, January 21, 1873, box 774, Signal Office Agricultural Correspondence.

57. Postmaster, Craigville, NY, to Chief Signal Officer, 1873, box 774, Signal Office Agricultural Correspondence.

difference, that the faith of today has been the outgrowth of intelligent observation and experience."[58]

Of course, incredulity was not conquered everywhere, and many postmasters observed skeptical and occasionally hostile reactions that questioned the value of the Synopsis and Probabilities. A Missouri postmaster reported, "Nobody ever looks at them; so far as we are concerned the whole thing is a Humbug," and a Wisconsin postmaster lamented that local farmers "smile and say your papers cannot change the weather." Another Wisconsin postmaster, "disgusted with the sharp remarks of the farmers," stopped displaying the Synopsis and Probabilities.[59] The most common complaint about the Synopsis and Probabilities was that it arrived too late (sometimes thirty-two or forty hours) to be of value, which a Pennsylvania postmaster pronounced "so manifestly useless." A Delaware postmaster wondered, "of what account is yesterday's Probabilities for today?"[60]

Aware of complaints like these, the Signal Service followed the 1872 congressional resolution expanding the Division of Telegrams and Reports for the Benefit of Commerce and Agriculture with a series of efforts to broaden the reach and variety of weather information available in the countryside. By fall 1872, nearly ninety agricultural societies and nearly forty boards of trade and chambers of commerce had established committees to serve as liaisons to the Signal Service.[61] In the mid-1870s, the Signal Service introduced and expanded river-

58. Chas. F. Vanderford to Chief Signal Officer, February 6, 1875, box 779, Signal Office Agricultural Correspondence. By the early nineteenth century, hundreds of various local "farmer's almanacs" were in publication but not in competition with each other, since most were limited to state or regional weather forecasts and to circulations well under one hundred thousand. These almanacs contained general long-range weather forecasts (based largely on planetary meteorology) to help readers anticipate optimum planting and harvesting times. Although nearly every almanac since the late seventeenth century, not only farmer's almanacs, included some type of long-range weather prediction, almanacs devoted entirely to the weather did not appear until the 1877 publication of Canadian geologist and ornithologist Henry Vennor's *Vennor's Weather Almanac.* On the role of almanacs in weather forecasting, see Robb Sagendorph, *America and Her Almanacs: Wit, Wisdom and Weather, 1639–1970* (Boston: Little, Brown, 1970), 25, 118–19.

59. C. M. Hines to Office of the Chief Signal Officer, June 2, 1874; A. S. Kelley to Chief Signal Officer, May 13, 1874, both in box 777, Signal Office Agricultural Correspondence.

60. For complaints, see Jas. T. Kenney to Chief Signal Officer, July 18, 1873; Jacob Vaughn to Acting Signal Officer, March 11, 1873; E. B. Pendleton to Chief Signal Officer, March 13, 1873, all in box 774, Signal Office Agricultural Correspondence. Quotations in Jonathan Jessup to Chief Signal Officer, January 24, 1873; D. H. Brereton to Chief Officer Signal Service, January 10, 1873, both in box 774, Signal Office Agricultural Correspondence.

61. Craft, "Private Weather Organizations," 1068.

level reports, frost warnings, and hurricane warnings, as well as a system of flag signals to relay forecasts and storm warnings. People who did not live near a Signal Service station or Western Union or newspaper office decoded geometric and colored flags signifying changes in temperature and conditions (e.g., a red sun indicated higher temperatures; a blue sun, general precipitation; a blue crescent, clear weather). Flag signals were accompanied by "The Weekly Weather Chronicle" (1872–1877) and the Farmers' Bulletin (1873–1881).[62] But the value of this information depended on its timely distribution, as a meteorological committee from Murfreesboro, Tennessee, made clear when it found "no practicable benefit" of the "Weekly Weather Chronicle" because it was "received several days after the reports are made up with no *probabilities* for the future."[63] The Farmers' Bulletin would have the same problem.

FARMERS' WEATHER INDICATORS

Beginning in 1873, the Signal Service sent reports based on the midnight forecast by morning express mail to over six thousand rural post offices, where each postmaster was required by order of the postmaster general to display the bulletin in a special frame and report the time of its receipt and posting each day.[64] The Farmers' Bulletin, not much different from the Synopsis and Probabilities, provided a summary of the previous day's weather and a one-sentence regional forecast of barometric pressure, winds, temperature, cloud conditions, and precipitation. For example, the New England Probabilities issued at 1:00 a.m. on April 3, 1876, read, "Falling barometer, southerly winds, warmer, hazy, followed by cloudy weather and falling barometer."[65] In some areas, postmasters judged the Farmers' Bulletin a timely, relevant, and trusted form of information. In West Northfield, Massachusetts, for example, the Farmers' Bulletins arrived daily on the 10:00 a.m. mail train "and [were] immediately posted in a conspicuous place. The *Farmers* (*particularly*) and others here have watched the publication with *much* interest and great curiosity which have been so *reliable* in *most* instances that *much confidence is being placed in*

62. Whitnah, *History of the Weather Bureau*, 31, 28; Bates and Fuller, *America's Weather Warriors*, 10–11.

63. Wm. P. Henderson to Chief Officer of the Army, January 2, 1873, box 774, Signal Office Agricultural Correspondence.

64. US Department of War, *Annual Report of the Secretary of War* (1876), 110.

65. "Farmers' Bulletin," April 3, 1876, box 782, Signal Office Agricultural Correspondence.

them."[66] In Streetsboro, Ohio, the postmaster reported that "a majority [of farmers] take great interest in the 'Probabilities' and many a plan for work on the morrow is laid out from it," and in Georgetown, Pennsylvania, "the citizens and farmers would like to see them every day."[67]

But in other areas, farmers complained that the Farmers' Bulletin, like the Synopsis and Probabilities, was late, redundant, or irrelevant. Rural postmasters pronounced the Farmers' Bulletin "trouble that does no one any good," "of no practical utility whatever," and "a useless waste of time and money."[68] A Michigan postmaster wrote that "the Bulletin is received so late in the day 1. o'clock that farmers say they can tell nearer by looking at the sky what the weather will be than to look at the report," and a Maryland postmaster observed, "Our mail does not arrive till from 2 to 3 PM so that the time Mr. Probabilities tells us story about is almost entirely past."[69] The Signal Service calculated that daily bulletins reached the countryside no later than 2:00 p.m., and on average by 11:00 a.m., but many communities reported receiving the Farmers' Bulletin much later, and many postmasters suggested that the bulletins be discontinued if they couldn't arrive sooner.[70] In some areas, Farmers' Bulletins merely duplicated the "probabilities" from daily newspapers, and in other areas, as an Ohio postmaster reported, "no attention is paid to them." In Montgomery Station, Indiana, the postmaster declared, "This is universally

66. E. E. Belding to Chief Signal Officer, February 22, 1876, box 782, Signal Office Agricultural Correspondence (emphasis in original). See also John E. Mander to War Department, Signal Service United States Army, Division of Telegrams and Reports for the Benefit of Commerce and Agriculture, November 17, 1877; G. B. Townley to War Department, Signal Service United States Army, Division of Telegrams and Reports for the Benefit of Commerce and Agriculture, April 21, 1877, both in box 785, Signal Office Agricultural Correspondence.

67. A. D. Peck to General Albert J. Myer, March 19, 1875, box 779; J. C. Trimble to Chief Signal Officer, December 19, 1877, box 785, both in Signal Office Agricultural Correspondence.

68. S. Bride to War Department, Signal Service United States Army, Division of Telegrams and Reports for the Benefit of Commerce and Agriculture, April 22, 1876, box 782; F. F. Robley to Chief Signal Officer, April 25, 1876, box 782; M. L. Bartlett to Office of the Chief Signal Officer, December 3, 1877, box 785, all in Signal Office Agricultural Correspondence.

69. Merritt Boyd to Chief Signal Officer, April 25, 1876; L. D. Henson to Chief Signal Officer, April 29, 1876, both in box 782, Signal Office Agricultural Correspondence.

70. US Department of War, *Annual Report of the Secretary of War* (1876), 110. For typical complaints, see Wm. F. Jones to H. M. Howgate, March 30, 1875, box 779; Wm. E. Ward to Postmaster General, June 1, 1874, box 777, both in Signal Office Agricultural Correspondence.

conceded to be a humbug., it is doing no good in this part of the country would like to have it discontinued."[71]

To others, the Farmers' Bulletin represented a broader failure of the Signal Service's attempts to serve both commerce and agriculture. In January 1878, farmers in St. Johns, Michigan, heard lectures on various topics, including an appraisal of the value of government weather forecasts for agricultural regions. Michigan agricultural scientist R. F. Kedzie declared that Farmers' Bulletins "do not measure up to the full requirements of the farmer" who lived so far from cities and Signal Service stations that he received outdated forecasts. Kedzie concluded that "the actual benefits derived from the signal service by the farmers are correspondingly small" and urged reform of the agency.[72] Calls for an expanded weather service persisted throughout the decade—between 1876 and 1877, for example, the chief signal officer received 294 separate appeals for the establishment of new weather reporting stations in 212 different locations—but limited congressional funding prevented the Signal Service from coming close to meeting this demand, despite its acknowledgement "that the public interests require that more stations should be provided for."[73]

In addition to Farmers' Bulletins and new observation stations, instrumentation offered another potential solution to the problem of forecasting in the countryside. In 1877, Myer wrote that "any intelligent farmer, supplied with the necessary simple instrument . . . and furnished with data" should be able to forecast, and by 1879 he had secured a patent on "The Weather Case, or Farmer's Weather Indicator."[74] (See figure 2.3.) About 2.5 feet high, just over

71. H. Beall to War Department, Signal Service United States Army, Division of Telegrams and Reports for the Benefit of Commerce and Agriculture, July 7, 1877, box 785; J. C. Montgomery to Chief Signal Officer, April 29, 1876, box 782, both in Signal Office Agricultural Correspondence.

72. "The Michigan Farmers," *Chicago Inter Ocean*, January 29, 1878. Farmers' institutes, public lectures on agricultural practices and research, were regularly held in Michigan, along with Nebraska, Pennsylvania, Rhode Island, Colorado, Ohio, Mississippi, Maine, Missouri, Wisconsin, and New York throughout the late nineteenth century, well before the federal funding of agricultural experiment stations by the 1887 Hatch Act. Jeffrey W. Moss and Cynthia B. Lass, "A History of Farmers' Institutes," *Agricultural History* 62, no. 2 (1988): 150–51.

73. US Department of War, *Annual Report of the Secretary of War* (Washington, DC: Government Printing Office, 1877), 162–67; quotation in Garrick Mallery to Charles Rogers and F. W. Balch, April 20, 1875, Miscellaneous Correspondence of the Signal Office, 1870–93, Agricultural Series, Letters Sent, 1873–78, vol. 2, RG 27, NACP.

74. US Department of War, *Annual Report of the Secretary of War* (1877), 115; Scheips, "'Old Probabilities,'" 39.

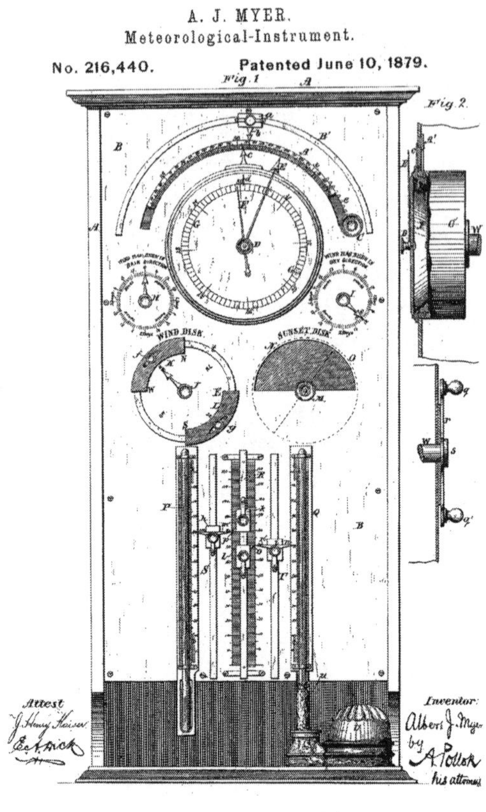

FIGURE 2.3. Chief Signal Officer Albert J. Myer was awarded a patent for "The Weather Case, or Farmer's Weather Indicator" in 1879. US Patent 216,440, issued June 10, 1879, United States Patent and Trademark Office, www.uspto.gov.

1 foot wide, and 4.5 inches thick, this clocklike instrument contained an aneroid barometer, a "Wind Disk" for measuring wind direction, a "Sunset Disk" for indicating a "fair" or "foul" sunset, and dry- and wet-bulb thermometers for comparing dry and moist weather conditions in order to predict temperature change. In 1880, the *American Agriculturalist* noted that the Signal Service distributed the instruments, which Myer planned to sell at cost, "so that farmers living beyond the reach of the daily weather reports may of themselves . . . make forecasts of the weather much more accurately than they could without the aid furnished by the Signal Service."[75] Yet Myer reminded farmers that his

75. "The U.S. Signal Service and Farmers," *American Agriculturalist*, June 1880, 231. This was not the first time that the federal government sought to provide meteorological instruments

invention, which was "not intended to be used independently of the official weather reports," would not eliminate uncertainty about the future or the need for subjective judgment: "The weather case is only to aid, sometimes, in making up one's mind as to what the weather of the next day will be. While it will often be very useful, there will be many instances in which everything will be left in doubt."[76]

Other technological solutions to the problem of disseminating forecasts in the countryside included railroad, cannon, and flag signals, which were adopted sporadically during the 1870s and 1880s with varying degrees of success. In 1881, the *Cultivator and Country Gentleman* proposed a combination of telegraph and cannon signals "intended as a universal system to give certain, instant and general warning of tornadoes, or rain storms, for the benefit of agriculture."[77] A Virginia farmers' club complained in 1885 that "the farmers, the class to be most benefited by the weather signal service, in reality derive the least benefit therefrom; only those living in proximity to large centers can be informed of the changes likely to occur, or in the case of extraordinary changes those living near telegraph offices," and recommended flag signals be displayed on railroad cars.[78] In general, Signal Service officials focused on increasing the number of observation stations and extending telegraph lines as the key to improving the weather service in the countryside, whereas farmers' groups often urged a combination of telegraphy and other signaling methods.

In the 1880s, Myer's successor, General William B. Hazen, oversaw a research-oriented era of professionalization, which included new forms of information tailored to agriculture. In addition to studies of tornadoes, analysis of climatological data, and an official meteorology textbook, the Signal Service issued crop-specific weather reports, including special frost warnings for sugar growers, special observations and reports for cotton-producing states, and tobacco, corn, and wheat region reports.[79] Cotton region reports added a

to farmers. In 1855, the agricultural branch of the US Patent Office had already distributed weather observation instruments, including two hundred rain gauges, for the purposes of climatological data collection and analysis. Whitnah, *History of the Weather Bureau*, 15; Fleming, *Meteorology in America*, 25–28, 109.

76. Albert J. Myer, "The Weather Case, or Farmers Weather Indicator," *Nature*, October 10, 1878, 624.

77. "Signal Service Warnings," *Cultivator and Country Gentleman*, March 3, 1881, 143.

78. "Weather Probabilities," *Cultivator and Country Gentleman*, December 24, 1885, 1038–39.

79. "The Work of the Signal-Office Under General Hazen," *Science* 5, no. 111 (1885): 237–38. A new series of Signal Service Professional Papers included Paper No. 7, *Report on the Character*

daily observation at 6:00 p.m. and the comparison of daily temperature highs and lows, while corn and wheat region reports had an additional 7:00 a.m. observation. The Signal Service also issued frost warnings for fruit growers whenever the temperature dropped below 40°F.[80]

Hazen's initiatives, like Myer's, met with modest appreciation, but the problem of distributing forecasts in the countryside persisted. The *Cultivator and Country Gentleman* cited the "great value" of frost warnings for Louisiana sugar growers and praised Signal Service publications on tornadoes, but it also noted complaints about inadequate frost warnings, which "le[d] many persons to question the benefits of this service to the cause of agriculture."[81] In 1886, the *Manufacturer and Builder* explained the limitations of the government weather reporting network just as Myer had a decade earlier:

> At present the means of disseminating the official daily predictions of the service are limited. The official predictions are telegraphed to the stations along the lines of a number of the principal railways, and are there posted up for the information of the agricultural community; but in most of the States this information reaches only a limited proportion of the population.[82]

"TO EVERY RURAL HOME"

Demands for improved weather forecasting in the countryside reached the national stage at the same time that government officials began to question the efficacy of military administration of a national weather service. Between 1881 and 1884, Secretary of War Robert Todd Lincoln, Senator John A. Logan,

---

of Six Hundred Tornadoes (1882) and Paper No. 9, *Charts and Tables Showing Geographical Distribution of Rainfall in the United States* (1883). For the Signal Service's meteorology textbook, see William Ferrel, *Recent Advances in Meteorology, Systematically Arranged in the Form of a Text-Book . . .* (Washington, DC: Government Printing Office, 1886).

80. Hughes, *Century of Weather Service*, 21–23; Whitnah, *A History of the Weather Bureau*, 33.

81. "Recent Publications: Destructive Storms," *Cultivator and Country Gentleman*, September 17, 1885, 768; "The Signal Service," *Cultivator and Country Gentleman*, November 18, 1880, 748; quotation in "The Signal Service—Wheat," *Cultivator and Country Gentleman*, September 27, 1883, 777–78. For additional criticism from farmers' groups, see "Weather Probabilities," *Cultivator and Country Gentleman*, December 24, 1885, 1038–39.

82. "Extending the Usefulness of the Signal Service," *Manufacturer and Builder* 18, no. 11 (1886): 243.

and a committee of the National Academy of Sciences each urged a transfer to civilian oversight.[83] From 1884 to 1886, Senator William B. Allison led a joint congressional investigation of federal scientific agencies, and in the case of the Signal Service, investigated whether its work was military or civilian in nature. Although General Philip Sheridan testified that the weather service "belongs to the business of the country" in advocating civilian oversight, the Allison Commission's recommendation—three members called for a gradual transfer to a civilian bureau and the other three for an immediate transfer—did not immediately come to fruition.[84] Perhaps the loudest calls for the transfer came from Massachusetts's Blue Hill Observatory meteorologist Henry Helm Clayton, who led a movement "among the land-grant colleges, experiment stations, state boards of agriculture, and the Grange to capture the weather bureau."[85]

National and state farmers' organizations increasingly turned to the federal government with their appeals for an improved weather service for agriculture in the mid-1880s. In 1884, the National Agricultural Congress petitioned Chief Signal Officer Hazen for "the extension of [weather service] benefits to every rural home however humble," but Hazen cited prohibitive costs of one hundred thousand dollars to send additional telegrams to roughly one-third of the lines operated by Western Union, seventy-five thousand dollars for providing signal flags, and seventy-five thousand for displaying them.[86] After the *Cultivator and Country Gentleman* printed Hazen's response, agrarian organizations intensified their efforts. In 1885, the Virginia Farmers' Assembly adopted a resolution "to memorialize the proper federal authorities in favor of extending the weather signal service to the rural districts," and at the 1886 National Grange meeting, the Committee on Agriculture approved the passage of a similar resolution:

83. Fleming, "Storms, Strikes, and Surveillance," 328–29.

84. Editorial, *NYT*, January 23, 1886, 4. For an account of the administrative delays and legislative technicalities that doomed an 1887 bill proposing the transfer of the weather service to the US Department of Agriculture, see [Henry Helm Clayton], *The Transfer of the United States Weather Service to a Civil Bureau* (Boston: Alfred Mudge & Son, 1889), 5.

85. Quotation in A. Hunter Dupree, *Science in the Federal Government: A History of Policies and Activities to 1940* (Cambridge, MA: Belknap Press, 1957), 192.

86. Robert Beverly, A. J. McWhitter, and M. C. Ellzly, "Petition to General W. B. Hazen," 1884, folder "Benefits to Agr. of the Signal Service Weather Reports"; W. B. Hazen, "Response to Petition from National Agricultural Congress," January 22, 1884, both in folder "Benefits to Agriculture of the Signal Service Weather Reports," box 131, HR 48A-H2.2, Committee on Agriculture, RG 233, NAB.

Whereas, It is of more importance to farmers than to any other class of citizens to know what changes are likely to occur in the weather; and,

Whereas, The farmers furnish three fourths of the exports of the country, and are the largest tax-payers in support of the General Government, therefore,

Resolved, That the Congress of the United Sates be respectfully requested to pass such laws as shall furnish the benefits of the Signal Service to the farmers of the country.[87]

These were not unusual demands. Farmers' calls for better access to modern, scientific weather information were consistent with the business- and market-oriented politics of the Farmers' Alliance and the Populist movement more broadly. The "general desire for information and almost universal effort at research" of the Farmers' Alliance was clearly shared by the National Grange's Committee on Agriculture and the many state agricultural societies and local farmers' clubs that petitioned the federal government for access to weather forecasts in the 1870s and 1880s.[88] In 1886, during House hearings on two bills proposing the extension of the weather service into the countryside, Robert Beverly, president of the National Agricultural Congress, pointed to widespread discontent with the current reach of the weather service, testifying that "there has not been an agricultural meeting of any society state or local in the last two years that has not called for this bill. Everyone has called for it." Beverly noted that trains didn't pass his own house, fifty miles from Signal Service headquarters in Washington, DC, until 5:00 p.m. each day, and even Farmers' Bulletins arriving at 10:00 or 11:00 a.m. were too late for farmers who began work before sunrise.[89]

By the end of the 1880s, few believed that the Division of Telegrams and Reports for the Benefit of Commerce and Agriculture was of much benefit

87. "Virginia Farmers' Assembly," *Baltimore Sun*, August 13, 1885; *Journal of Proceedings: Twentieth Session of the National Grange of the Patrons of Husbandry, 1886* (Philadelphia: J. A. Wagenseller, 1886), 133.

88. Notes of J. W. H. Davis, John B. Rushing Papers, Briscoe Center for American History, University of Texas at Austin, as quoted in Postel, *Populist Vision*, 138.

89. For the Relief of Farmers of the United States by Extending to Them the Benefits of the Signal Service, H.R. 2318, 49th Cong., 1st sess. (1886); For Extending the Benefits of the Signal Service to Farmers of the United States, H.R. 2506, 49th Cong., 1st sess. (1886); Unpublished Hearings on H.R. 2318 and H.R. 2506 Before the House Committee on Agriculture, 49th Cong., 1st sess., 22, quotation at 30 (May 5, 1886), in folder "Extending the Benefits of the Signal Service to the Farmers of the United States," box 16, Committee Papers, RG 233, NAB.

to farmers. In 1889, US Army Signal Corps lieutenant John Walsh wrote to the secretary of agriculture to argue that the transfer of the weather service to the US Department of Agriculture would be a boon for farmers, predicting, like many others, a more efficient weather service "unhampered by military red tape."[90] In 1890, the secretary of the Michigan State Board of Agriculture complained of the Signal Service's "apparent lack of vital interest in serving the welfare of the people," noting that farmers were particularly disadvantaged by a lack of timely storm warnings.[91] But residents of Lenawee County, a farming region in southeastern Michigan, did not face this problem, given that they had operated their own local telegraph network throughout the decade.[92] The Lenawee farmers' information network began locally and then expanded, unlike the government weather service, which began as a national network and then endeavored to reach into the countryside.

"EVERY FARMER IS HIS OWN OPERATOR"

In June 1881, northwest of Tecumseh, Michigan, (population 2,100), farmer Raymond J. W. Bowen strung a telegraph line between his house and a neighboring farmer's. By December, the line extended half a mile west to a third house, and by fall 1882, three miles into the center of Tecumseh. In January 1883, this private cooperative was incorporated as the Commercial Telegraph Company, and it grew rapidly thereafter.[93] By 1888, the company operated

90. John Walsh to Jeremiah Rusk, December 28, 1889, box 1, Records Relating to the Transfer of Meteorological Functions from the War Department to the Agriculture Department, 1887, Meteorological Correspondence of the Signal Office, 1870–1893, Correspondence Series, RG 27, NACP.

91. Henry Reynolds, "Petition from Michigan State Board of Agriculture to Transfer the Weather Service to the USDA," March 20, 1890, folder "Transfer of the Weather Bureau to the Dept. of Agr.," box 93, HR 51A-H1.8, Committee on Agriculture, RG 233, NAB.

92. Lenawee County boasted numerous farmers' organizations, including the Farmers' Social Club (est. mid-1870s) and the Tecumseh Grange (est. 1875). Agricultural production in Lenawee County in the 1880s, particularly around the town of Tecumseh, centered on flour, wool, dairying, and celery farming. Clara Waldron, *One Hundred Years, a Country Town: The Village of Tecumseh, Michigan, 1824–1924* (1974; repr., Tecumseh, MI: T. A. Riordan, 1968), 129, 143–45.

93. Waldron, *One Hundred Years*, 146. On the tradition of setting up rural telephone cooperatives (sometimes with barbed wire fences), see Ronald R. Kline, *Consumers in the Country: Technology and Social Change in Rural America* (Baltimore: Johns Hopkins University Press, 2000), 26–40.

sixty-five miles of wire linking ninety telegraph instruments, two-thirds of which were located in farmhouses, with the remaining third located, according to the *New York Times*, "in stores where these farmers do their trading."[94] By 1890, the line that had originally connected two farmers three hundred feet apart "strictly as a social affair" had grown to include one hundred miles of wire and 170 instruments, 95 of which belonged to farmers in Tecumseh and surrounding towns. The farmers' informal telegraph line had become a commercial network. The Commercial Telegraph Company, along with the Tecumseh and Macon Telegraph Company's six miles of wire linking eleven more operators, "form[ed] a continuous line spreading in almost every direction from the village of Tecumseh."[95]

The cost of operating and maintaining Commercial Telegraph Company lines was shared by its members, who purchased their own telegraph keys, erected their own poles (readily available at a nearby tamarack swamp), and strung their own wires. The telegraph's wet-cell batteries were clustered in designated houses along the line, with fifteen to twenty batteries per house. As the *New York Times* reported, "Every farmer is his own operator, battery man, and line repairer."[96] The by-laws of the Commercial Telegraph Company stipulated that shareholders would pay twenty cents to send a message of ten words or fewer, with no charge for the reply, with each word over ten incurring an additional two cents (a fee not often collected). Members engaged in social exchanges were required to yield to business-related messages, and only "necessary business" was permitted on the wires between 10:00 p.m. and 5:00 a.m.[97]

Farmers outside Tecumseh, because of their distance from city and village telegraph offices (twenty-five miles southwest of Ann Arbor and over fifty miles west of Detroit), would not have had access to the nearly seven hundred thousand miles of Western Union telegraph wire that covered the country by the end of the 1880s.[98] By 1890, the Commercial Telegraph Company connected its ninety-five farmers to nineteen stores, ten doctors' offices, seven railroad offices, six post offices, seven mechanics, and two grain dealers, in addition to a

94. "A Unique Telegraph Line," *NYT*, May 11, 1888.

95. "Historian Reviews Telegraph," *Tecumseh Herald*, January 20, 1966, folder "Commercial Telegraph Company," Clara Waldron Historical Room Archives, Tecumseh District Library, Tecumseh, MI (hereafter Commercial Telegraph Company records).

96. "Historian Reviews Telegraph," *Tecumseh Herald*, January 20, 1966, Commercial Telegraph Company records; quotation in "A Unique Telegraph Line," *NYT*, May 11, 1888.

97. "The Commercial Telegraph Co. By-Laws," n.d.; "Tecumseh and Macon Telegraph Company By-Laws," n.d., Commercial Telegraph Company records.

98. Field, "The Magnetic Telegraph," 402.

couple of teachers and a so-called pleasure resort.[99] Linked to the network in Tecumseh were druggists, a hardware store, a veterinarian, a painter, a banker, a miller, an undertaker, a jeweler, and a Western Union telegraph office, with the proprietor of the village store serving as the area's "telegraph operator, express agent, and postmaster."[100]

The Commercial Telegraph Company's network connected Lenawee County farmers to their local economy as well as to the weather forecasts telegraphed daily from Signal Service headquarters and then relayed by signal flags at the Western Union office.[101] Farmers could receive daily weather reports on their own instruments or read the signal flags in time to cover crops from unexpected frost or adjust their schedules to beat a coming storm. The *Christian Advocate* cited the Michigan farmers' telegraph network as "evidence that ruralists are making progress . . . and are not so far behind the so-called business classes as many suppose," and the *New York Times* saw a promising solution to the problem of rural information delay:

> When a piece of important news is received by the station officers, and at the newspaper office some event of great national or State importance, it is sent over the farmers' lines, and by this means the farmers, who are regarded as slow and behind the times, are often several hours ahead of the reportedly faster denizens of the cities, who are waiting for their afternoon papers to appear. There are two or three independent systems of these lines in operation in the county, arranged so that they can be connected with each other at intersecting points, and the whole system is being worked very cheaply and successfully.[102]

These other systems were likely set up with the help of Raymond J. W. Bowen, creator of the original telegraph line between his house and a neighbor's, who served as general manager of the Commercial Telegraph Company as well as a

99. "Historian Reviews Telegraph," *Tecumseh Herald*, January 20, 1966, Commercial Telegraph Company records.

100. "List of Instruments for Commercial Telegraph Company," n.d., Commercial Telegraph Company records; quotation in "The Farmers' Telegraph," *Chicago Daily Tribune*, November 16, 1892.

101. For a mid-twentieth-century account of the Michigan farmers' telegraph network citing the dissemination of weather information first—before social connection—in its list of the Commercial Telegraph Company's various functions, see "Historian Reviews Telegraph," *Tecumseh Herald*, January 20, 1966, Commercial Telegraph Company records.

102. "Hints to the Farmer," *Christian Advocate*, December 10, 1891; "A Unique Telegraph Line," *NYT*, May 11, 1888.

consultant and supplier of telegraph keys. Indeed, Bowen's obituary recalled his experience setting up local telegraph networks around the state.[103]

The Lenawee farmers' telegraph network was eventually shut down by the arrival of the telephone. In 1895, Bell Telephone poles linked the towns of Tecumseh and Saline, and long-distance service was available between Adrian and Detroit. The Commercial Telegraph Company was bought by the Tecumseh Telephone Company in 1898, by which time the telegraph wires were falling into disrepair, and Bowen went to work for the telephone company after the buyout.[104] The farmers' telegraph ceased operation as telephone service became more widely available, although, as one resident lamented, "it was now necessary to drop what you were doing and take down a receiver in order to listen in."[105] The telegraphic transmission of weather forecasts, news, prices, orders, and gossip that lasted nearly two decades in Lenawee County was, however successful, an unusual circumstance, and the problem of forecasting in "the distant parts of the country" remained.

"A HIGHER DEGREE OF LOCAL ADAPTATION"

Throughout the 1880s, critics charged that the weather service was costly, inefficient, and inaccurate—and would be better run as a civilian institution.[106] Although the Signal Service boasted verification percentages between 85 and 90 percent in the mid-1880s, the *Chicago Journal* complained that "the signal-service weather predictions are as faulty as ever. It is safer to bet on the election than to bet on the signal service."[107] Proponents of a civilian weather service eagerly awaited the arrival of the "new meteorology": a "free and rational"

103. "The Steiner Telegraph Key," n.d., advertisement; "R. J. W. Bowen Dies Here on Wednesday," *Tecumseh Herald*, March 31, 1932, both in Commercial Telegraph Company records.

104. Waldron, *One Hundred Years*, 147.

105. On the introduction of rural telephony, see Kline, *Consumers in the Country*, chap. 1; Claude S. Fischer, *America Calling: A Social History of the Telephone to 1940* (Berkeley: University of California Press, 1992). Quotation in Waldron, *One Hundred Years*, 135.

106. On criticisms of the Signal Service, see Clayton, *Transfer of the Weather Service*, 11–29.

107. On verification percentages, see W. B. Hazen, *History of the Signal Service, with Catalogue of Publications, Instruments, and Stations* (Washington, DC: US Army Signal Service, 1884), 10. Quotation in Clayton, *Transfer of the Weather Service*, 28.

science administered by a decentralized agency.[108] At the end of the decade, T. C. Chamberlin, president of the University of Wisconsin, writing on behalf of members of a farmers' institute to Senator John C. Spooner, identified the problem and its solution:

> It is obvious to all . . . that [the weather service] needs reorganization . . . if it is to . . . become in the highest practical degree subservient to the interests of agriculture. It must be rendered more exact, and be made more specifically applicable to the peculiarities of the different regions and to the needs of different agricultural industries. In other words, there must be a higher degree of local adaptation.[109]

Chamberlin recognized the challenge that the Signal Service had faced since the early 1870s: making weather information local.

By the late 1880s, the combination of state weather agencies, agricultural colleges, and agricultural experiment stations offered a promising approach to decentralized local forecasting. A state weather service project envisioned by the nation's leading meteorologist, Cleveland Abbe, grew to include over two thousand volunteer observers, who correlated weather data with local agricultural production, and state weather agencies were hailed as "a more comprehensive and thorough system of disseminating the information" that could "place within reach of every farming community" Signal Service forecasts.[110]

108. Thomas J. Brown, "The Present Crisis in Meteorology; and the Opportunity and Duty of the Agricultural Interest at this Juncture," undated pamphlet, box 1, Records Relating to the Transfer of Meteorological Functions from the War Department to the Agriculture Department, 1887, Meteorological Correspondence of the Signal Office, 1870–1893, Correspondence Series, RG 27, NACP.

109. Quotation in Clayton, *Transfer of the Weather Service*, 16.

110. State weather agencies cropped up sporadically between the 1840s and the 1880s, but no sustained state system survived without Signal Service oversight. The most successful early state system was in New York, with thirty stations networked in a system for the Board of Regents from 1825 to the early 1850s. The Pennsylvania state legislature approved four thousand dollars for stations that paled in comparison to New York's, and in Massachusetts, twelve stations were established by 1850 but were short-lived. Abbe's project covered Illinois, Indiana, Kansas, Mississippi, Missouri, Nebraska, New Jersey, and Ohio, which joined Iowa and Nevada, which had created state weather agencies in 1878 and 1881, respectively. Whitnah, *History of the Weather Bureau*, 11–12, 32–33; Hughes, *Century of Weather Service*, 27; "The Task of State Weather Services," *Science* 13, no. 310 (1889): 20; quotation in "Auxiliary Weather Bureaux in the States," *Manufacturer and Builder* 19, no. 6 (1887): 123.

Agricultural colleges, which received monthly data from state weather agencies before it was forwarded to Signal Service headquarters, were a key part of local weather information networks. The president of one state agricultural college observed in the late 1880s that "agricultural colleges and the newly formed or forming experiment stations constitute a *cadre* organization upon which the new system [a civilian weather service] can be soon developed and perfected at a very moderate cost," noting that state agricultural colleges in Alabama, Ohio, Tennessee, Wisconsin, Massachusetts, and New York had already established their own county- or state-level weather bureaus.[111]

While state weather services, agricultural colleges, and experiment stations were building decentralized information networks that represented a departure from the military's centralized system, resistance to the growing consensus favoring a civilian weather service came from Chief Signal Officer Adolphus W. Greely.[112] Greely and his allies argued that military discipline was necessary to coordinate synchronous observations, that a civilian agency would be far more expensive, that military personnel would be more suited to frontier or wartime service, and that a civilian organization would quickly become stocked with political appointees.[113] But such protests by Greely, whose organization had been derided as "the highly ornamental, highly expensive, and all but useless Drum-Major of Meteorology," proved fruitless.[114] Sufficient momentum for civilian oversight had been generated by bills and hearings in the 1880s, the Signal Service's tattered public image after an embezzlement scandal in 1881, and Greely's own acknowledgement of widespread inefficiencies.[115]

In 1890, Congress finally approved the transfer of the weather service to the US Department of Agriculture, which formally took place on July 1, 1891,

111. Whitnah, *History of the Weather Bureau*, 33; Dupree, *Science in the Federal Government*, 167; Clayton, *Transfer of the Weather Service*, 21–22, quotation on 21.

112. On state weather services, agricultural colleges, and experiment stations as an "an alternative model of organization" to a centralized weather information network, see Fleming, "Storms, Strikes, and Surveillance," 330.

113. Clayton, *Transfer of the Weather Service*, 6–7.

114. Thomas J. Brown, "The Present Crisis in Meteorology; and the Opportunity and Duty of the Agricultural Interest at this Juncture," undated pamphlet, box 1, Records Relating to the Transfer of Meteorological Functions from the War Department to the Agriculture Department, 1887, Meteorological Correspondence of the Signal Office, 1870–1893, Correspondence Series, RG 27, NACP.

115. Whitnah, *History of the Weather Bureau*, 46–48; *Condition of Signal Corps*, H.R. Doc. No. 50-143 (1889).

with the creation of the US Weather Bureau. Signal Corps observers would no longer telegraph their daily weather observations to Washington, DC, but they did lend their signaling expertise during the labor conflict that rocked the Carnegie steel town of Homestead, Pennsylvania, in 1892 and the riots in the wake of Chicago's Pullman strike in 1894, helping National Guardsmen establish signaling networks to relay details of the violence.[116] The Panic of 1893 and the ensuing depression meant budget-cutting measures for the new Weather Bureau. Secretary of Agriculture J. Sterling Morton reduced the number of telegraphed forecasts and curtailed theoretical research in favor of improving the accuracy of short-term forecasting.[117] Still, by 1894, the Weather Bureau had forty forecasters on staff (as compared to four in 1891) and a more competitive system for promotion. Weather Bureau chief Mark W. Harrington noted that civilian forecasters would have "more freedom in . . . verbal expression" without being constrained by the "hard and fast lines" of standardized military reports, and, as a result, USDA forecasters would "express their forecasts for the benefit of the public rather than for the benefit of their official records."[118]

Yet despite the civilian reconstitution of the weather service, the proliferation of state weather agencies, and changes in forecasting protocol, the problem of distributing forecasts in the countryside remained. In the mid-1890s, the Weather Bureau's twenty-four- to thirty-six-hour forecasts were telegraphed from new district centers in Washington, Chicago, San Francisco, and Portland to substations and then to hundreds of towns and newspapers nationwide. Telephone companies in major cities displayed the forecast in their offices and relayed it to their subscribers, and Weather Bureau maps printed in newspapers found a yearly circulation of close to fifty million in 1894.[119] But Harrington, a University of Michigan meteorologist who had grown up on a farm in Illinois, acknowledged that "the problem of reaching, effectively and in time, the more sparsely settled districts is still unsolved. For instance, how

116. Fleming, "Storms, Strikes, and Surveillance," 331.

117. Whitnah, *History of the Weather Bureau*, 65–66. On Morton's controversial reductions in funding requests, personnel, and the seed distribution program that served nearly two million farmers, see Daniel P. Carpenter, *The Forging of Bureaucratic Autonomy: Reputations, Networks, and Policy Innovations in Executive Agencies, 1862–1928* (Princeton, NJ: Princeton University Press, 2001), 199–200.

118. Mark W. Harrington, "What Meteorology Is Doing for the Farmer," in *Yearbook of the United States Department of Agriculture* (Washington, DC: Government Printing Office, 1894), 117–18, quotation on 118.

119. Robert DeC. Ward, "The Newspaper Weather Maps of the United States," *American Meteorological Journal* 11, no. 3 (1894): 106.

can we tell the isolated overseers of cattle ranges in Wyoming of the approach of a blizzard?"[120]

PROBABILITIES BY PHONE

Harrington's answer came at the turn of the century, when Rural Free Delivery (RFD) and rural telephone networks carried forecasts far into the countryside. Calls for RFD echoed throughout the late nineteenth century, from the National Grange in the 1870s, postmaster general and department store magnate John Wanamaker in 1889, and the People's Party in 1892 and 1896.[121] Georgia Populist Tom Watson worked tirelessly for the cause, but RFD was not exactly a product of grassroots politics. Rather, it was an experimental program made permanent by "postal officials [who] conceived, planned, lobbied for, and secured the passage of RFD as did no other actor or force in the United States," calling on agrarian organizations to pressure Congress into funding it.[122] The Post Office implemented eighty-two experimental RFD routes in 1896 (following a short-lived experiment in 1891–1892) and several hundred more over the next two years. In 1900, RFD boasted over 2,500 routes serving over 1.8 million Americans, with 15,000 routes in 1902 and nearly 25,000 the following year. The partisan nature of RFD expansion in areas that were politically valuable to Republicans in congressional and presidential elections at the turn of the century meant that the distribution of RFD routes disproportionately favored the Northeast and Midwest.[123]

When RFD was permanently established in 1902, periodicals, mail-order catalogues, farm journals, crop reports, and personal correspondence arrived in the mailboxes of rural Americans nationwide, along with the Weather Bureau's forecast cards, produced in Weather Bureau stations across the nation

120. Harrington, "What Meteorology Is Doing," 118.

121. Samuel Kernell and Michael P. McDonald, "Congress and America's Political Development: The Transformation of the Post Office from Patronage to Service," *American Journal of Political Science* 43, no. 3 (1999): 796–97. For a comprehensive history of Rural Free Delivery, see Wayne E. Fuller, *RFD: The Changing Face of Rural America* (Bloomington: Indiana University Press, 1964).

122. Postel, *Populist Vision*, 145; Carpenter, *Forging of Bureaucratic Autonomy*, 124–27, quotation on 124.

123. Charles H. Greathouse, "Free Delivery of Rural Mails," *Yearbook of the Department of Agriculture, 1900* (Washington, DC: Government Printing Office, 1900), 513; Carpenter, *Forging of Bureaucratic Autonomy*, 128, 134 (figure 4.4), 138; Kernell and McDonald, "Congress and America's Political Development."

by local officials and observers who used rubber type sets to stamp the daily forecast on postcards.[124] Weather Bureau officials had envisioned that RFD would solve the problem of distributing forecasts in the countryside, a prediction confirmed by farmers' testimonials. In 1900, Iowa farmer E. D. Nauman believed that RFD

> will greatly assist the farmer in a material and practical way by giving him the markets and United States weather forecasts daily. Of the two, I regard the weather forecasts fully as important as the markets. Under the old system the farmers, for whom to a large extent the weather bulletins are intended, do not see them with sufficient regularity to be of much value to them. And at that season of the year when the weather forecasts are of most value to the farmer (harvest time) he is too busy to visit the post office to either see the bulletin there displayed or to get his daily paper.[125]

A California farmer explained that RFD "brings our daily papers promptly, so that saves us time and anxiety. The weather report is dropped in our box, and that is the first thing I look at, to see what it says about the weather tomorrow."[126] And in areas that did not yet have RFD routes, residents wrote to their congressmen requesting the delivery of daily weather reports.[127]

RFD expanded alongside rural telephone networks, and by 1902, many Midwestern and Plains farmers subscribed to a telephone news bureau, whose operators called each evening to read a summary of news, weather, and market reports to all the households sharing a single line.[128] Over one hundred thousand Midwestern farms received telephone forecasts in 1902, and in 1904, some fifty thousand Ohioans received the morning forecast by telephone minutes after it had been issued by the Weather Bureau's district forecasting center.[129] That year, the USDA declared that "the rural telephone lines are now

124. "What the Government Is Doing to Aid the Farmer," *Chicago Daily Tribune*, July 18, 1909; E. B. Calvert, "How the Weather Bureau Disseminates Forecasts and Warnings," in *Report of the Chief of the Weather Bureau, 1895–96* (Washington, DC: Government Printing Office, 1896), xxii.

125. Greathouse, "Free Delivery of Rural Mails," 524.

126. Greathouse, "Free Delivery of Rural Mails," 524.

127. See, for example, Harvey Bassett to J. W. Bailey, February 1, 1904, box 1871, Weather Bureau Correspondence 1870–1912, RG 27, NACP.

128. Kline, *Consumers in the Country*, 29, 42–43.

129. States with telephoned weather reports in 1904 included Iowa, Ohio, Wisconsin, Oklahoma, Illinois, Missouri, Indiana, Arkansas, Massachusetts, New York, and Nebraska. USDA,

the best and most economical means of distributing weather information. The forecasts are quickly disseminated, covering a large territory with little or no expense to the Government."[130] As the Weather Bureau shifted its focus to rural telephone lines, it reduced the number of forecasts mailed through RFD. By mid-1905, the bureau reported over one million subscribers to its daily telephone forecasts, and as the *Southern Farm Magazine* noted approvingly, "Far removed from the daily paper, with its weather forecasts, the farmer can thus keep informed by phone of probabilities, and thus escape loss in crop time."[131] By 1909, the *Chicago Tribune* considered "telephones . . . handmaids to the weather bureau."[132]

Although the spread of rural telephone lines had solved the problem of access to weather forecasts in the countryside, debates over forecasts' accuracy persisted. Despite the work of decentralized state weather services to make local forecasts for agricultural regions, controversies over forecasts' relative accuracy and value, as well as the value of government weather forecasting, would continue.

In 1909, Iowa writer Emerson Hough articulated these uncertainties regarding the nature of weather forecasting in a muckraking essay on the Weather Bureau in which he repeatedly asked, "Now what does the farmer *get* when he gets a forecast?"[133] After reciting a litany of the bureau's shortcomings—its cost, its lack of theoretical innovations, its avalanche of lengthy publications—Hough arrived at this conclusion for an imagined Minnesota farmer:

> It is good advice to John Smith of Sleepy Eye, to trust in the Weather Bureau, but to keep a rheumatic toe. He can further serve himself by observation of the stripes on caterpillars, the spots on the breastbone of a goose, by observations

---

Report of the Secretary of Agriculture, 1904 (Washington, DC: Government Printing Office, 1904), 15.

130. USDA, *Yearbook of the United States Department of Agriculture, 1904* (Washington, DC: Government Printing Office, 1905), 16.

131. USDA, *Report of the Secretary of Agriculture, 1906* (Washington, DC: Government Printing Office, 1906), 21; quotation in "The Telephone on Farms," *Washington Post*, November 28, 1903. On the enthusiastic reception of rural telephone weather forecasts, see "Farmers to Get Daily Bulletins," *Atlanta Constitution*, February 16, 1906; "Weather Over Telephone," *Washington Post*, February 16, 1906; "Clearing and Colder Today," *Atlanta Constitution*, March 8, 1906; "Farmers Given Weather News," *Atlanta Constitution*, June 25, 1906.

132. "Telephones Aids to Weather Bureau," *Chicago Daily Tribune*, November 21, 1909.

133. Emerson Hough, "Does the Weather Bureau Make Good?," *Everybody's Magazine*, May 1909, 613.

> of the phenomena shown by domestic animals, wild fishes, tame pigeons, observations of the silk on the corn, of the fur on the muskrat.[134]

Hough advocated an epistemic compromise between competing systems of knowledge production: the standardized weather forecasts produced by professional meteorologists in a government bureaucracy and the local weather knowledge based on the experience and empirical observation of farmers, rural Americans, and other self-taught forecasters. But arriving at such a compromise between the Weather Bureau and weather prophets was not as simple as Hough imagined it would be.

134. Hough, "Does the Weather Bureau Make Good?," 620.

CHAPTER 3

Weather Prophecies

Long before the Signal Service brought the "Probabilities" into daily life, private forecasters employed a range of methods to predict the weather a month, a season, and even a year ahead. In June 1905, the *Raleigh Morning Post* cited "the infinite desirability of foreknowing the seasons for the benefit of husbandmen [as] at once the opportunity of charlatans and the justification of national weather services." The article further suggested that the successful distribution of the federal government's short-term weather predictions had led to a boom in commercial long-range forecasting:

> The rapid strides of the United States Weather Bureau in recent years toward popular favor through the widespread dissemination of the forecasts—a service made possible largely by the phenomenal spread of the telephone and the development of the rural delivery service—has apparently given a new impetus to unscientific, not to say unscrupulous, forecasts, based upon some theory of cycles or of planetary control. And the chief of the weather bureau is believed to be not only justified, but morally enjoined to counteract as far as possible the mischievous effects of the work of astrologers, who pretend to foretell the character of coming seasons or the progress of storms and ordinary weather conditions for a month or a year in advance, and whose unfounded and unreliable forecasts are too often given undue circulation.[1]

1. "Long Range Weather Forecast," *Raleigh Morning Post*, June 18, 1905, box 1913, Weather Bureau Correspondence 1870–1912, Letters Received, 1896–1912, Records of the Weather Bu-

Commercial weather forecasting, as the *Morning Post* recognized, posed an epistemic threat to government meteorology, and newspapers regularly dramatized a confrontation between Weather Bureau forecasters, who provided fairly accurate twenty-four- to thirty-six-hour projections of empirical surface observations, and private forecasters who peddled speculative long-range predictions.

This conflict between competing modes of knowledge production about the natural world intensified at the turn of the century, when the Weather Bureau came under the leadership of Willis L. Moore, who oversaw a vigorous campaign to discredit commercial long-range forecasters as purveyors of "meteorological soothsaying."[2] Soon after Moore took the helm in 1895, he launched a "war on the weather prophets" aimed at distancing forecasting from prophecy, professional meteorological expertise from lived experience and informal observation, and modern scientific knowledge from superstition and fraud. Yet these distinctions were not as clear as newspapers and government officials imagined them to be, and the Weather Bureau's construction of its public reputation revealed the fragility of its scientific authority and the durability of time-honored ways of knowing the weather.[3]

In the 1890s and the early twentieth century, as the Weather Bureau fielded a steady stream of requests for long-range forecasts, it sought to discredit those who provided them, while simultaneously working to build the reputation of a civilian weather service and develop a scientific method for long-range weather prediction. This was no easy task, as the bureau had to convince the public to trust in a government bureaucracy—but accept the limitations of its short-term weather forecasts—and compete with the popular and ubiquitous commercial "weather prophets" who met the considerable demand for long-range forecasts. Through the turn of the twentieth century, the Weather Bureau denounced long-range forecasters as unlearned or opportunistic frauds who trafficked in

reau (RG 27), National Archives College Park (hereafter NACP), College Park, MD (hereafter WB Correspondence 1870–1912).

2. "How 'Fake' Weather-Forecasters Fool Farmers," *New York Times* (hereafter *NYT*), December 11, 1904.

3. For brief discussions of the Weather Bureau's quest for scientific authority in this period, see Donald R. Whitnah, *A History of the United States Weather Bureau* (Urbana: University of Illinois Press, 1961), 87–88; Bernard Mergen, *Weather Matters: An American Cultural History Since 1900* (Lawrence: University Press of Kansas, 2008), 4, 9–12. This chapter draws on an archival record that reveals a more nuanced and tenuous constitution of federal scientific authority in everyday life.

uncertainties while it focused on improving the certainties of short-term government forecasts. But as it defined weather forecasting as a modern scientific practice, the bureau came to accept the very uncertainty it had formerly condemned as the failing of weather prophecy: following its "war on the weather prophets," the Weather Bureau entered into long-range forecasting with the introduction of weekly forecasts in 1908, signaling an institutional and cultural shift in how foreknowledge of the weather was produced and understood. The bureau's acceptance of uncertainty involved a reconceptualization of forecasting itself: in the late nineteenth century, the bureau viewed the uncertainty of long-range forecasting as a liability in a science of accuracy and a hallmark of amateur meteorology, vernacular weather wisdom, and, at its worst, fraud. However, by 1910, the Weather Bureau had accepted uncertainty as an inescapable feature of long-range forecasting and indeed a fundamental aspect of what would become a probabilistic form of knowledge.[4]

The vocabulary of weather forecasting shifted as well from the late nineteenth to the early twentieth century. As Katharine Anderson has noted, nineteenth-century weather prediction was "poised between divination, opinion, and calculation," and "weather forecasts were both experiment and prophecy."[5] The

4. For sociological studies of the conceptualization and communication of uncertainty in the National Weather Service, see Gary Alan Fine, *Authors of the Storm: Meteorologists and the Culture of Prediction* (Chicago: University of Chicago Press, 2007); Phaedra Daipha, *Masters of Uncertainty: Weather Forecasters and the Quest for Ground Truth* (Chicago: University of Chicago Press, 2015).

5. Katharine Anderson, *Predicting the Weather: Victorians and the Science of Meteorology* (Chicago: University of Chicago Press, 2005), 15, 19. Most work in the history of meteorology in the United States has focused on scientific and institutional contexts for the production of knowledge, although recent studies in the history of meteorology have acknowledged a broader cultural context for ways of knowing weather and climate. This chapter combines both approaches to examine how the US Weather Bureau and commercial long-range forecasters competed for scientific and cultural authority while responding to evident public demand for long-range weather prediction in the 1890s and early twentieth century. For institutional histories and histories of science, see Whitnah, *History of the Weather Bureau*; Patrick Hughes, *A Century of Weather Service: A History of the Birth and Growth of the National Weather Service, 1870–1970* (New York: Gordon and Breach, 1970); James Rodger Fleming, *Meteorology in America, 1800–1870* (Baltimore: Johns Hopkins University Press, 1990); Kristine C. Harper, *Weather by the Numbers: The Genesis of Modern Meteorology* (Cambridge, MA: MIT Press, 2008). On the cultural production of weather knowledge, see William B. Meyer, *Americans and Their Weather*, 2nd ed. (New York: Oxford University Press, 2014); Vladimir Jankovic, *Reading the Skies: A Cultural History of English Weather, 1650–1820* (Chicago: University of Chicago Press, 2001); Sarah Strauss and Benjamin S. Orlove, eds., *Weather, Climate, Culture* (New York: Berg, 2003);

Weather Bureau endeavored to separate calculation and divination by establishing and policing a rigid boundary between scientific forecasting and premodern prophecy. Throughout the nineteenth century, the term *weather prophet* commonly referred to military and civilian personnel trained in government and academic settings as well as the self-taught private prognosticators who based their forecasts on natural signs and weather folklore. But at the turn of the twentieth century, the meanings of *forecaster* and *prophet* began to diverge as the Weather Bureau sought to redefine professional scientific forecasting—not always successfully—as the antithesis of vernacular prophecy.[6] This chapter focuses on how historical actors at the turn of the twentieth century understood the categories of forecasting and prophecy and how they distinguished—or did not—between scientific and ostensibly unscientific ways of knowing the weather. By not imposing on the past present-day judgments about what counts as legitimate scientific meteorology, this chapter follows Anderson's important observation that "splitting popular from elite knowledge, sound from unsound science, or even successful from unsuccessful theories . . . seems problematic as a way of understanding the impact of nineteenth-century meteorology."[7]

Anderson, *Predicting the Weather*; James Rodger Fleming, Vladimir Jankovic, and Deborah R. Coen, eds., *Intimate Universality: Local and Global Themes in the History of Weather and Climate* (Sagamore Beach, MA: Science History Publications, 2006); Deborah R. Coen, *Vienna in the Age of Uncertainty* (Chicago: University of Chicago Press, 2007); Jan Golinski, *British Weather and the Climate of Enlightenment* (Chicago: University of Chicago Press, 2007); Mergen, *Weather Matters*; James Rodger Fleming, *Fixing the Sky: The Checkered History of Weather and Climate Control* (New York: Columbia University Press, 2010), chaps. 2 and 3. For an overview of the history and historiography of meteorology in the United States, see James Bergman, "Meteorology and Atmospheric Science," in *A Companion to the History of American Science*, ed. Georgina M. Montgomery and Mark A. Largent (West Sussex, UK: Wiley Blackwell, 2016), 160–73.

6. In her analysis of the public construction of individual and institutional scientific reputation in early Victorian weather prophecy and the subsequent administration of weather science in Britain, Katharine Anderson argues that categories of "respectability" and "propriety" are more historically useful for understanding the production of scientific knowledge in the public sphere than are rigid retrospective distinctions between science and pseudoscience, or between popular and institutional contexts for scientific work. Like Anderson, I am concerned with the public reception of meteorology and the nineteenth-century "public theatre of science," but here I focus explicitly on the US Weather Bureau's construction and policing of a rhetorical boundary between science and quackery as a tool for building its institutional reputation in the late nineteenth and early twentieth centuries. Katharine Anderson, "The Weather Prophets: Science and Reputation in Victorian Meteorology," *History of Science* 37 (1999): 180, 203.

7. Anderson, "Weather Prophets," 181.

"GOOSEBONE PROPHETS"

At 7:00 a.m. on Sunday, March 11, 1888, Chief Signal Officer Adolphus W. Greely issued the official daily weather map and twenty-four-hour forecast: "Fresh to brisk easterly winds, with rain, will prevail to-night, followed on Monday by colder brisk westerly winds and fair weather throughout the Atlantic states."[8] But the following morning brought a powerful blizzard that buried the east coast in four feet of snow and turned New York City into a "wilderness" with seventy-mile-per-hour winds and snow drifts towering thirty-five feet high. Trains in New Jersey, New York, Connecticut, and Massachusetts stopped running, and businesses, schools, and the New York Stock Exchange shut down. High winds and heavy ice pulled down telegraph lines, plunging Hartford, Boston, and Providence into unfamiliar isolation. As the *New York Times* remarked, "It is hard to believe in this last quarter of the nineteenth century that for even one day New York could be so completely isolated from the rest of the world as if Manhattan Island was in the middle of the South Sea."[9] As New York City dug itself out, reports surfaced of corpses huddled in doorways. In all, over three hundred storm-related deaths were reported on land and almost one hundred at sea. The storm ended on March 14, 1888, but its legacy as the most severe blizzard in American history has endured.[10]

Greely explained in *National Geographic* why the storm's course and intensity had defied prediction, and elsewhere he classified the blizzard as "a somewhat unusual class of storm on a very grand scale."[11] On Sunday, March 11, Boston weather officials had displayed cautionary signals, but, as the *Boston Globe* put it, "the thought of the morrow brought no suggestion of such a storm." In November, a *New York Times* editorial faulted the Signal Service for allowing the storm to arrive "absolutely unheralded," arguing that "the Weather Bureau predictions led the people of New-York to expect 'clearing

8. National Oceanic and Atmospheric Administration, "U.S. Army Signal Service Daily Weather Map, March 11, 1888, 7 A.M.," NOAA Central Library US Daily Weather Maps Project, http://docs.lib.noaa.gov/rescue/dwm/data_rescue_daily_ weather_maps.html.

9. "Crushed Under the Snow," *NYT*, March 14, 1888; "Cut Off," *Boston Daily Globe*, March 13, 1888; A. W. Greely, "Great Storm off the Atlantic Coast of the United States, March 11–14," *National Geographic* 1, no. 1 (1888): 38–39; "Blasted," *Boston Daily Globe*, March 14, 1888; "In a Blizzard's Grasp," *NYT*, March 13, 1888. Quotations in "Cut Off"; "In a Blizzard's Grasp."

10. "Blasted," *Boston Daily Globe*, March 14, 1888; Mary Cable, *The Blizzard of '88* (New York: Athaneum Books, 1988), 1–2; Mark Monmonier, *Air Apparent: How Meteorologists Learned to Map, Predict, and Dramatize the Weather* (Chicago: University of Chicago Press, 1999), 2.

11. Quoted in Cable, *Blizzard of '88*, 178.

weather' on the 12th of March last, the day of the memorable blizzard." The editorial concluded that government forecasts for the Atlantic states were either inadequate for storms of such magnitude or based on a careless synthesis of weather observations.[12]

Credit for accurately predicting the blizzard of 1888 went instead to Horace Johnson, a prosperous Connecticut farmer whose career as a private forecaster was launched by his statement that "a disastrous blizzard would occur between March 12 and 15."[13] Johnson, who earned "instant prominence" and "considerable notoriety" thereafter, would be anointed the "Sage of Middle Haddam" and "oracle of Connecticut." "Uncle Horace," self-taught in meteorology and astronomy, earned a loyal following among local farmers for his "prophecies [that were] the result of years of scientific research" into planetary motion and its possible effects on atmospheric conditions. Although Johnson would later characterize his work as similar to that of government forecasters, his prediction of the blizzard of March 12, 1888, was fundamentally different in that it was a long-range forecast issued well in advance of the storm (from three weeks to six months ahead of time, according to varying reports).[14]

Long-range weather forecasting was hardly a new phenomenon when Horace Johnson predicted the blizzard of 1888, as American almanacs from the late seventeenth century onward had regularly featured seasonal and yearly predictions. But the first almanac focusing entirely on the weather, *Vennor's Weather Almanac*, appeared in 1877. Canadian geologist and ornithologist Henry Vennor rose to fame on the basis of the long-range forecasts he published in almanacs and newspapers between 1877 and 1885, and he came to embody vernacular weather expertise in American culture.[15] (See figure 3.1.) The *Times*

12. "Cut Off," *Boston Daily Globe*, March 13, 1888; Editorial, *NYT*, November 26, 1888.

13. "Another Blizzard Predicted," *NYT*, March 19, 1888. Many newspaper accounts credited Horace Johnson with forecasting the blizzard of March 12, 1888. See, for example, "'Old Reliable' up in Connecticut," *Baltimore Sun*, December 16, 1890; "Big Blizzard Coming," *Boston Daily Globe*, January 28, 1892; "Will Destroy New York," *NYT*, May 30, 1907; "Sage of Middle Haddam," *Boston Daily Globe*, June 2, 1912; "Horace Johnson, Noted Weather Sharp, Dead," *Boston Daily Globe*, January 21, 1917.

14. "A Visit to the Prophet Who Predicted the 1888 Blizzard," *NYT*, July 16, 1916; "Sage of Middle Haddam," *Boston Daily Globe*, June 2, 1912; "Horace Johnson, Noted Weather Sharp, Dead," *Boston Daily Globe*, January 21, 1917; "Another Blizzard Predicted," *NYT*, March 19, 1888; "Editorial Points," *Boston Daily Globe*, September 17, 1903; "Will Destroy New York," *NYT*, May 30, 1907.

15. Robb Sagendorph, *America and Her Almanacs: Wit, Wisdom and Weather, 1639–1970* (Boston: Little, Brown, 1970), 156; Scott Somerville, "A *Vennor*able Weather Prophet," *Chinook*

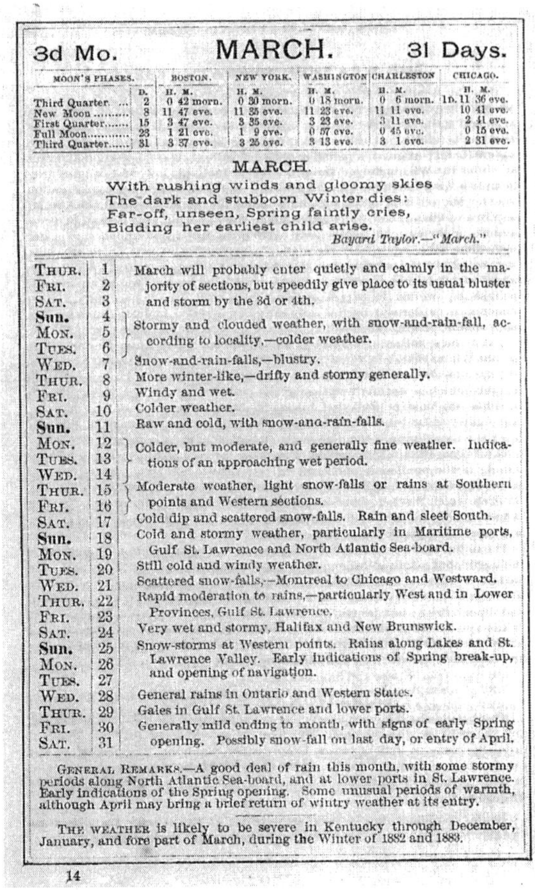

3d Mo. **MARCH.** 31 Days.

MOON'S PHASES.	D.	BOSTON. H. M.	NEW YORK. H. M.	WASHINGTON H. M.	CHARLESTON H. M.	CHICAGO. H. M.
Third Quarter ...	2	0 42 morn.	0 30 morn.	0 18 morn.	0 6 morn.	1D. 11 36 eve.
New Moon	8	11 47 eve.	11 35 eve.	11 23 eve.	11 11 eve.	10 41 eve.
First Quarter.......	15	3 47 eve.	3 35 eve.	3 23 eve.	3 11 eve.	2 41 eve.
Full Moon...........	23	1 21 eve.	1 9 eve.	0 57 eve.	0 45 eve.	0 15 eve.
Third Quarter......	31	3 37 eve.	3 25 eve.	3 13 eve.	3 1 eve.	2 31 eve.

MARCH.

With rushing winds and gloomy skies
The dark and stubborn Winter dies:
Far-off, unseen, Spring faintly cries,
Bidding her earliest child arise.

Bayard Taylor.—"March."

THUR.	1	March will probably enter quietly and calmly in the majority of sections, but speedily give place to its usual bluster and storm by the 3d or 4th.
FRI.	2	
SAT.	3	
Sun.	4	Stormy and clouded weather, with snow-and-rain-fall, according to locality,—colder weather.
MON.	5	
TUES.	6	
WED.	7	Snow-and-rain-falls,—blustry.
THUR.	8	More winter-like,—drifty and stormy generally.
FRI.	9	Windy and wet.
SAT.	10	Colder weather.
Sun.	11	Raw and cold, with snow-ana-rain-falls.
MON.	12	Colder, but moderate, and generally fine weather. Indications of an approaching wet period.
TUES.	13	
WED.	14	
THUR.	15	Moderate weather, light snow-falls or rains at Southern points and Western sections.
FRI.	16	
SAT.	17	Cold dip and scattered snow-falls. Rain and sleet South.
Sun.	18	Cold and stormy weather, particularly in Maritime ports, Gulf St. Lawrence and North Atlantic Sea-board.
MON.	19	
TUES.	20	Still cold and windy weather.
WED.	21	Scattered snow-falls,—Montreal to Chicago and Westward.
THUR.	22	Rapid moderation to rains,—particularly West and in Lower Provinces, Gulf St. Lawrence.
FRI.	23	
SAT.	24	Very wet and stormy, Halifax and New Brunswick.
Sun.	25	Snow-storms at Western points. Rains along Lakes and St. Lawrence Valley. Early indications of Spring break-up, and opening of navigation.
MON.	26	
TUES.	27	
WED.	28	General rains in Ontario and Western States.
THUR.	29	Gales in Gulf St. Lawrence and lower ports.
FRI.	30	Generally mild ending to month, with signs of early Spring opening. Possibly snow-fall on last day, or entry of April.
SAT.	31	

GENERAL REMARKS.—A good deal of rain this month, with some stormy periods along North Atlantic Sea-board, and at lower ports in St. Lawrence. Early indications of the Spring opening. Some unusual periods of warmth, although April may bring a brief return of wintry weather at its entry.

THE WEATHER is likely to be severe in Kentucky through December, January, and fore part of March, during the Winter of 1882 and 1883.

14

FIGURE 3.1. The 1883 edition of Henry Vennor's famous almanac included a general seasonal forecast of "the Probabilities for the Autumn and Winter of 1882–1883" followed by more specific weekly and daily forecasts for each month in Canada and certain regions of the United States. Reprinted from Henry Vennor, *Vennor's Weather Almanac, 1883* (Baltimore: A. Vogeler, 1882), 14, quotation on 8.

of London saw potential in Vennor's seasonal forecasts, noting, "The power of foreseeing the weather of the next few days would do much, the power of foreseeing the weather of the next season would do almost everything, to take

(Spring 1979): 36–37. On almanacs as a popular form of weather prediction in colonial and early America, see Arwen P. Mohun, *Risk: Negotiating Safety in American Society* (Baltimore: Johns Hopkins University Press, 2013), 54–57.

away from agriculture the uncertainty which is now its greatest hindrance."[16] By the end of the century, Vennor had a host of well-known colleagues, including John H. Tice, W. T. Foster, and Rev. Irl R. Hicks, all of St. Louis; A. J. De Voe of Hackensack; Henry C. Maine of Rochester; W. W. Marsh of Cincinnati; and former Signal Service and Weather Bureau employee Lawrence Dunne, who employed various forecasting methods, including periodicity, planetary meteorology (which the Weather Bureau dismissed as "one of the relics of astrology"), lunar phases, patterns based on the weather of saints' days and other holidays, and observable changes in flora and fauna.[17]

In addition to regularly printing forecasts of the "long-rangers," newspapers frequently mentioned a menagerie of nonhuman prophets in articles that invoked a scientist-seer rivalry: "A Sort of Gopher Weather Bureau at Santa Ana," "Woodchuck as a Seer," "Tree Frog a Weather Sharp; One Animal Whose Meteorological Reputation Science Has Not Damaged."[18] In addition to the groundhog whose shadow foretold a longer winter, other animals predicted—through their behavior and appearance—long-term trends and short-term changes in weather. The ranks of animal forecasters included moles (the depth of their holes indicated the severity of winter), frogs (who sought refuge in water before inclement weather), yellow-billed cuckoos (whose cry announced a coming storm), caterpillars (whose coloring in late fall revealed the pattern of the coming winter), and fish (who refused to bite before a storm and retreated to deeper water in advance of cold weather).[19] Weather and climate could also be read in animal bodies, as evidenced by the ubiquity of "goosebone weather

16. Henry Vennor, *Vennor's Almanac and Weather Record for 1878–9* (Montreal: "Witness" Printing House, 1879), 2, https://books.google.com/books?id=lycXAAAAYAAJ.

17. E. B. Garriott, *Long-Range Weather Forecasts* (Washington, DC: Government Printing Office, 1904), 36; Lawrence Dunne, *Practical Weather* 5, no. 4 (1900), pamphlet, box 1493, General Correspondence of the Weather Bureau, 1894–1942, Letters Received, 1894–1911, Records of the Weather Bureau (RG 27), National Archives College Park, College Park, MD (hereafter WB Correspondence 1894–1942). For lists of long-rangers well known to the federal government, see "Fake Forecasts," *Monthly Weather Review* 32, no. 7 (1904): 322; Willis L. Moore to H. M. Watts, May 11, 1904, box 1809, WB Correspondence 1870–1912.

18. "Southern California News," *Los Angeles Times*, January 30, 1896; "Starts His Winter Nap," *Chicago Daily Tribune*, December 17, 1899; "Tree Frog a Weather Sharp," *Washington Post* (hereafter *WP*), June 9, 1907.

19. "Caterpillar as a Weather Prophet," *Atlanta Constitution*, January 29, 1905; "The Mole as Weather Prophet," *WP*, December 24, 1905; "Frogs as Weather Prophets," *Los Angeles Times*, July 30, 1899; "Tree Frog as Prophet," *Boston Daily Globe*, July 14, 1907; "Frog Is a Weather Prophet," *WP*, August 9, 1908; "Signs of an Early Spring," *Chicago Daily Tribune*, April 9, 1899; "Picture to Paint," *Boston Daily Globe*, August 21, 1904; "Caterpillar as a Weather Prophet,"

prophets," who examined the breastbone of a goose born the previous spring: dark spots presaged cold weather, lighter spots milder, with sections of the bone corresponding to December, January, February, and March. There was more than one way to read a goosebone, however, and its prophecies varied:

> If the November goose bone be thick,
> So will the winter weather be;
> If the November goose bone be thin,
> So will the winter weather be.
>
> If the breastbone of a goose is red, or has many red spots, expect a cold and stormy winter; but if only a few spots are visible the winter will be mild.
>
> The whiteness of a goose's breastbone is superstitiously thought to indicate or foreshow the amount of snow during winter.[20]

Such weather lore included plant as well as animal life: the thickness of corn husks foretold the severity of winter, a thin onion skin indicated a mild winter, the appearance of skunk cabbages announced the start of spring, and late flowering of goldenrod signaled late frost.[21]

Newspaper accounts of forecasters who saw future weather in the present state of nature commonly depicted an epistemic confrontation between rural weather prophecy and scientific expertise. In 1908, the *New-York Tribune* dramatized a debate between a "rural interpreter of portents" and a skeptical "city man" who dismissed natural weather signs as superstitious nonsense.[22] The *Tribune*'s imagined conflict—between superstition and science, subjective interpretation and rational calculation, and rural and urban modes of apprehending the natural world—had indeed occurred during the Weather Bureau's "war

---

Atlanta Constitution, January 29, 1905; "Fishes as Barometers," *WP*, September 16, 1906; "Fishes Know Weather," *Boston Daily Globe*, November 25, 1906.

20. "Goosebone Markings," *WP*, December 30, 1906; "Hard Winter, Says Goose," *Atlanta Constitution*, December 26, 1911; Edward B. Garriott, *Weather Folk-Lore and Local Weather Signs* (Washington, DC: Government Printing Office, 1903), 40.

21. "Vegetable Weather Prophet," *WP*, October 3, 1897; "Signs of an Early Spring," *Chicago Daily Tribune*, April 9, 1899; "The Goldenrod as a Weather Prophet," *Atlanta Constitution*, September 13, 1910.

22. "Some Popular Weather Signs," *WP*, November 1, 1908.

on the goosebone prophets."[23] At stake in this conflict over the production of meteorological foreknowledge was the value of weather forecasting as a public good and the authority of a branch of government science that figured prominently in public life throughout the late nineteenth century. As one journalist noted in 1893,

> there is not a bureau in the national government whose maxims and procedure are not better established, nor, when one considers the immense and varied interests—railway, shipping, agricultural, commercial and individual—which are affected by the weather, is there any branch of the service which affects so many people, and affects them so directly, as this, unless we except the postal business?[24]

Just as the postal service was the face of the federal government in the early nineteenth century, the Weather Bureau was the face of federal government science in the late nineteenth century.[25] Thus the bureau's daily maps and forecasts became the basis for public assessment of the efficacy and legitimacy of weather forecasting as a government science as well as a mechanism for the bureau's production of accuracy and policing of uncertainty.

FORECASTER ON TRIAL

Weather Bureau officials in the 1890s, wary of criticism and reluctant to make inaccurate long-range forecasts—despite evident demand—cultivated a culture of certainty to which all forecasters were expected to conform. In 1891, Chief Mark Harrington, upon learning of a San Francisco official's publicly stated intention to make monthly forecasts, warned that "such advertisement should not be made."[26] In 1893, Secretary of Agriculture J. Sterling Morton shut down the "wholly theoretical" work of Weather Bureau meteorologist F. H. Bigelow, a proponent of solar radiation theories who envisioned monthly

23. "War on the Goosebone Prophets," undated newspaper clipping enclosed in A. I. Root to Willis L. Moore, May 25, 1904, box 1805, WB Correspondence 1870–1912.

24. James P. Hall, "Our Crippled Weather Service," *Science* 22, no. 547 (1893): 44.

25. Richard R. John, *Spreading the News: The American Postal System from Franklin to Morse* (Cambridge, MA: Harvard University Press, 1995). On post offices and weather stations similarly embodying the presence of the federal government in everyday life, see Mergen, *Weather Matters*, 26.

26. Mark Harrington to R. E. Kerkane, December 4, 1891, Letters Sent by the Chief of the Bureau, 1891–1895, 1897–1911, vol. 1, RG 27, NACP.

and perhaps yearly forecasts in the bureau's future. Such forecasts would, Morton exclaimed to Harrington in June 1893, "degenerate, so far as precision and certainty is concerned, into the style of the ancient almanacs, wherein we read 'about this time expect rain,' running down the column and covering several days, and even weeks." "The real object of the Weather Bureau work," Morton declared, "is to state with more certainty what the weather will be tomorrow, or the next day."[27]

Less than six months later, Harrington traveled to Boston to preside over an unusual public trial involving New England district forecaster Henry Helm Clayton, a well-known meteorologist who had campaigned for the transfer of the military weather service to a civilian agency and earned widespread recognition for his pioneering work in upper-air kite and balloon observations and local weekly forecasting at Blue Hill Meteorological Observatory outside of Boston. But the Weather Bureau put Clayton under scrutiny not for his experimental work in local long-range forecasting but rather for fifteen errors in his daily maps between June and September 1893.[28] The US Department of Agriculture notified Clayton of the official charges: "gross carelessness and irregularities in the printing of the maps."[29] Assistant Bureau Chief Henry Harrison Dunwoody had conducted his own investigation in September and October 1893, and, discovering what he called "gross irregularities . . . such as the absolute omission of figures and the isothermal lines," reported his findings to the secretary of agriculture.[30] As a result, Clayton was suspended for the month of November, after which time he told a *Boston Globe* interviewer that "it was

27. Sterling Morton to Mark Harrington, June 29, 1893, Letters Sent by the Chief of the Bureau, 1891–1895, 1897–1911, vol. 3, RG 27, NACP; "Weather Bureau Reforms," *NYT*, July 13, 1893.

28. H. Helm Clayton to Willis Moore, December 10, 1904, folder 18, box 5, Henry Helm Clayton Papers, 1877–1949 and undated, Smithsonian Institution Archives, Washington, DC; "Forecasting the Weather," *WP*, July 20, 1894; Sterling P. Fergusson and Charles F. Brooks, "Henry Helm Clayton: 1861–1946," *Science*, n.s., 105, no. 2723 (March 7, 1947): 247–48; "Carelessness Alleged," *Boston Daily Globe*, November 7, 1893.

29. "Off for a Month," *Boston Daily Globe*, November 6, 1893.

30. "Carelessness Alleged," *Boston Daily Globe*, November 7, 1893. Dunwoody was not Clayton's only adversary within the bureau. Chief Signal Officer A. W. Greely unsuccessfully opposed Clayton's bureau appointment in late 1891, likely because of Clayton's calls to transfer the weather service from military to civilian oversight in the late 1880s, which Greely staunchly opposed. Mark Harrington to A. W. Greely, October 5, 1891; Mark Harrington to H. H. Clayton, October 5, 1891, Letters Sent by the Chief of the Bureau, 1891–95, 1897–1911, vol. 1, RG 27, NACP. On Clayton's efforts to reconstitute the weather service as a civilian agency, see [Henry Helm Clayton], *The Transfer of the United States Weather Service to a Civil Bureau* (Boston:

nothing unusual to make such a mistake, and the mistakes in question occurred when hurricanes were prevalent in different parts of the country, and such an error was liable to happen at such a time."[31]

Clayton's downplaying of his mapping inaccuracies, perhaps more than the inaccuracies themselves, drew Harrington to Boston to conduct a formal inquiry. Bureau officials would have been aware of Clayton's comments, since all newspaper articles mentioning the bureau were to be submitted to headquarters using a standardized clipping form.[32] As the *Boston Globe* reported, the bureau "did not want the impression to be in circulation that the weather department was making mistakes every day."[33] Not surprisingly, Clayton's nonchalance did not sit well within this culture of certainty.[34]

Clayton's trial, conducted with "all the ceremony incident to a court-martial," considered each individual inaccuracy in his maps—in June he had made fourteen errors; in July, thirteen; in August, eight; and in September, fifteen. The "station map"—the local version of the national weather map—varied a bit from station to station, but weather maps in the 1890s typically indicated, for each major city, temperature and air pressure, wind direction, and current conditions, as well as isotherms and isobars, hand-drawn curves that smoothly delineated areas of similar temperature and pressure.[35] (See figure 3.2.) The flaws in Clayton's maps varied. The fifteen errors itemized for Clayton's September work fell into six categories: isobars not marked at all, isobars not marked at one end, isobars marked incorrectly, complete absence of temperature lines, isotherms marked incorrectly at one end, and isotherms drawn but not labeled with degree marks. Given the prominence of the Boston station and the fact that it was second only to Washington in the number of maps produced, Clayton's inaccuracies were highly visible: his map was

---

Alfred Mudge & Son, 1889). On Greely's opposition to the transfer, see "The Caucus on the Tariff," *NYT*, May 28, 1888.

31. "Mr. Clayton Returns," *Boston Daily Globe*, December 2, 1893.

32. "To Question Clayton," *Boston Daily Globe*, December 15, 1893.

33. "Chief Harrington Is Coming," *Boston Daily Globe*, December 14, 1893.

34. Following Clayton's trial, mapmaking errors drew increased scrutiny within the Weather Bureau, as Milwaukee local forecast official Willis Moore began to record the errors made by each of his mapmakers. W. L. Moore to Mark Harrington, December 9, 1893 and December 19, 1893, box 883, Forecast Division Letters Received, 1893–1894, RG 27, NACP.

35. Frank H. Bigelow, *Storms, Storm Tracks, and Weather Forecasting* (Washington, DC: Government Printing Office, 1897), 5–9; Mark Monmonier, "Telegraphy, Iconography, and the Weather Map: Cartographic Weather Reports by the United States Weather Bureau, 1870–1935," *Imago Mundi* 40 (1988): 18.

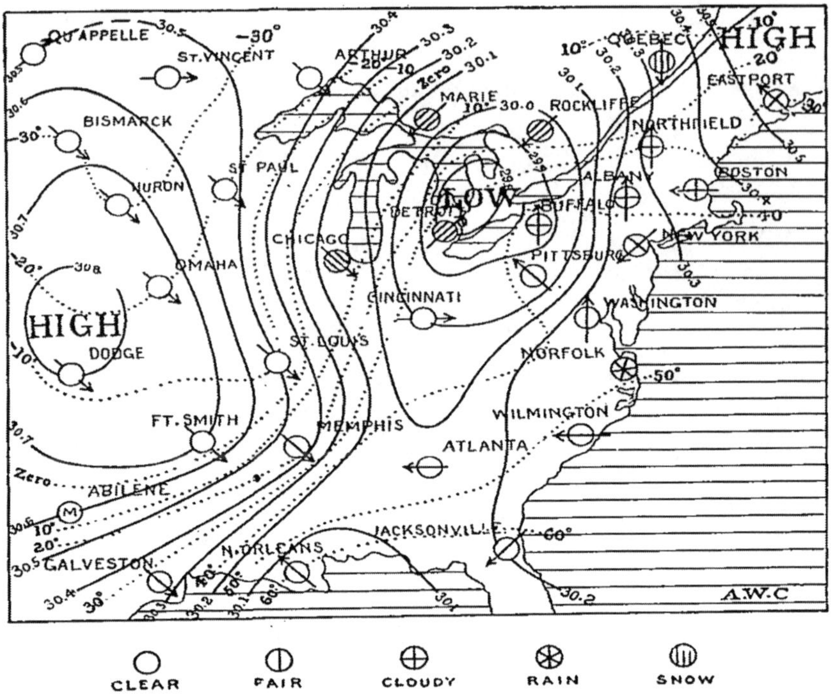

THE BOSTON "HERALD" WEATHER MAP.

FIGURE 3.2. This station map for 8:00 a.m. on January 14, 1894, made by Boston Weather Bureau observer A. W. Crosby, appeared on the back page of the *Boston Herald*, where it had regularly appeared since September 1892. The isobaric lines delineate areas of similar barometric pressure, and areas of high and low pressure are labeled accordingly. Reprinted from Robert DeCourcy Ward, "The Newspaper Weather Maps of the United States," *American Meteorological Journal* 11, no. 3 (1894): 99–100, illustration following p. 100.

printed on the back page of the *Boston Herald* and displayed in all major post offices throughout New England and in hundreds of educational institutions. As Clayton's replacement in the Boston office remarked, weather maps were "the basis daily for the transaction of business interests of great value, and often when human life is a part."[36]

36. Quotation in "An Important Station," *Boston Daily Globe*, December 6, 1893. Weather maps commonly appeared in Weather Bureau publications throughout the late nineteenth century, but in 1894, only four newspapers (*Boston Herald*, *New Orleans Times-Democrat*, *Cincinnati Tribune*, *San Francisco Examiner*) regularly printed them. The Boston station published a day and a night map each day during an era when many stations were moving toward publishing

The hearing ultimately focused on the circumstances of mapmaking in the Boston office and the extent to which Clayton was responsible for having produced uncertainties in the minds of those who consulted his maps. Clayton, represented by Boston attorney Charles S. Rackeman, depicted a distracting atmosphere conducive to inaccuracy: a rushed production process in which he made his map while a clerk was still receiving data telegraphed from Washington headquarters; "almost constantly occurring" errors in ciphered telegraphic communication; and frequent interruptions from journalists, navigators, and other forecast seekers. In closing, Clayton's counsel argued that despite Clayton's mistakes, a trained reader would have easily corrected the errors based on other parts of the maps.[37] And he was right: there is no evidence that Clayton's mislabeled maps had any serious consequences for their readers, none of whom wandered unprepared into a storm or ventured out on dangerous waters as the result of unmarked isobars and isotherms.

Many in attendance at Clayton's trial were trained weather map readers—"scientific men who had come to say a good word for Mr. Clayton," as the *Boston Globe* called them—and the general tenor of their testimony was that Clayton's considerable talents in meteorological research and forecasting outweighed his mapping errors, which were not the "gross irregularities" Dunwoody had made them out to be.[38] In fact, argued one professor, inaccuracy was an inevitable feature of weather forecasting. "It is impossible," he said, "to conduct the work of a bureau of this kind without errors; it is no use to expect it."[39]

But expect it the Weather Bureau did. According to Harrington, who presided as judge over the proceedings, "Clayton's plea was that of confession and avoidance," and in the bureau's culture of certainty in the 1890s, to confess

---

only one map a day. "Friends Rallied," *Boston Daily Globe*, December 17, 1893; Robert DeC. Ward, "The Newspaper Weather Maps of the United States," *American Meteorological Journal* 11, no. 3 (1894): 96–107.

37. "Friends Rallied," *Boston Daily Globe*, December 17, 1893.

38. Those speaking on Clayton's behalf included Abbot Lawrence Rotch, director of the Blue Hill Observatory where Clayton worked before joining the bureau; Professor W. M. Davis of Harvard; Professor S. C. Chandler of Cambridge; E. G. Preston, secretary of the Boston Chamber of Commerce; M. A. Ballou, secretary of the Boston stock exchange; Clarence W. Barron of the Boston news bureau; Arthur E. Sweetland, a volunteer observer from Winthrop, Massachusetts, who had benefited from Clayton's tutelage; and numerous residents of the Boston area and southeastern Massachusetts. "Friends Rallied," *Boston Daily Globe*, December 17, 1893.

39. "Friends Rallied," *Boston Daily Globe*, December 17, 1893.

to mapping inaccuracies rendered one unfit to forecast.[40] Immediately after the trial, Clayton requested a transfer to "more strictly scientific work," promising to resign otherwise. Harrington acknowledged Clayton's capacity as a researcher but noted that Clayton's failure to meet the standards of "the accuracy essential in this work" meant that his resignation would be accepted effective December 31, 1893.[41] The trial resolved what Clayton's superiors understood as a personnel problem but at the same time revealed a more formidable epistemic problem for the Weather Bureau: how to account for uncertainty in the practice of daily forecasting while building the public reputation of government weather science on a foundation of accuracy and certainty.

Clayton's maps posed a threat to the bureau's public reputation as a symbol of inaccuracy in an institution that defined itself by the opposite. The symbolic meaning of the late nineteenth-century weather map, which "inspire[d] a mixture of admiration, appreciation and awe among the general public," was complicated by Clayton's trial.[42] The proceedings revealed the weather map as contested terrain: not a representation of empirical certainty and unquestioned scientific authority but rather a site for public debate over the professional credibility of the Weather Bureau and its vision of forecasting as an exact science. Clayton's maps represented not only the achievements of government weather science but also the very uncertainty the bureau's forecasters sought to conquer.

"AN EXPERT AND NOT A PROPHET"

Although Clayton was the only government forecaster to face a public trial, his colleagues cited an unprecedented degree of public accountability by the end of the century.[43] The first convention of Weather Bureau officials, held in Omaha in 1898, focused on the challenge of public relations. One speaker, G. N. Salisbury of Seattle, argued that the bureau's visibility had resulted in a more knowledgeable yet more demanding public, noting that "more and more are the general public, even those of only ordinary education and intelligence, becoming quite familiar with what the Bureau does, and what it aims to do."

40. Harrington to Morton, December 23, 1893, Letters Sent by the Chief of the Bureau, 1891–95, 1897–1911, vol. 4, RG 27, NACP.

41. Harrington to Clayton, December 21, 1893, Letters Sent by the Chief of the Bureau, 1891–95, 1897–1911, vol. 4, RG 27, NACP.

42. Monmonier, "Telegraphy, Iconography, and the Weather Map," 15.

43. "To Question Clayton," *Boston Daily Globe*, December 15, 1893; Weather Bureau, "Clippings, Suggestions, etc.," December 23, 1893, box 883, Forecast Division Letters Received, 1893–1894, RG 27, NACP.

But given the demand for more precise two-day or even weekly forecasts, Salisbury advocated a conservative approach to forecasting that hinged on reckoning with—and avoiding—uncertainty:

> If we do not know, with considerable certainty, what the weather will be two or more days in advance its prediction should not be attempted. It savors of charlatanism; it begets lack of confidence in us; our information becomes unreliable, and whom can we blame save ourselves, if we offer the odium or oblivion assigned to all false prophets?[44]

Salisbury warned that inaccurate long-range forecasts would damage the bureau's standing, declaring that "each correct statement and each accurate prediction adds to the grandeur and beauty of its temple of reputation; while every misstatement, and every gross failure, is a stone taken from its foundation, lessening the stability and threatening the overthrow of the structure."[45] Maintaining the "temple of reputation" demanded accuracy in file keeping as well as forecasting, according to speaker E. A. Beals of Cleveland. He observed that unseasonable weather invariably brought newspaper reporters to the local bureau office in search of climatological data and that only offices with well-maintained records would be able to comply. Otherwise, Beals warned, journalists would turn to the authority of the "oldest inhabitants," thereby ensuring that "the public mind receives further confirmation as to . . . the infallibility of the goose bone and the ground hog in foretelling all sorts of calamities."[46]

Beals's scenario—the well-organized bureaucracy and scientific authority of the professional Weather Bureau office trumping weather lore and local history—embodied the bureau's efforts to establish a boundary between modern scientific forecasting and premodern (and fraudulent) prophecy. At the Omaha convention, bureau official J. Warren Smith praised Chief Willis Moore, who "went out among the people . . . [and] showed them that he was an expert and not a prophet," and a subsequent comment by bureau observer A. F. Sims exhorted the expert to "lead the people out of the darkness of ignorance into the light of intelligent meteorology."[47] Sims's metaphor of

44. US Weather Bureau, *Proceedings of the Convention of Weather Bureau Officials, Held at Omaha, Nebr., October 13–14, 1898* (hereafter *Omaha Convention*) (Washington, DC: Weather Bureau, 1899), 12, 15–16.

45. *Omaha Convention*, 16.

46. *Omaha Convention*, 71.

47. *Omaha Convention*, 73, 74.

post-Enlightenment science vanquishing the darkness of weather prophecy underscored the incompatibility of these two ways of knowing the future, and his conversion imperative evidently resonated with Moore, who thereafter set out on a proselytizing mission, which was, by some accounts, a quick success. By 1902, Moore had earned the *Washington Post*'s distinction as the government scientist who had done the most "to combat superstition and ignorance" in the broader ideological project of "civilizing the masses and bringing them in touch with modern science."[48]

Nearly thirty years before the *Post*'s acclaim, the seventeen-year-old Moore was working as a reporter for New York's *Binghamton Republican*, and then as a printer for the *Hawk-Eye* in Burlington, Iowa, where he landed after his gold-seeking expedition in the Black Hills went bust. As Moore later recalled, his housemate at a Burlington boardinghouse had just set up a local signal station, and the young Moore decided that he too would be well suited for "the weather business." At his boss's urging, Moore ventured to Washington in 1876 to take the entrance exam for the Signal Service's training school at Fort Myer, Virginia, and was subsequently promoted to Milwaukee district forecaster in 1891, Chicago forecast official in 1894, and chief of the Weather Bureau from 1895 to 1913. (See figure 3.3.) As bureau chief, the erstwhile printer and newspaperman improved the production of weather maps and popularized the work of government meteorological science, and at the center of this public-oriented work was his attack on weather prophecy.[49]

At the turn of the century, Moore acknowledged both the economic potential and the improbability of long-range forecasting. In a *Forum* essay, he envisioned that accurate weekly and monthly forecasts indicating when to expect rain in corn and wheat regions or ideal planting weather in the cotton belt would mitigate farmers' risk in a national market and ultimately create an agricultural utopia. Such foreknowledge, Moore speculated, would yield "a wonderful conservation of human energy" and national efficiency with "effort . . . withheld in one part of the country, and prodigious energy exerted in another." But such a scenario was still a distant "dream," Moore warned, and he sought to "especially caution the public against the imposture of charlatans and as-

48. "Scientist as Prophet," *WP*, August 3, 1902. On Moore's efforts to convey the value of the bureau's work to the public, see Mergen, *Weather Matters*, 13.

49. "Scientist as Prophet," *WP*, August 3, 1902; James B. Morrow, "The Man Who Chases Cyclones," *Boston Daily Globe*, September 5, 1909; quotation in James B. Morrow, "Blame Moore for Hot Wave," *Los Angeles Times*, September 5, 1909; Monmonier, "Telegraphy, Iconography, and the Weather Map," 19–23.

FIGURE 3.3. Willis L. Moore, chief of the US Weather Bureau from 1895 to 1913. NOAA Photo Library, http://www.photolib.noaa.gov/htmls/pers0040.htm, National Oceanic and Atmospheric Administration/Department of Commerce.

trologists, who simply prey upon the credulity of the people." The range of a legitimate forecast, he emphasized, was two to three days in summer and one to two days in winter, when rapidly changing atmospheric conditions could make even a one-day forecast a significant challenge. The "Weather Bureau . . . does not claim to be able to do more than it is possible to accomplish," he wrote.[50]

50. Willis L. Moore, "Weather Forecasting," *Forum* 25 (1898): 351–52. Government forecasters were permitted to make extended forecasts (beyond twenty-four hours in the 1880s, forty to forty-eight hours in the 1880s, and forty-eight to seventy-two hours in the 1890s) only in special circumstances. On these policies and the internal agency history of long-range forecasting, see Robert Reeves, *The Development of Operational Long-Range Weather Prediction in the United States* (Boston: American Meteorological Society, forthcoming), chap. 2. For a newspaper

Government officials had a precise understanding of what they could accomplish given that the Weather Bureau, like the Signal Service before it, conducted systematic verifications to assess forecasters' accuracy and to quantify the value of short-term forecasting in general. In 1873, two years after the introduction of the daily "Probabilities," regional verification percentages ranged from 72 to nearly 82 percent, and by 1883, "predictions of barometric, thermometric, wind-direction, and general weather changes" had improved to between 84 and 90 percent accuracy.[51] In the 1890s, Weather Bureau forecasters earned promotions and awards based on verifications of their daily forecasts, but by the turn of the century, with averages between 82 and 83 percent, the verification process had become the subject of confusion and disagreement within the bureau.[52] It was difficult to compare the accuracy of general, state, and local forecasts made across the country, in different localities, some of which had significant climatological diversity within the geographic boundaries of a single forecasting district and some of which used different methods of calculating verification percentages. Bureau officials discussed the technical nuances of how to quantify failure, given that, according to the standards used in Washington, DC, "a forecast of fair weather counts nothing, even if the forecast proves correct, while, on the other hand, it counts 5 failure for each period it is in error," whereas in the different system used in the Chicago district, "failures of a positive character—that is, forecasts of weather which does not come—are separated from failures of a negative character, or weather happenings not predicted, and thus the relative values of the two are clearly perceived."[53] Officials also debated whether to continue verifying morning and evening temperatures rather than maximum and minimum, where to set the threshold for indicating warmer or colder weather, and how to verify qualitative terms like "warmer," "slightly warmer," "slightly cooler," and "cooler," which had different implications for different regional climates.[54]

---

announcement of government "long-time predictions," see "Getting Very Bold," *Boston Daily Advertiser*, April 30, 1889.

51. W. B. Hazen, *History of the Signal Service, with Catalogue of Publications, Instruments, and Stations* (Washington, DC, 1884), 5, 10, quotation on 10.

52. Willis L. Moore to Col. Wm. H. Powell, January 28, 1901, box 1510, WB Correspondence 1894–1942.

53. Ferdinand J. Walz, "The Relation between General and Local Forecasts," in US Weather Bureau, *Proceedings of the Second Convention of Weather Bureau Officials, Held at Milwaukee, Wis., August 27, 28, 29, 1901* (hereafter *Milwaukee Convention*) (Washington, DC: Government Printing Office, 1902), 122, 124.

54. *Milwaukee Convention*, 131.

The vocabulary of verification was crucial in bridging the disjuncture between quantitative and qualitative assessments of forecasting. As bureau officials and farmers well knew, a forecast that was technically verified was not always locally accurate, and vice versa. As an Atlanta forecaster noted, "It frequently happens that the observer while sitting at his office window can count several showers within a radius of 1 mile of his station, while not a drop falls in his gage. So far as the public is concerned his forecast was correct, while his marking by the official verifier will be zero."[55] The qualitative condition forecast was also subject to differing interpretations that could undermine the public perception of accuracy regardless of verification percentages. As a Savannah forecaster reminded his colleagues, "With us a 'fair' day means a day without rain or even with rain if not in excess of a certain small amount, whereas the public, when the term 'fair' is used, expect a day not only without rain, but with more or less sunshine; they can not reconcile the words 'fair' and 'cloudy.'"[56] Some bureau officials wondered if their internal standards for quantitative verification failed to capture the value of forecasts to the public in the event of "a good forecast with a bad verification."[57]

Although blizzards, hurricanes, and severe weather presented the most obvious occasions for public evaluation of government and private forecasting, routine daily forecasts had a far wider influence. As a Georgia Weather Bureau official observed,

> While the first and most important duty of the Weather Bureau is the forecasting of severe storms, tropical hurricanes, cold wave, and frosts, it should be borne in mind that these conditions are comparatively infrequent, whereas the daily weather forecasts, meeting the public eye on every hand and at all times, are assuredly the most potent agency by which a community judges our efficiency.[58]

55. J. B. Marbury, "Verification of Weather Forecasts," in *Milwaukee Convention*, 157.

56. H. B. Boyer, "Weather Forecasts and the Public," in *Milwaukee Convention*, 151.

57. H. J. Cox, "Should the Verifying Change of Temperature Be Smaller, and Should Not the Terms 'Slightly Warmer' or 'Slightly Cooler' Be Credited?," in *Milwaukee Convention*, 128. On consistency, quality, and value as measures of "forecast goodness," see Allan H. Murphy, "What Is a Good Forecast? An Essay on the Nature of Goodness in Weather Forecasting," *Weather and Forecasting* 8 (June 1993): 281–93. For a recent account of National Weather Service forecasters grappling with the complexities of "forecast goodness," see Daipha, *Masters of Uncertainty*, 40–41.

58. Boyer, "Weather Forecasts and the Public," in *Milwaukee Convention*, 152.

One noteworthy attempt to shape public understanding of the weather of everyday life came in the form of an unusual almanac issued in 1901 by Willis Moore, former printer: *Moore's Meteorological Almanac and Weather Guide, for The Farmer, the Horticulturalist, the Shipper, the Mariner, the Merchant, the Tourist, the Health-Seeker, and for those who Wish to Learn the Art of Weather Forecasting.* Rather than showcasing the seasonal predictions of a typical almanac, Moore's contained tables of climatological data listing the highest and lowest monthly temperatures on record at each Weather Bureau station, along with significant historical events in place of forecasts for the days of each month.[59] The bulk of Moore's 128-page almanac, which began with the customary signs of the zodiac along with the dates of the year's eclipses and equinoxes, consisted of articles depicting weather forecasting not as an art but as a technological and scientific practice driven by the bureaucratic orchestration of simultaneous and standardized weather observations, their telegraphic transmission, and the professional expertise of government meteorologists who skillfully translated observational data into short-term forecasts. The rest of the almanac contained maps of blizzards, hurricanes, hot and cold waves, tornadoes, snowfall, and monthly precipitation averages—intricate visualizations of storm tracks or climatological data. Moore described his almanac as an "effort . . . to correct many popular but erroneous impressions relative to climate and weather," and indeed, the *Literary World* praised Moore for "admit[ting] the curious investigator to a participation in all [the Weather Bureau's] secrets" and judged the almanac "extremely useful for all whose vocations are at all conditioned upon weather."[60]

Moore's Meteorological Almanac and Weather Guide also contained an extended version of what was by now his standard argument and boilerplate denouncing long-range weather forecasters, with "BEWARE OF CHARLATANS!" in bold print at the top of the page, lest any reader misunderstand Moore's assessment of the long-rangers as "imposters" who "market their fraudulent wares" or otherwise "ignorant and misguided persons" of whom an educated public should be either wary or dismissive. Moore criticized long-range forecasters'

59. On Moore's almanac as a "demonstration of the transformative progression from weather lore to meteorology," see Sarah Strauss, "Weather Wise: Speaking Folklore to Science in Leukerbad," in Strauss and Orlove, *Weather, Climate, Culture*, 43. On Moore's almanac as "also a demonstration of how weather lore and meteorology coexist," see Mergen, *Weather Matters*, 13.

60. Willis L. Moore, *Moore's Meteorological Almanac and Weather Guide* (Chicago: Rand, McNally, 1901), 3; "New Publications," *Literary World* 32, no. 2 (1901): 26.

lack of specificity regarding dates or locations of predicted storms while positioning government forecasting as at once a more certain yet still uncertain practice. He noted at the outset that

> practical meteorology is to some extent a tentative work. It may be placed upon a plane with the theory and practice of medicine and surgery. The forecaster is in a degree guided in his calculations by symptoms, and he is able to diagnose the atmospheric conditions as accurately as the physician is able to determine the bodily condition of the patient. He is able to forecast changes in the weather more certainly than the skilled physician can predict the course of a well-defined disease.[61]

By delineating the scientific knowledge behind the art of weather prediction and by describing forecasting in the vocabulary of both indeterminacy and precision, *Moore's Meteorological Almanac and Weather Guide* embodied the ongoing epistemic compromise between forecasting and prophecy evident in the interactions between government and commercial forecasters.

"THE MOST RELIABLE WEATHER PROPHET OF HIS GENERATION"

Although Weather Bureau officials and the press commonly depicted scientific forecasters and weather prophets in opposition, some commercial long-rangers sought a partnership, along with recognition and a salary, from the government. Weather Bureau officials, the secretary of agriculture, and occasionally the president received unsolicited letters offering various systems of long-range forecasting for sale, like the one W. W. Marsh, who had been publishing monthly forecasts in the *Cincinnati Enquirer* since the mid-1890s, sent to Theodore Roosevelt in 1906. Marsh billed his "simple and almost accurate" system of forecasting one to two years in advance as superior to "the government forecasts being only 24 hours in advance and based principally on uncertainty." A railroad telegrapher and former newsboy and bootblack who earned fifty dollars a month on which to support his six young sons, the "Winton Place Prophet" explained to Roosevelt, "I wish to advance my family and myself and have not the means to do so—my weather system is for sale—a government almanac issued one or two years in advance containing my weather

61. Moore, *Almanac*, 53, 54, 28.

forecast."[62] But Secretary of Agriculture James Wilson declined since Marsh had no scientific training and did not produce accurate forecasts "based upon definite principles." "The fact that your forecasts have been extensively published," Wilson replied, "does not necessarily imply that they have a value."[63]

One of Marsh's fellow long-rangers, W. T. Foster of St. Louis, was far more persistent in his attempts to collaborate with the federal government, but in so doing inadvertently escalated Moore's war on the weather prophets. A Union Army captain and newspaper editor who studied astronomy, Foster became a planetary meteorologist who first published newspaper forecasts in the 1880s and then expanded his business to a full-time enterprise in the early 1890s, supporting his family on the $2,500 his forecasts brought in annually.[64] Foster sold his weekly, monthly, and yearly forecasts in a variety of forms, including weekly weather bulletins, monthly weather charts, monthly weather maps, monthly rainfall and temperature forecasts, and newspaper bulletins, counting close to three hundred daily and weekly newspapers among his subscribers.[65] Believing that a complex system of magnetic relationships between the sun, moon, and planets governed atmospheric changes, Foster worked for nearly thirty years to discover "a physical basis for meteorology, the real causes of all weather changes." (See figure 3.4.) But to complete his calculations, Foster needed Moore's help in obtaining complete sets of daily temperature and precipitation records, given his conviction that "we can know the future of the weather by the past."[66]

Imagining a harmonious relationship between short-range and long-range forecasting, Foster began in 1896 what would become an ongoing struggle with Moore and the federal government to obtain historical weather data. Moore initially took Foster's forecasting acumen seriously enough to invite him to submit for verification morning temperature forecasts for the next three months in either Kansas City or Chicago, although not without reservations over Foster's

62. Willis L. Moore to H. M. Watts, May 11, 1904, box 1809, WB Correspondence 1870–1912; W. W. Marsh to Theodore Roosevelt, December 16, 1906, box 2027, WB Correspondence 1870–1912.

63. James Wilson to W. W. Marsh, December 20, 1906, box 2027, WB Correspondence 1870–1912.

64. Frederick Clifton Peirce, *Foster Genealogy* (Chicago: W. B. Conkey, 1899), 688; W. T. Foster to W. L. Moore, May 19, 1896, box 1153, WB Correspondence 1870–1912.

65. Foster to Moore, March 28, 1905, box 1906; W. T. Foster, "War on the Long-Range Forecasters," June 1904, box 1814, both in WB Correspondence 1870–1912.

66. Foster to Moore, May 19, 1896, box 1153; quotation in W. T. Foster, "Long-Range Weather Forecasts," n.d., pamphlet, box 1906, both in WB Correspondence 1870–1912.

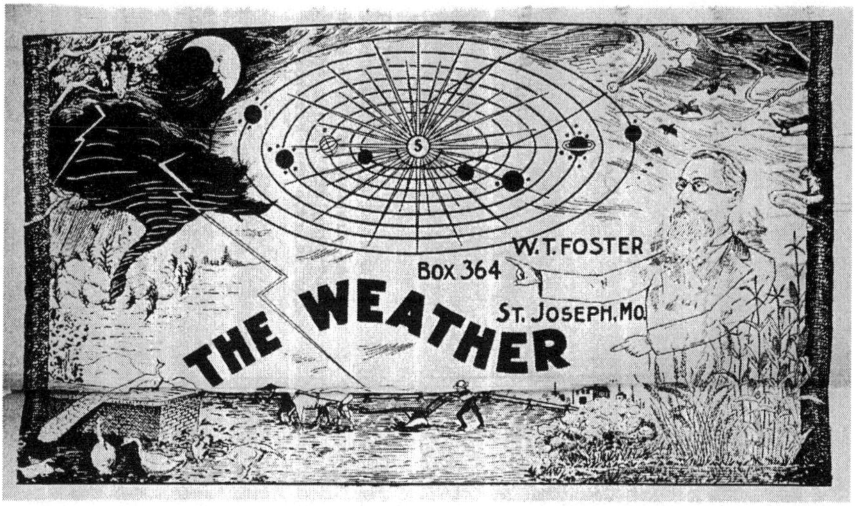

FIGURE 3.4. In his letterhead depicting the interconnected elements in the science of planetary meteorology (the solar system, weather, agriculture, animals, and humans), W. T. Foster looms largest as the learned figure linking the celestial and terrestrial worlds. Weather Bureau Correspondence 1870–1912, Letters Received, 1896–1912, box 1416, Records of the US Weather Bureau, National Archives College Park.

"making what appear to be sensational forecasts for personal gain."[67] After reviewing Foster's mean daily temperature forecasts for Chicago for April, May, and June of 1900, Moore found only "a few coincidental verifications" and thereby refused to furnish copies of the data to Foster for fear of contributing to fraud.[68] In response, Foster submitted a chart comparing his monthly temperature forecasts with the recorded mean daily temperature for Chicago in April 1900 to support his claim of a 66 percent verification percentage, which Moore dismissed as revealing "practically no agreement" between predicted and recorded temperatures. (Still, Moore complained, "Mr. Foster has made a great deal of trouble for the Weather service . . . I have given a great deal of time to the investigation of his work.")[69]

67. Foster to Moore, March 24, 1900, box 1651; Moore to Foster, March 3, 1900, box 1437, both in WB Correspondence 1870–1912.

68. Moore to Foster, March 13, 1900; Moore to Foster, May 4, 1900, both in box 1416, WB Correspondence 1870–1912.

69. W. S. Boyd to W. L. Moore, May 15, 1900, box 1416; Willis L. Moore to William Loeb, Jr., October 10, 1904, box 1651, both in WB Correspondence 1870–1912.

But Foster argued that the bureau's verification standards were an inadequate measure of the value of his forecasts, which predicted—with 66 to 75 percent accuracy—significant changes in the weather within a day or two of their occurrence.[70] Foster explained to Moore that his subscribers valued his forecasts despite—or perhaps because of—their imprecision: "A variation of one day, sometimes two, from the forecast will be accounted good by those most interested and is so claimed in my publications. A difference of a few degrees between the forecast and the temperature does not vitiate the forecast in the minds of those who are watching them closely." Foster went on to insist that "the public cares nothing for technicalities; the people want general results."[71] Foster's monthly forecasts were qualitatively probabilistic in that they hinged on a vocabulary of the conditional:

> Last bulletin gave forecasts of disturbance to cross continent June 2 to 6, warm wave one to 5, cool wave 4 to 8. Next disturbance will reach Pacific coast about 8, cross west of Rockies by close of 9, great central valleys 10 to 12 . . . Warm wave will cross west of Rockies about 8 . . . Dangerous storms may be expected not far from June 19. . . . I expect that disturbance to be one day east of meridian 90 on June 19. That will place it in the Ohio valley and on the lower lakes. But it may be a little behind that or a little east of these points.[72]

Foster's monthly forecast actually relied on the Weather Bureau's daily forecasts, noting that "the weather maps will locate the storm center on that or any other day for their hind-casts are good. But wherever that storm center is on June 19th moderately dangerous storms will occur 100 to 300 miles southeast of its center as it moves eastward."[73] Despite these uncertainties of storms' timing, location, and duration—which failed the Weather Bureau's verification standards—Foster claimed "the only method by which the weather can be forecast with certainty for long periods in advance."[74]

Indeed, testimonials from across the Midwest and the Plains, which Foster reprinted in a promotional pamphlet, lauded the "unerring accuracy" and "remarkable correctness" of his forecasts, which many farmers depended on to

70. Foster, "Long-Range Weather Forecasts."

71. Foster to Moore, May 11, 1900, box 1416, WB Correspondence 1870–1912.

72. W. T. Foster, "Weather Bulletin," *Whitney Point (NY) Reporter*, June 2, 1904, box 1813, Weather Bureau Correspondence 1870–1912.

73. Ibid.

74. Foster, "Long-Range Weather Forecasts."

time their planting and harvesting. The Augusta, Kansas, *Gazette* hailed Foster as "the most reliable weather prophet of his generation," and many newspapers deemed him more reliable than the US Weather Bureau. And Foster would ultimately thank Weather Bureau head forecaster E. B. Garriott and Moore for "hav[ing] done more toward popularizing my weather forecasts than any other two men in the world."[75] For his part, Foster, like other long-rangers before him, sought a public competition with the federal government, and in fall 1900, through his attorney, W. S. Boyd, challenged the secretary of agriculture and Willis Moore to make three months' worth of temperature forecasts to be compared with Foster's, which the secretary declined on the grounds that monthly or seasonal forecasts were not scientifically feasible.[76]

Undeterred, Foster and another ally, a Los Angeles attorney named Townsend, turned to Congress for recognition and funding, and in March 1904, Republican senator Thomas Bard of California proposed a bill to "promote further discovery and research in meteorology" through a public long-range forecasting contest. Bard put forth the bill on behalf of Townsend, known to the Weather Bureau as "the friend of a well known long-range planetary forecaster" (none other than W. T. Foster) but did not consider the contest a threat to the bureau's professional scientific authority, suggesting to Moore that "it will prove effective even if the result shall be to demonstrate that many unofficial Weather Forecasters (so-called) have no scientific basis for their assumptions."[77] The contest would award prizes for the most accurate temperature and rainfall forecasts over a six-month period.[78] Contestants who achieved a certain verification percentage in their daily temperature forecasts for six months or their monthly rainfall totals between March and August would then be asked to articulate the scientific principles underlying their methods to a three-person jury of university experts in the meteorological sciences, with a "temperature prize" of one hundred thousand dollars and a "rainfall prize" of fifty thousand dollars to the forecasting system deemed most

75. Foster, "Long-Range Weather Forecasts"; W. T. Foster to E. B. Garriott, March 28, 1905, box 1906, WB Correspondence 1870–1912.

76. W. S. Boyd to Secretary of Agriculture, October 16, 1900; James Wilson to W. S. Boyd, October 19, 1900, both in box 1416, WB Correspondence 1870–1912.

77. "The Promotion of Meteorology," *Monthly Weather Review* 32, no. 5 (May 1904): 220; Thomas R. Bard to Willis Moore, February 25, 1904, box 1786, WB Correspondence 1894–1942.

78. Contestants would be required to forecast the temperature for any three states from a predetermined list: Massachusetts, New York, Pennsylvania, District of Columbia, Georgia, Ohio, Minnesota, Missouri, Kansas, Texas, and California. "The Promotion of Meteorology," *Monthly Weather Review* 32:220–21.

scientifically sound and replicable.[79] Bard's competition sought to uncover "the physical basis of meteorology," which would mitigate the uncertainty of forecasting, as the *Los Angeles Times* acknowledged when it reported that "the government weather experts do not know [the physical basis for meteorology]. Nobody knows what causes weather till after it has happened, at any rate, and then the knowledge is worthless." The *Times* did allow that telegraphic communication made the Weather Bureau's twenty-four-hour forecasts relatively accurate but concluded that the bureau was unable "to analyze the weather and say what is going to happen next with certainty."[80]

Willis Moore, who was "unable to see any merit in the bill," opposed Bard's "ridiculous" contest on the grounds that it would perpetuate the very uncertainties it proposed to eliminate.[81] Moore warned Bard that the reputation of government science would be undermined by "a guessing contest in which an ignorant or fraudulent forecaster would be as liable to win as the most reputable scientific forecaster." Uncertainty was the key to long-rangers' stature, Moore argued, noting that their imprecise predictions of storms' timing and location only improved their chances of verifying. And Moore denounced their dire predictions of disaster, complaining that "most long-range weather forecasters are of that pernicious class of people that predict swarms of locusts, wars, famines, and other scourges. They perpetrate a positive injury upon the community at large." In 1902, Atlantic City hotel managers asked Moore to discredit a privately issued hurricane warning that had compelled frightened guests and hotel workers to flee, but his attempt to reassure the public failed to stop the panic or the exodus.[82]

Moreover, what Moore called "long range calamity forecasters" like the one who emptied Atlantic City also disrupted an epistemic boundary between the scientific and the supernatural, thereby challenging the positivist certainties of

79. *A Bill to Promote Further Discovery and Research in Meteorology*, S. 5277, 58th Cong., 2d sess. (1904).

80. "Big Prizes Offered Weather Prophets," *Los Angeles Times*, March 27, 1904.

81. "The Promotion of Meteorology," *Monthly Weather Review* 32:220; "Forecasting the Weather," *Los Angeles Times*, August 10, 1904; Willis L. Moore to William Loeb, Jr., October 10, 1904, box 1651, WB Correspondence 1870–1912.

82. Moore to Bard, February 29, 1904, folder 18, box 5, Clayton Papers; US Weather Bureau, *Proceedings of the Third Convention of Weather Bureau Officials Held at Peoria, Ill., September 20, 21, 22, 1904* (hereafter *Peoria Convention*) (Washington, DC: Government Printing Office, 1904), 45.

scientific knowledge production itself.[83] As Moore wrote to Bard, "It is wrong for them to tell the public that there is something occult and mysterious in science." Of course, Moore also objected over concern that Bard's contest would garner more public attention for private forecasters, and indeed, mere rumors of the competition drew inquiries from would-be contestants in the United States, Canada, and France eager to claim prizes and prestige.[84] Moore's worst fears were never realized, as Bard's bill never made it out of committee. But suggestions of "something occult and mysterious in science" fueled Moore's early twentieth-century crusade against long-range forecasting and his vigilant policing of a boundary between science and superstition, forecasting and prophecy.

"A MATTER OF SCIENCE AND NOT OF RELIGION"

When Weather Bureau officials gathered for a convention in Peoria in September 1904, Moore exhorted his colleagues to denounce fraudulent long-range forecasting more aggressively. He urged the bureau to act against the long-range forecasters who had recently begun "to spring up like a mushroom growth, prosper, and grow rich and fat on the proceeds of their work."[85] The purpose of government weather forecasting—originally "for the benefit of commerce and agriculture"—was to protect public safety and mitigate weather-related risk, and Moore maintained that such risk was only exacerbated by the uncertainties of long-range forecasting. He cited the losses incurred by farmers, manufacturers, shippers, merchants, and businesses who orchestrated their activities according to inaccurate long-range forecasts. An Ohio manufacturer wrote of a friend in Florida who wasted over one hundred dollars protecting orange trees from a frost that weather prophet Irl Hicks erroneously predicted: "I was told of other instance where foolish or ignorant men expended a good deal of money in preparing for Hicks's forecasts. From my point of view it

83. Quotation in Willis L. Moore to William Loeb, Jr., October 10, 1904, box 1651, WB Correspondence 1870–1912.

84. Moore to Bard, February 29, 1904, folder 18, box 5, Clayton Papers; W. A. Bowles to Weather Bureau, April 11, 1904, box 1786, WB Correspondence 1894–1942; E. Stone Wiggins to the Secretary of War, June 24, 1904, box 1818, WB Correspondence 1870–1912; Albert Berthot to Willis Moore, January 19, 1905, box 1786, WB Correspondence 1894–1942.

85. *Peoria Convention*, 43.

seems as if the Weather Bureau should do something to protect people from this particular quackery."[86]

Of all those who might make ruinous economic decisions based on long-range forecasts, farmers had the most to lose, according to Weather Bureau officials. In 1904, chief forecaster E. B. Garriott instructed section directors to protect ostensibly naïve farmers from fraudulent forecasters by urging rural editors to "print articles that will teach people the truth about forecasts and storm warnings, instead of misleading their readers by publishing the predictions of fakirs."[87] Editors like J. H. Davis of the *California Agricultural Journal* agreed that the war on the weather prophets should target the rural press: "Eliminate the fake reports *from the country papers*, and you have gained the victory."[88] The *Washington Post* lauded Moore's efforts to banish weather lore from the countryside:

> By making science bear directly on the most prevalent superstitions, by showing that weather forecasting is a matter of science and not of religion, by convincing the rural population that when the bureau predicts rain for the day following the fact that all the roosters in the country are perching upon the top fence rail will not change it. . . . Prof. Moore has done more for the advancement of intelligence than any other scientist in Washington.[89]

And some producers, like Illinois nurseryman and fruit grower George Foster, supported the Weather Bureau's campaign against time-honored methods of anticipating planting and harvesting times: "These weather prophets I have no use for and never had yet they have a fast hold upon many of the people who believe what they say just as conscientiously as they plant their potatoes in the dark of the moon and other things in like manner."[90]

In an attempt to teach farmers to transform their faith in natural signs into trust in a federal government bureaucracy, the Weather Bureau published a bulletin titled *Weather Folk-Lore and Local Weather Signs*, which denounced a range of long-range forecasting methods, including projections based on climatological data and seasonal averages, planetary meteorology (including

86. A. I. Root to Willis L. Moore, May 25, 1904, box 1805, WB Correspondence 1870–1912.

87. "Not Official Sharps," *WP*, July 12, 1904.

88. J. H. Davis to Willis L. Moore, April 18, 1905, box 1916, WB Correspondence 1870–1912.

89. "Scientist as Prophet," *WP*, August 3, 1902.

90. George J. Foster to William G. Burns, May 31, 1904, box 1813, WB Correspondence 1870–1912.

HOW "FAKE" WEATHER-FORECASTERS FOOL FARMERS

Chief Moore of the Government Weather Bureau Endeavoring to Break Up the Illegitimate Business of the Pseudo Prophets Who are Growing Rich Through the Credulity of Our Agricultural Population—The Humbug Exposed in Diagrams.

FIGURE 3.5. The montage beneath the headline positions Willis L. Moore between the well-equipped forecaster's office and the imposing edifice of Weather Bureau headquarters. The weather prophets and almanac readers discredited by Moore and the Weather Bureau are depicted, literally and figuratively, on the margins of professional scientific meteorology. Reprinted from "How 'Fake' Weather-Forecasters Fool Farmers," *New York Times*, December 11, 1904.

planetary motion, lunar phases, and "stellar influences"), and natural observation of plants and animals. *Farming* magazine called the bulletin "iconoclastic," declaring that "this all means that the traditional ground hog, goose bone, changes of the moon, and other time-honored weather indicators as a matter of fact have nothing to do with the weather, and therefore must be eliminated from the calculations of the farmer who wants to be up-to-date."[91]

To demonstrate that government weather forecasting was a modern, rational, scientific practice, Moore conducted public verifications of the commercial forecasters he sought to expose as unscientific yet calculating frauds. Moore made his argument through a series of charts published in the Weather Bureau's 1903–1904 annual report and in a 1904 *New York Times* article that celebrated his crusade against the self-taught weather forecasters who operated outside the boundaries of professional scientific meteorology. (See figure 3.5.) Moore's verifications, which the *Times* billed as the "Humbug Exposed in Diagrams," superimposed the bureau's recorded temperature data for a given

91. Garriott, *Weather Folk-Lore and Local Weather Signs*; "Good-by to the Ground Hog," *WP*, November 4, 1906.

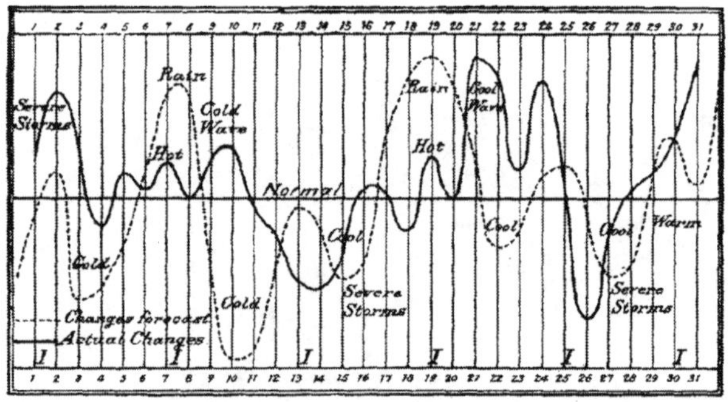

MARCH, 1904, WEATHER CHART: The solid black wavy line, the word "normal," and the explanation in the lower left-hand corner were put upon the above chart by the Weather Bureau. All other matter, including the words "severe storms," "cold," "rain," "cold wave," "hot," "cool," etc., were copied from the original chart issued by this particular long-range forecaster. The dotted wavy line represents the temperatures predicted by him; the solid black wavy line the actual temperatures as recorded at St. Louis, Mo., a city on the ninetieth meridian.

FIGURE 3.6. In his annual report for 1903–1904, Willis L. Moore published a more detailed version of the verification charts that would appear in the *New York Times* in 1904 as an illustration of weather prophets' consistent inaccuracy. On this chart the dotted line represents the temperature forecasts of a well-known St. Louis long-range forecaster, and the solid line represents the temperature changes in March 1904 as recorded by the US Weather Bureau. Reprinted from Weather Bureau, *Report of the Chief of the Weather Bureau, 1903–1904* (Washington, DC: Government Printing Office, 1905), xv.

month on a popular long-range forecaster's chart in order to illustrate the consistent inaccuracies of weather prophecy.[92] (See figure 3.6.)

Moore itemized the faults of the unnamed forecaster's work: a vertical axis without temperature increments labeled, a "normal" horizontal axis originally printed as three separate lines (which Moore called "another artifice of the long-range forecaster to avoid being specific"), and warnings of storms and high winds that never materialized. Moore based his charges on head forecaster E. B. Garriott's denunciation of "storm periods," two- or three-day intervals in which storms were forecast in some locality. As Garriott observed, such an imprecise method "admits no failures," since a storm was bound to occur somewhere and verify the forecast.[93] Moore's charts depicted not only the failures but also the distinct boundary between short- and long-range forecasting, legitimizing

92. "How 'Fake' Weather-Forecasters Fool Farmers," *NYT*, December 11, 1904. Moore was not the first to publish weather verification charts. In 1864, the British Board of Trade requested verification charts of former *H. M. S. Beagle* captain Robert FitzRoy's weather forecasts. Anderson, *Predicting the Weather*, 151.

93. US Weather Bureau, *Report of the Chief of the Weather Bureau, 1903–1904* (Washington, DC: Government Printing Office, 1905), xv–xvi; Garriott, *Long-Range Weather Forecasts*, 9.

the former by discrediting the latter, thereby representing in visual terms the bureau's construction of scientific authority.

Yet not all bureau officials agreed that its "relentless warfare" on long-rangers was best waged in the press.[94] In 1904, J. Warren Smith told his colleagues that their "campaign of education" in periodicals and scientific journals was less effective than marching into a newspaper office with weather maps and long-range forecasts in hand in order to demonstrate the superior accuracy of short-term government forecasts and convince the editor to discontinue long-range predictions. When Smith performed these verifications, he not only discredited long-range forecasting but also followed the bureau's edict that its officials "will best service the public interest when they teach the communities they serve the limitations of weather forecasting."[95] And if this strategy did not work, Willis Moore, having "fl[own] into an obvious and futile rage," threatened to escalate the "war on the goosebone prophets" by shutting down Weather Bureau stations in communities where editors continued to publish long-range forecasts.[96]

Moore never followed through, but the government's conservative logic of forecasting persisted from the 1890s into the early years of the twentieth century: emphasizing the accuracy of short-term forecasting while refusing to make more uncertain long-range predictions, permitting local forecasts forty-eight or seventy-two hours in advance only in special circumstances.[97] Still, despite bureau officials' attempts to educate the public accordingly, demand for long-range forecasts would not subside. As Smith admitted, newspaper editors often told him that "the public wants something of this kind; the Bureau does not furnish it and these long-range fellows do."[98] And a Toledo editor lamented that when he followed the bureau's instructions and discontinued long-range forecasts, "the public clamored for them so . . . we resumed them."[99]

94. William G. Burns to Clara S. Reagan, May 5, 1905, box 1916, WB Correspondence 1870–1912.

95. *Report of the Chief of the Weather Bureau, 1903–1904*, xiii.

96. *Boise Evening Capital News*, October 12, 1904, box 1876, WB Correspondence 1870–1912; "War on the Goosebone Prophets," n.d., newspaper clipping, enclosure in A. I. Root to Willis Moore, May 25, 1904, box 1805, WB Correspondence 1870–1912; "Must Muzzle Papers or Lose Forecasts," *Washington Times*, September 26, 1904, box 1651, WB Correspondence 1870–1912.

97. Reeves, *Development of Operational Long-Range Weather Prediction*, chap. 2.

98. *Peoria Convention*, 43, 249, 248; *Report of the Chief of the Weather Bureau, 1903–1904*, xiii.

99. H. P. Crouse to Willis L. Moore, August 26, 1904, box 1834, WB Correspondence 1870–1912.

Such demand was evident in the piles of letters requesting long-range forecasts that arrived at Weather Bureau headquarters. Police superintendents, fire insurance companies, chimney builders, Florida citrus growers, the Heinz pickle company, a lawyer planning a fishing trip, and a new mother concerned about taking her infant on a train from New Orleans to San Francisco in a heat wave—all needed to know the weather weeks or months ahead.[100] In this way, the archival record reveals an epistolary reckoning with the uncertainties of long-range weather prediction, in which individual citizens and government officials wrote back and forth regarding to what extent and with what degree of certainty it was possible to achieve foreknowledge of the weather. And some letter writers, mindful of the seasonal temporalities that governed agriculture, noted that long-range forecasts had the greatest potential economic value to farmers. As a Nebraskan grateful for the recent arrival of Rural Free Delivery observed, "It is roumered here that the govt is going to give us the Weather Forecasts of course that would be all right but 24 or 48 hours is not just what the Farmer needs. He needs Long Time Forecasts."[101]

But rural producers benefited from short-term forecasts and storm warnings along with the manufacturers, shippers, and businesses whose testimonials to the value of government forecasting were regularly collected by the Weather Bureau and its predecessor, the Signal Service, and cited in government reports and published circulars to justify their congressional funding. Shipping firms and railroad corporations adjusted their deployment of vessels and freight trains in anticipation of fog or snow storms, and canal shippers used daily forecasts and river reports to steer clear of ice in river junctions. Before the introduction of refrigerated railroad cars in the 1880s, shippers of perishable goods consulted forecasts of extreme temperature changes when scheduling their shipments, and merchants awaiting delivery used forecasts to anticipate delays. The economic value of short-term forecasts and storm warnings was cited by agricultural producers, including fruit growers in California, farmers in corn and wheat regions and southern sugar cane growers who heeded frost warn-

100. Richard Sylvester to Willis L. Moore, August 30, 1898, box 1315; O. J. Shannon to Weather Bureau, June 30, 1903, box 1719; H. R. Heinicke to US Weather Bureau, November 30, 1903, box 1750; Robert W. Davis to Willis L. Moore, February 23, 1899, box 1356; H. J. Heinz Co. to Weather Bureau, August 10, 1903, box 1728; Frank Talbot to Supt. Weather Bureau, April 25, 1903, box 1704; Nellie P. Rosenberg to Weather Bureau, August 20, 1903, box 1730, all in WB Correspondence 1870–1912.

101. Joseph Pittam to E. H. Hinshaw, March 16, 1904, box 1792, WB Correspondence, 1870–1912.

ings, and ranchers who sheltered their cattle after receiving cold-wave warnings from the Denver office. As the secretary of the Sacramento Valley Development Association wrote to Moore, "the good that has resulted from this efficient handling of storm, flood and frost notices can hardly be measured in dollars and cents."[102]

The Weather Bureau's aggregated monthly observations had economic value for looking forward and looking backward. Real-estate speculators and prospective homesteaders used climatological history to project the future value of land in a distant part of the country. Manufacturers of agricultural machinery and fertilizer, along with retailers and traveling salesmen, used the bureau's climatological data to anticipate demand, and engineers building bridges, sewers, and irrigation systems consulted rainfall data to determine necessary capacity of water flow. Climatological records also proved indispensable to railroad agents, insurance companies, and lawyers in assessing claims of weather-related financial loss and personal injury. Railroad companies checked rebate claims for perishable goods spoiled in the heat or damaged by freezing temperatures against the bureau's climatological data, and courts often heard testimony of bureau officials regarding icy sidewalks, damaging winds, and lightning strikes before awarding damages.[103]

The bureau's short-term forecasts and long-term climatological reports had significant economic value to networks of commercial exchange that relied on knowledge of the weather of the immediate future or the distant past to inform business decisions. But those who "keep pace with modern progress," Moore argued, did not put their faith in long-range forecasts, and according to his logic, the uncertainty of long-range forecasts had no place in the marketplace of modern capitalism—itself as unpredictable as weather.[104] In response to all the letters that arrived in Washington seeking long-range forecasts for business or personal interests, Weather Bureau officials returned standardized

102. "Scientist as Prophet," *WP*, August 3, 1902; "Chief in Reading Weather," *WP*, August 27, 1905; quotation in O. H. Miller to Willis Moore, February 5, 1909, box 2, Reports on Value of Climatological Publications, 1903–1908, RG 27, NACP.

103. T. B. Jennings to Willis Moore, June 18, 1908; G. Howland to Willis Moore, June 17, 1908; A. J. Mitchell to Willis Moore, June 5, 1908; Levi A. Judkins to Willis Moore, June 5, 1908, all in box 1, Reports on Value of Climatological Publications, Administrative and Fiscal Records, RG 27, NACP; Henry J. Cox, "Use of Weather Bureau Records in Court," in *Yearbook of the United States Department of Agriculture, 1903* (Washington, DC: Government Printing Office, 1904), 303–9.

104. "How 'Fake' Weather-Forecasters Fool Farmers," *NYT*, December 11, 1904; *Report of the Chief of the Weather Bureau, 1903–1904*, xiii–xvii.

explanations of their institutional commitment to the certainties of relatively precise short-term forecasting, emphasizing "that a definite calculation of the weather is not possible for so distant a date" and "that it is not possible to make forecasts, with any degree of certainty, for periods greater than two or three days in advance."[105]

"THE SCIENTIFIC BASIS OF LONG-RANGE WEATHER FORECASTING"

On March 27, 1906, Willis Moore announced at a Maritime Association dinner at the Waldorf-Astoria that the Weather Bureau was about to begin issuing monthly forecasts. Moore, who had just delivered an address on the history of the government weather service, was mocked by a man at his table over the bureau's accuracy record, whereupon Moore stood up and retorted that "the Weather Bureau believes that for the first time in the history of meteorological science it has within its grasp the scientific basis of long-range weather forecasting—that is forecasting of the character of the month to come." The forecasting system was still months from operational, Moore allowed, but the bureau had committed itself to an experimental program of monthly forecasts based on new sources of meteorological data: isobaric charts spanning the northern hemisphere (based on reports from an extensive network of stations, including new stations in Alaska), upper-air kite and balloon observations, and possibly solar radiation studies. Observations of temperature, moisture, pressure, and wind direction and velocity at an altitude of one mile were telegraphed daily from the Mount Weather meteorological research facility in Virginia to Washington beginning in June 1907.[106] Thus began the bureau's shift from ground to air, away from an empirical focus on tracking and projecting observable surface conditions and toward a more theoretical investigation of the atmosphere for the secret to long-range forecasting.

105. H. E. Williams to M. E. Finnegan, August 10, 1900, box 1471, WB Correspondence 1870–1912; Forecast Official to Thomas E. Williams, August 19, 1902, box 1641, WB Correspondence 1870–1912.

106. "Forecasts for a Month," *NYT*, March 28, 1906; "Long-Range Weather Forecast," *Macon (GA) Telegraph*, March 30, 1906; "Rival of Hicks," *Aberdeen (SD) Daily News*, April 5, 1906; Whitnah, *History of the Weather Bureau*, 89; US Weather Bureau, *Report of the Chief of the Weather Bureau, 1907–1908* (Washington, DC: Government Printing Office, 1909), xiv–xv. Quotation in "Forecasts for a Month."

Moore surprised the Waldorf-Astoria crowd, some of his colleagues startled by "the new departure in forecasting," and a skeptical public.[107] The *New York World* remarked that "Prof. Moore has the proud distinction of having missed the weather more often than any man of his inches. His guesses have kept the entire nation guessing."[108] Moore fielded a number of inquiries after the dinner but did not elaborate on his announcement other than to express his reluctance "to widely herald our plans or to make predictions as to their success." As the *Boston Herald* reported,

> Mr. Moore said there would be no effort to predict the exact character of the weather on a certain day a month in advance. "We will not attempt to predict, for instance," said Mr. Moore, "that there will be rain on April 21 or that April 30 will be bright and fair. The effort will be merely to forecast the general character of the weather for that week's period, as, for example, that it will be cold and wet, or hot and dry."[109]

This approach resonated with well-known long-ranger A. J. De Voe, who read of Moore's announcement and then submitted his own monthly forecasts to Secretary of Agriculture James Wilson with a letter asking how much the Weather Bureau would pay him for monthly forecasts that were 90 percent accurate. Wilson declined his services.[110]

But De Voe and others surprised by Moore's announcement did not realize that behind the bureau's "war on the goosebone prophets" was a history of quiet experimentation with the theory and practice of long-range forecasting dating back to the end of the nineteenth century. After several days of heavy rain in fall 1897, a local bureau observer in Walla Walla, Washington, offhandedly predicted seven to ten more days of rain, which caused great "anxiety" among the area's wheat farmers. The Spokane bureau observer feared the economic implications of his colleague's statement, given the attention of mortgage

107. Norman B. Conger to Chief of US Weather Bureau, April 3, 1906, box 1962, WB Correspondence 1870–1912.

108. "Foster's Weather Bulletin," *Los Angeles Times*, April 16, 1906. For criticism of the bureau's long-range forecasting project, see "Long-Range Weather Forecast," *Macon (GA) Telegraph*, March 30, 1906; Frank Waldo, "Long-Range Prediction Impossible," *Boston Daily Globe*, May 26, 1907.

109. "Long Range Forecasts of Weather Conditions," *Boston Herald*, April 3, 1906, box 1962, WB Correspondence 1870–1912.

110. A. J. De Voe to James Wilson, December 3, 1906; Wilson to De Voe, December 12, 1906, both in box 2025, WB Correspondence 1870–1912.

brokers to both the crop yield and the weather forecast: "Hundreds of farms are heavily mortgaged in this portion of the country and the prospective large yield has been looked upon by both the farmer and the Mortgage Company to obliterate these incumbrances this season, and any information looking to a failure or injury of crops may lead to the injury of farmers." Moore, obviously displeased with the Walla Walla observer, issued a stern reminder that "he must confine himself to forecasts as prescribed by instructions."[111] At the 1898 meeting of Weather Bureau officials in Omaha, Patrick Connor of Kansas City, Missouri, a neighbor of renowned weather prophets W. T. Foster and Rev. Irl Hicks, reported his occasional success in long-range forecasting based on a theory of solar magnetism. His colleague B. S. Pague of Portland, Oregon, was more successful, having spent the last two years regularly making three- to seven-day forecasts for farmers, fruit growers, boatmen, and shippers. Pague attributed his accuracy to the regional geography that yielded clearly delineated movements of high and low pressure areas across the Pacific Northwest. A. B. Crane of Pensacola, Florida, ventured that local periodicities—repetitions in weather patterns over time in a particular place—occasionally enabled long-range predictions. But such a method was hit or miss, Crane stressed, and according to his positivist scientific framework, meteorology was not sufficiently advanced to produce reliable long-range forecasts: "Science is defined as consisting simply of the systematic arrangement of facts, and more facts are needed before the artist, however energetic or skillful, can unfold that intricate study, long-range weather predictions."[112]

The science and art of long-range forecasting were combined in the popular turn-of-the-century theories of solar radiation that combined elements of planetary meteorology with astrophysics. The logic behind all sunspot and solar radiation theories was one of correlation: assuming that terrestrial and solar conditions were interrelated, analyzing periodicity in sunspot cycles would enable long-range forecasting of terrestrial phenomena. Sunspot periodicity was met with great enthusiasm in scientific and popular publications, some of which envisioned a meteorological utopia in which sunspot cycles would enable forecasting of not just temperature and precipitation but natural, economic, and political events, "depth and quantity of discharge of rivers, retreat and ad-

111. "Walla Walla Feels Gloomy," *Spokane Spokesman-Review*, September 8, 1897; H. E. Wilkinson to Willis L. Moore, September 8, 1897, both in box 1244, WB Correspondence 1870–1912.

112. *Omaha Convention*, 43–48, 157–60.

vance of glaciers, number of shipwrecks, bank failures and commercial crises, the crops, prices of grain, famines, wars, and even flights of butterflies."[113]

The Weather Bureau's vision was more modest, and, no doubt due to its vociferous critique of weather prophets, its entry into long-range forecasting was relatively quiet. The bureau first published experimental weekly (not monthly, as Moore had suggested) forecasts in April, May, and June 1908 and then made them a regular feature in March 1910, after the success of special forecasts like the one for Sunday, February 13, 1910:

> During the present week a general storm, followed by a cold wave, will cross the United States. The center of this storm will appear over the Pacific States within the next two days, cross the Rockies, Plains States, and central valleys during the middle days of the week, and reach the Atlantic seaboard by Friday. The cold wave promises to be rather severe. It will overspread the North Pacific States by Tuesday morning, the middle and northern Plains States and Central Valleys by Thursday, and reach the Atlantic seaboard by Friday or Friday night.[114]

The bureau's forecast of this blizzard met a far more favorable public reception than Chief Signal Officer Greely's had in March 1888, and one Oklahoma newspaper celebrated "the remarkable accuracy" of the weekly forecast and deemed it "the story of a prediction and its fulfillment."[115] Another notable success came in 1910, when the Weather Bureau issued a weekly forecast on August 21 that predicted a cold wave for the second half of the coming week, which indeed brought record low temperatures in the Rockies and the Plains, along with snow in Wyoming. Newspapers across the nation echoed this praise from the *Oklahoman*: "The day was a great triumph for the weather man. The prophesy was on long time, as weather forecasts go. . . . It was accurate to the hour, and to distance, direction, and temperature; geographically correct—absolutely correct. . . . No one but the doubter was disappointed." And the *Charlotte Observer* reported, "This forecast was read by many but most people

113. C. G. Abbot, "The Relation of the Sun-Spot Cycle to Meteorology," *Monthly Weather Review* 30, no. 4 (1902): 178–81, quoted in Garriott, *Long-Range Weather Forecasts*, 45.

114. *Report of the Chief of the Weather Bureau, 1907–1908*, xiv; Hughes, *Century of Weather Service*, 39–41; quotation in US Weather Bureau, *Report of the Chief of the Weather Bureau, 1909–1910* (Washington, DC: Government Printing Office, 1911), 16.

115. Quoted in *Report of the Chief of the Weather Bureau, 1909–1910*, 16.

straightway dismissed it from their minds. During the week, however, there followed such a remarkably accurate verification of the prediction made days before the cold started that the public sat up and took notice."[116]

Despite such accolades, the bureau's weekly forecasts were not all that different from those Moore and his colleagues had denounced as fraudulent prophecies. As climatologist Robert DeCourcy Ward later observed, the bureau's weekly "weather outlook" was not a precise forecast but rather "a broad generalized statement," typified by a sample outlook for the North and Middle Atlantic states that predicted "period of snows over North and rains and snows over South portion about middle of week; temperatures near or below normal."[117] The bureau's weekly forecasts were characterized by the same uncertainties of time and place the bureau had formerly condemned in long-rangers' vague "storm periods."[118]

Although the Weather Bureau's weekly outlooks signaled a formal institutional acceptance of long-range forecasting, some of its personnel still refused to clear a space for chance. In January 1906, two months before Willis Moore promised his Waldorf audience monthly forecasts, a government meteorologist from western Australia made a similarly surprising announcement about short-term forecasts. W. Ernest Cooke revealed in the pages of the *Monthly Weather Review* his new forecasting system that not merely accepted but actually quantified uncertainty. The problem with daily weather forecasting, he believed, was that forecasters had no way to convey to the public their degree of faith in their own predictions. The "condition of confidence," as Cooke termed it, needed to be quantified on a scale of "states of doubt or certainty."[119]

Cooke's forecasts included, for general conditions, temperature, and precipitation, a weighted confidence score ranging from five to one:

5. We may rely upon this with almost absolute certainty.
4. We may rely upon this with tolerable certainty, but may be wrong about once in ten times.

116. US Department of Agriculture Weather Bureau, *Report of the Chief of the Weather Bureau, 1910–11* (Washington, DC: Government Printing Office, 1912), 16–17, quotations on 17.

117. Robert DeC. Ward, "The Present Status of Long-Range Weather Forecasting," *Proceedings of the American Philosophical Society* 65, no. 1 (1926): 6.

118. Garriott, *Long-Range Weather Forecasts*, 8.

119. W. Ernest Cooke, "Forecasts and Verifications in Western Australia," *Monthly Weather Review* 34, no. 1 (1906): 23.

3. Very doubtful. More likely right than wrong, but probably wrong about four times out of ten.
2. Just possible, but not likely. If showers are indicated, for example, they will not be heavy even if they occur at all.
1. The barest possibility. Not at all likely.[120]

Cooke had designated most of the past year's forecasts as fours (48 percent) and fives (34 percent), with ones (1 percent) and twos (3 percent) rare occurrences, and the "serious doubt" of the three (15 percent) almost never applied to the overall forecast but rather was reserved for specific elements. Cooke's self-calculated verification percentages for his confident forecasts (threes, fours, and fives) revealed that he was indeed a good judge of his own certainty: his fives were accurate almost 99 percent of the time, his fours 94 percent, and his threes 79 percent. The incorporation of a forecaster's subjective judgment and the inscription of the word *uncertainty* into daily weather predictions signaled, according to Cooke, not a failure but rather an advance in forecasting. "It is more scientific and honest," he declared, "to be allowed occasionally to say 'I feel very doubtful about the weather for to-morrow, but to the best of my belief it will be so-and-so.'"[121]

But more honesty in weather forecasting was not the best policy, according to the bureau's chief forecast official, E. B. Garriott, who issued a sharp rejoinder pronouncing Cooke's plan "impracticable" for the United States. Garriott's objections were logistical—the prohibitive cost of telegraphing wordy forecasts and the impossibility of fitting them on forecast cards and weather maps—and epistemological, given his conviction that the American public would not accept uncertainty in daily forecasts. Garriott did not think Americans who "insist upon having . . . forecasts expressed concisely and in unequivocal terms" could navigate "the bewildering complication of uncertainties" posed by Cooke's forecaster confidence ratings.[122]

Cooke, one of the first to systematize probabilistic weather forecasting, maintained that his forecasts did not add costly verbiage but rather introduced numerical probabilities that would clarify rather than confuse. Drawing a comparison to astronomical observations, he argued that information that revealed its "probable error" was more precise and thus more valuable. Ultimately the

120. Ibid.

121. Ibid., 24, 23.

122. "Note by Prof. E. B. Garriott," *Monthly Weather Review* 34, no. 1 (1906): 24.

value of any forecast was determined by public perception of its accuracy, and a forecast that conveyed the probability of its verification would, in many cases, forestall criticism. Therefore, Cooke concluded, quantifying the uncertainty in a forecaster's judgment would only increase the value of weather forecasts. Unlike Willis Moore, who continually labored to translate the value of Weather Bureau forecasting to a skeptical public after the fact, Cooke's weighting system was designed to make forecasts more valuable to the public by making them probabilistic from the very start.[123] Both men took public opinion very seriously, but for Moore, it loomed as a threat to the glory and grandeur of the bureau's "temple of reputation," whereas for Cooke it was the very foundation of forecasting practice.

WILLIS MOORE, LONG-RANGER

A system like Cooke's might have been useful in producing the forecast for President Taft's inauguration on March 4, 1909. The night of March 3 brought a messy mix of heavy rain, moderate snow, and high winds to Washington, DC, but according to the Weather Bureau forecast, the storm would be over in time for Taft's swearing in. Willis Moore confidently predicted "a clear day, with plenty of sunshine and invigorating air. The temperature will range between 35 and 40 degrees, and every indication points to the best weather conditions."[124] But the next day brought the worst Inauguration Day weather conditions in nearly a century, leading some newspapers to ask, "Is Bureau Chief Willis L. Moore a false prophet?"[125] "Blizzard Isolates Capital and Oath Is Administered in the Senate Chamber," read the *Washington Post* headline, and some compared the severity of the "furious blizzard" to the historic blizzard of March 1888.[126] And just as had happened in 1888, government forecasters were wrong and long-range forecaster Horace Johnson, "a thorn in the

123. W. Ernest Cooke, "Weighting Forecasts," *Monthly Weather Review* 34, no. 6 (1906): 274–75, quotation on 275. On Cooke's contributions to the development of probabilistic weather forecasting, see Allan H. Murphy, "The Early History of Probability Forecasts: Some Extensions and Clarifications," *Weather and Forecasting* 13 (March 1998): 8–9; Richard Jeffrey, *Probability and the Art of Judgment* (New York: Cambridge University Press, 1992), 46–47.

124. "A Fair Inauguration Day," *NYT*, March 4, 1909.

125. "Post Keeps Tab on Nation's Weather for a Day," *Cincinnati Post*, April 30, 1909, box 1, Records Relating to Opinions Concerning Emerson Hough's Article on the Weather Bureau, RG 27, NACP.

126. "William H. Taft Inaugurated President," *WP*, March 5, 1909; "Topics of the Times," *NYT*, March 6, 1909. Quotation in "William H. Taft Inaugurated President."

flesh of Moore," was right.[127] On February 12, Johnson had predicted—with less precision—stormy weather for the days surrounding Taft's inauguration:

> The indications are that there will be a disturbance in the atmosphere about March 4. The atmospheric and tidal waves, in connection with the orbital condition of the planets, give certain indications of an intermediate disturbance to take place from the 3d to the 6th. Let us hope it will not take place until after the inauguration service is completed. . . . P.S.—Willis L. Moore, have your raincoat and rubbers ready for the occasion.[128]

Two days after the blizzard, the *New York Times* faulted the Weather Bureau's handling of the aftermath of its "prophecy go[ne] utterly wrong" as well as the certitude with which it was issued: "It was a mistake, when so many people's happiness depended on the state of the sky, to make a prophecy unqualified by a warning that unforeseen elements might disarrange all calculations, and it is another mistake to go on stolidly as if nothing had happened."[129] Moore's own unequivocal "condition of confidence," as Cooke would have termed it, perhaps more than the blown forecast itself, undermined the reputation of the Weather Bureau as Taft took office. Yet the bureau could not win: a subsequent critic would ridicule the uncertainties produced by its "nice use of the words 'probably,' 'possibly,' and 'occasional'" and its "general disingenuous ambiguity."[130]

The snow storm lasted two days, but public critique of government forecasting lingered. Former Weather Bureau employee turned private forecaster Elias B. Dunn second-guessed Moore in the pages of *Everybody's Magazine*, asserting that someone with even a basic knowledge of meteorology should have known, based on synoptic maps tracking the low-pressure center at that time of year, that snow would likely continue even after the center had passed through Washington, DC. Moore's statement to the *New-York Tribune* that "no forecaster could foresee" the continuing precipitation did not sit well with Dunn, and Moore came under more intense critique from Iowa journalist and

127. "Sage of Middle Haddam," *Boston Daily Globe*, June 2, 1912. On the Moore-Johnson rivalry, see "Spurns Weather Test," *NYT*, October 21, 1907; "A Cold Winter Prophesied," *WP*, October 2, 1911; Editorial, *WP*, March 3, 1913; "Horace Johnson, Noted Weather Sharp, Dead," *Boston Daily Globe*, January 21, 1917.

128. "Sage of Middle Haddam," *Boston Daily Globe*, June 2, 1912.

129. "Topics of the Times," *NYT*, March 6, 1909.

130. Emerson Hough, "Our Foozling Weather Bureau," *Everybody's Magazine*, April 1910, 569.

novelist Emerson Hough, whose widely read muckraking essay in *Everybody's Magazine* derided an inefficient and bloated Weather Bureau, "one of the most beautifully bureaucratic bureaus known in this land of the free," as "a weather trust" that failed to adequately serve the public with timely, accurate, and locally relevant forecasts.[131] The Weather Bureau responded to "Hough's anti-weather bureau article" by gathering endorsements from producers, shippers, trade associations, and commodity exchanges across the nation and then distributing a rebuttal pamphlet to approximately two hundred Weather Bureau stations.[132]

Moore managed to bounce back from the public embarrassments of his Inauguration Day forecast and Hough's harsh treatment, but he could not overcome increasingly negative perceptions of his leadership, and his tenure as bureau chief ended in political disgrace. Congressional critics had alleged improper spending on the Mount Weather research facility in Virginia after its opening in 1904 (which some referred to as Moore's "mountain resort," complete with fine table linens and glassware), but by 1907 investigations had found no evidence of the misuse of public funds.[133] Complaints from former subordinates about Moore's strict management style—what one critic called "a venomous and vindictive nature and responsibility for misuse of power"—led to investigations by the Civil Service Commission in 1906 and the House Committee on Expenditures in the Department of Agriculture in 1912.[134] Moore's undoing, however, resulted from his overt political maneuvering for the office of secretary of agriculture under the incoming Wilson administration, an ambition that caused him to use the Weather Bureau reporting network as a campaign network. Bureau employees fanned out across the country to rally support in their local communities and in labor unions, and Moore promised increased appropriations for the Weather Bureau and pay raises for bureau personnel upon his appointment. Theron Akin of New York, who de-

131. Emerson Hough, "Does the Weather Bureau Make Good?," *Everybody's Magazine*, May 1909, 610.

132. Quotation in "A Case in Point," *Wichita Beacon*, May 8, 1909, Administrative and Fiscal Records, box 1, Records Relating to Opinions Concerning Emerson Hough's Article on the Weather Bureau, 1909, RG 27, NACP. For the Weather Bureau's pamphlet and its distribution, see box 4, Records Relating to Opinions Concerning Emerson Hough's Article on the Weather Bureau, 1909, RG 27, NACP.

133. Whitnah, *History of the Weather Bureau*, 117–19, quotation on 118.

134. Charles Burleigh Murray, *Life Notes of Charles B. Murray, Journalist and Statistician* (Cincinnati: Charles B. Murray, 1915), 161. I am grateful to John R. Stremikis for pointing me to this source.

manded a congressional investigation into the Weather Bureau, authored a House resolution that called for Moore to "at the proper time receive the toe-end of Woodrow Wilson's copper-toed boot and be relegated to the political scrap heap."[135] Akin got his wish. Moore submitted his resignation after meeting with Wilson in March 1913, but the president dismissed Moore on April 16 for "irregularities and misuse of orders in conduct of business of Bureau" well before his resignation was scheduled to go into effect on July 31. The Department of Justice subsequently launched an investigation but found no grounds for prosecution.[136]

After leaving government service, Moore took an academic position as professor of practical meteorology and head of the meteorological department at George Washington University in Washington, DC, approximately twenty miles from Mooreland Farm, his summer home in Rockville, Maryland, where he raised horses and hogs and sold peaches, apples, and, ironically, long-range weather forecasts.[137] In early 1916, Moore entered into a partnership with W. F. Carothers, a Houston attorney and long-range forecaster who had long sought, to no avail, paid work from the Weather Bureau.[138] Using Mooreland Farm as their business address, Moore and Carothers offered two-week forecasts on a subscription basis to newspaper editors for five dollars per month, including two articles each month written by Moore on "the wonders of natural science."[139] Moore exulted that Carothers "has undoubtedly discovered the central law of the weather" that was "the most important discovery in meteorological science since the act of Franklin in drawing down the lightning of the clouds and identifying it with the electricity of the laboratory." Carothers posited a correlation between observable solar activity—specifically the creation of "heat rifts" that burst through the photosphere—and the motion

135. *Resolution Calling upon Employees of the Weather Bureau to Give Evidence as to Certain Alleged Irregular Conditions Existing Therein and So Forth*, H.R. 858, 62d Cong., 3d sess. (February 21, 1913), 5.

136. On the series of Moore's misadventures, see Whitnah, *History of the Weather Bureau*, 117–28.

137. W. J. Maxwell, ed., *General Alumni Catalogue of George Washington University* (Washington, DC: George Washington University, 1917), 26; "G. W. Paper Uses New Weather Forecasts," *Washington Herald*, April 1, 1916, box 2704, WB Correspondence 1912–42; Willis L. Moore to Chief of Weather Bureau, March 28, 1916, box 2704, WB Correspondence 1912–42.

138. W. F. Carothers to Morris Sheppard, May 30, 1916; Weather Bureau, memorandum, December 16, 1916, both in box 2704, WB Correspondence 1912–42.

139. Willis Luther Moore and W. F. Carothers, "Weather Forecasts for Two Weeks," box 2704, WB Correspondence 1912–42.

of equatorial air masses that would enable the forecasting of storms and cold waves. As Moore explained, "having located a heat rift on the sun, . . . we may look with certainty for its return at the end of twenty-five days and for the cool weather conditions that in each case will follow in due time."[140] This method would, Moore claimed, produce two- or three-week forecasts that, "while not perfect as to the exact time of arrival of weather changes," would still be more accurate than the Weather Bureau's short-term forecasts and thereby yield "a saving of hundreds of millions to the farmer and to the man who works out of doors."[141] Moore and Carothers issued weekly and two-week forecasts based on a series of cool-wave movements—usually six occurring simultaneously—indicated by letters in the alphabet for each of their seven forecast districts across the nation as well as three separate districts for the west, central, and east sections of the cotton belt. Their "general weather movements" could vary within twenty-four to thirty-six hours, and they did not include hurricanes or other "phenomenal local weather features."[142] Furthermore, Carothers's method might soon produce seasonal forecasts and thereby create the kind of meteorological utopia Moore had envisioned back at the turn of the century when speculating about future advances in long-range forecasting, a prediction Moore cited "with some feelings of satisfaction, showing that I anticipated the result now accomplished."[143]

Moore and Carothers's forecasts were published in their hometowns—in the student newspaper at George Washington University and in southern Texas—but their business venture was short-lived, partly because of a lack of funding and partly because the Weather Bureau, under the direction of Moore's successor, continued the war on the weather prophets that Moore himself had begun over a decade earlier.[144] In speeches and newspaper interviews, Bureau Chief Charles F. Marvin denounced Moore and Carothers's forecasts as "fake"

140. Willis Luther Moore, "Newly Discovered Central Law of Weather Will Save Millions," *Washington Herald*, March 12, 1916; "Claims Discovery in Weather System," *Washington Star*, March 11, 1916, both in box 2704, WB Correspondence 1912–42.

141. Moore and Carothers, "Weather Forecasts for Two Weeks"; Moore, "Newly Discovered Central Law."

142. Carothers Observatory, "Key to Grain and Cotton Weather Forecasts," box 2704, WB Correspondence 1912–42.

143. Willis Luther Moore, "Newly Discovered Central Law of the Weather Will Save Millions," n.d., pamphlet, box 2704, WB Correspondence 1912–42.

144. B. Bunnemeyer to Chief US Weather Bureau, April 25, 1917; "G. W. Paper Uses New Weather Forecasts," *Washington Herald*, April 1, 1916, both in box 2704, WB Correspondence 1912–42.

and devoid of "scientific value," drawing Moore into a contentious public exchange.[145] Although Marvin dismissed their forecasts as having "practically no sound scientific foundation," the bureau nonetheless published a circular referring to Carothers's method titled "Warning Against Weather Forecasting Fallacies," which relegated the sun to the realm of planetary meteorology and astrology, reminding readers that "specious references to the moon, to the planets, and to the spottedness of the sun and its shafts of radiation alleged to dominate terrestrial weather are but picturesque frameworks upon which to display weather forecasts for sale." The circular lamented that so many people "still cling to . . . the idea that the weather conditions depend upon planetary and astrological combinations" but noted that weather services worldwide "all condemn and disapprove the methods and theories of the long-rangers."[146] The irony of Willis Moore's entry into the ranks of the long-rangers was not lost on Weather Bureau officials, one of whom remarked that Moore was "for many years, an ardent protagonist against the very principle which he now endorses."[147]

Although Moore's war on the weather prophets had relied on and indeed perpetuated battle lines dividing a public good from commercial fraud, in the end the distinction—and the outcome of the war itself—was not so clear. Both the Weather Bureau and Willis Moore had entered the formerly discredited world of long-range forecasting, and although the bureau's weekly forecasts would become more refined with the midcentury advent of computational meteorology and Moore's would be shortly discontinued, in the 1910s both signaled a turn away from an exclusive focus on the certainties of short-term forecasting and a new acceptance of unpredictability in weather forecasting.

In the first third of the twentieth century, solar radiation theories passed in and out of vogue, and debates over the legitimacy of long-range forecasting

145. "Weather Bureau Chief Hits New Discovery," *Washington Herald*, March 17, 1916; "Weather Bureau Condemns Plan," *Washington Herald*, March 28, 1916; "Weather Chief Hit by Moore," *Washington Herald*, March 26, 1916, all in box 2704, WB Correspondence 1912–42.

146. "Marvin Takes Issue with Willis Moore," *Washington Times*, March 17, 1916; Office of the Chief of the Weather Bureau, "Warning Against Weather Forecasting Fallacies," March 24, 1916, circular, both in box 2704, WB Correspondence 1912–42. The Weather Bureau's announcement was widely noted in the press. See, for example, "Forecasting of Weather Is 'Faked,'" *Idaho Daily Statesman*, March 30, 1916; "Warning Against Forecasters," *Aberdeen (SD) Daily News*, March 30, 1916; "Fake Weather Prophets and Long Prognostications Taboo," *Columbus (GA) Ledger*, April 19, 1916.

147. W. G. Burns to Chief US Weather Bureau, March 24, 1916, box 2704, WB Correspondence 1912–42.

waxed and waned, but the scope of the bureau's weekly forecast remained essentially unchanged until 1940, which saw the introduction of a five-day forecast that was more specific than the previous weekly outlook. In 1950, the bureau published its first thirty-day outlooks, which hinged on major advances in twentieth-century meteorology: the emergence of polar front theory from Norway's Bergen School after World War I, Carl-Gustav Rossby's discoveries about the movement of waves in the upper atmosphere in the 1930s, intensified long-range forecasting research during World War II, and innovations in computational atmospheric modeling in the mid-1960s.[148] Nonetheless, the problem of the long-range forecast—termed a "will-o'-the-wisp" in the late nineteenth century and an "academic problem" in the early twentieth century—persisted, and Moore's war on the weather prophets would occasionally echo during the twentieth century, when government, academic, and private forecasters squared off in debates over accuracy, authenticity, and authority.[149]

In 1904, at the height of the "war on the weather prophets," Weather Bureau professor C. M. Woodward issued a refutation of St. Louis long-ranger John H. Tice's theory of planetary equinoxes to "help clear the way for the coming of the true science of meteorology which the future certainly has in store for us."[150] Woodward's statement neatly encapsulated Willis Moore's aims during his eighteen years as chief of the Weather Bureau, during which time the institution sought to eliminate popular long-range forecasters in order to establish its professional reputation and the scientific authority of short-term forecasting. The bureau's culture of certainty in the 1890s—in which improving the accuracy of short-term forecasting was paramount and forecaster Henry Helm Clayton was put on trial for mapping mistakes—faded in the early twentieth century with the Weather Bureau's entry into the world of long-range forecasting. The advent of the Weather Bureau's weekly forecasts in 1908 acknowledged that the "true science of meteorology," based on data from the upper atmosphere and throughout the northern hemisphere, accommodated more uncertainty than did the empiricism of late nineteenth-century forecasting,

148. Hughes, *Century of Weather Service*, 39–41, 69, 71, 138. On the emergence of numerical weather forecasting in the mid-twentieth century, see Harper, *Weather by the Numbers*; Frederik Nebeker, *Calculating the Weather: Meteorology in the Twentieth Century* (San Diego, CA: Academic Press, 1995).

149. "Forecasting the Weather," *Omaha World Herald*, September 25, 1892; Charles Fitzhugh Talman, "Weather Forecasters Take on New Duties," *NYT*, November 21, 1926.

150. C. M. Woodward, "The Planetary Equinoxes—An Examination of Mr. Tice's Theory," in Garriott, *Long-Range Weather Forecasts*, 31.

which based its short-term predictions on geographical projections of observable surface conditions. Ultimately the Weather Bureau came to accept the very indeterminacy it had formerly denounced as the defining characteristic of long-range weather prophecy, thereby incorporating uncertainty into its weekly outlooks and its vision of scientific forecasting.

The influence of long-range weather forecasting in the late nineteenth century ranged far beyond the world of meteorology to reach capitalism's "economic chance-world." Long before they were discredited by the Weather Bureau, the seasonal predictions of John H. Tice directly inspired an Ohio farmer named Samuel Benner to publish, beginning in 1876, what became a popular series of commodity price forecasts called *Benner's Prophecies of Future Ups and Downs in Prices*.[151] Benner, in response to what he and many others considered a crisis of economic certainty in the wake of the Panic of 1873, seized upon the popular theory of periodicity in meteorology and aimed it at the market, which he believed was determined by predictable price cycles. Like the late nineteenth-century weather prophets whose work influenced his own, Benner believed that he could predict, with certainty, the far-off future.

151. Samuel Benner, *Benner's Prophecies of Future Ups and Downs in Prices* (Cincinnati: Samuel Benner, 1876), 119.

CHAPTER 4

Economies of the Future

In April 1894, in the midst of the United States' worst economic depression to date, with unemployment approaching 20 percent and armies of the unemployed massing and marching in protest, Samuel H. Terry, retailer and author of a popular guidebook for storekeepers, analyzed the causes of financial panics in the pages of the *Banking Law Journal*.[1] Terry ended his indictment of excessive speculation, inflation, and debt on a note of resignation. "It seems now a certainty," he wrote, "that from some cause we have to face these recurring panics every seventeen to twenty years; and it is wise to make an effort to be prepared for them, and the preparation should not be left until another one comes. Forewarned should be forearmed." Terry, who had lived through the panics of 1857, 1873, and now 1893, acknowledged their inevitability as well as their predictability in his call for sounder methods of anticipating the downturns of the boom-and-bust Gilded Age economy.[2]

1. On Samuel H. Terry's *The Retailer's Manual* (1869), the first of its kind, see Anne Krulikowski, "The Shop Around the Corner: Change, Continuity, and the Independent Neighborhood Grocer," in *Shopping: Material Culture Perspectives*, ed. Deborah C. Andrews (Newark: University of Delaware Press, 2015), 91–92.

2. Samuel H. Terry, "Causes of the Late and Previous Panics, and How to Avoid Them," *Banking Law Journal* 10, no. 8 (1894): 327–30, quotation on 330. On the late nineteenth-century discourse on moral responsibility for unforeseen financial gain or loss, see Jonathan Levy, "The Freaks of Fortune: Moral Responsibility for Booms and Busts in Nineteenth-Century America," *Journal of the Gilded Age and Progressive Era* 10, no. 4 (2011): 435–46.

Terry's observations would not have surprised nineteenth-century readers familiar with a discourse on business cycles, which posited that analysis of historical price data would reveal recurring patterns in markets and prices. As one historian has noted, "From the early decades of the century, observers of American business pondered the length of the business cycle, and offered myriad suggestions for its length, ranging from three to twenty years."[3] Although few late nineteenth-century economists took such theories seriously, well-known English economist William Stanley Jevons popularized cycles through his efforts in the late 1870s and early 1880s to demonstrate that sunspots, weather, harvests, and prices were linked in a sequential causal chain. And in 1914, economist Henry Ludwell Moore announced that "the law of the cycles of rainfall is the law of the cycles of the crops and the law of Economic Cycles."[4] The concept of business cycles gained momentum through the work of the professional economic forecasters who rose to prominence in the wake of the Panic of 1907—Roger Babson, Warren Persons, and Wesley Clair Mitchell chief among them—and would become central to business and government by the end of the 1920s.[5]

But some of the most widely read predictions of the economic future in the late nineteenth and early twentieth centuries originated outside of the worlds of business and economics, from farmer turned commodity price forecaster Samuel Benner, best-selling utopian novelist Edward Bellamy, and meteorologist Henry Helm Clayton, none of whom had formal academic training in economics. They produced and sold forecasts that earned them numerous adherents, some detractors, and popular acclaim as prophets of prices, political economy, and meteorological and market cycles, respectively. Although Benner, Bellamy, and Clayton had very different methods of anticipating the economic conditions of the future, they shared, in addition to their status as white, middle-class men based in the northeastern United States, faith in

3. Peter Eisenstadt, "The Origins and Development of Technical Market Analysis," *Essays in Business and Economic History* 15 (1997): 336–37, quotation on 337.

4. Mary S. Morgan, *The History of Econometric Ideas* (New York: Cambridge University Press, 1990), 18–20, 25; Mary Poovey, *Genres of the Credit Economy: Mediating Value in Eighteenth- and Nineteenth-Century Britain* (Chicago: University of Chicago Press, 2008), 275–83; Henry Ludwell Moore, *Economic Cycles: Their Law and Cause* (New York: Macmillan, 1914), 135, quoted in Morgan, *History of Econometric Ideas*, 28.

5. Walter A. Friedman, *Fortune Tellers: The Story of America's First Economic Forecasters* (Princeton, NJ: Princeton University Press, 2014), 22–23, 137–38, 171–74, 181; Morgan, *History of Econometric Ideas*, 39.

predictive certainties that defied the late nineteenth-century epistemic shift toward an acceptance of chance and contingency. These counterrevolutionaries in the probabilistic revolution reckoned with the "economic chance-world" by envisioning worlds of economic certainties.[6]

In historical accounts of American capitalism, American literary utopianism, and meteorological science, these individuals generally appear, if at all, as marginal figures. Benner has been characterized as an unscientific precursor to the legitimate business-forecasting profession that emerged in the wake of the Panic of 1907, and Bellamy as a mediocre writer who managed to seduce thousands of readers with his "silk-hat socialism."[7] Clayton is virtually absent from histories of meteorology, perhaps dismissed as a meteorological iconoclast devoted to unscientific theories of solar radiation as the elusive key to long-range weather forecasting and correlating cycles of weather and business.[8] But the production of knowledge about the future in the late nineteenth and early twentieth centuries defies present-day academic distinctions between marginal and mainstream ways of knowing, and this chapter considers all three as forecasters whose predictions reached national and international audiences and popularized their theories of futurity. The significance of Samuel Benner, Edward Bellamy, and Henry Helm Clayton lies not in particular forecasts, their relative accuracy, or whether their methods were scientific by the standards of the present day but in their ideas about predictability and uncertainty that shaped their visions of the economic future.

6. For a theoretical discussion of the resistance to probabilism in economic thought, see Claude Ménard, "Why Was There No Probabilistic Revolution in Economic Thought?," in *The Probabilistic Revolution*, vol. 2, ed. Lorenz Krüger, Gerd Gigerenzer, and Mary S. Morgan (Cambridge, MA: MIT Press, 1987), 139–46.

7. Eisenstadt, "Origins and Development of Technical Market Analysis," 338–39; Jonathan Levy, *Freaks of Fortune: The Emerging World of Capitalism and Risk in America* (Cambridge, MA: Harvard University Press, 2012), 181–82. For an important exception and the most thorough consideration of Benner's legacy, particularly his influence on twentieth-century business forecaster Roger Babson, see Friedman, *Fortune Tellers*, 23–27. On Bellamyite politics as "put[ting] 'the silk hat on socialism,'" see Sylvia E. Bowman, *The Year 2000: A Critical Biography of Edward Bellamy* (New York: Bookman Associates, 1958), 128.

8. For a recent analysis of Clayton's work as a local forecaster at Blue Hill Observatory, see James Bergman, "Knowing Their Place: The Blue Hill Observatory and the Value of Local Knowledge in an Era of Synoptic Weather Forecasting, 1884–1894," *Science in Context* 29, no. 3 (2016): 305–46.

"THE PANIC-PROPHET"

In 1876, in the first edition of what would become the popular series *Benner's Prophecies of Future Ups and Downs in Prices*, Ohio farmer and long-range commodity price forecaster Samuel Benner (1832–1913; see figure 4.1) lamented "all kinds of business at loose ends—astray, tossed on the tempestuous sea of uncertainty—from our imperfect knowledge of natural causes and the laws by which they operate; and our lack of accurate statistics of production and prices."[9] Benner identified the weather as the fundamental element of uncertainty in the agricultural economy and called for improved forecasting methods and more comprehensive weather and climate data to "form a system of probabilities that can be useful."[10] The premise of his *Prophecies* was that the "sea of uncertainty" demanded, with both economic and moral imperatives, a new form of knowledge that promised future certainties.

Benner, whose parents came to the United States from Germany in 1803, worked as a bookkeeper and then a superintendent in his father's ironworks before joining the Twelfth Ohio Cavalry Regiment in the Civil War and becoming a sergeant in 1865. After the war, he married and became a hog and corn farmer in Ohio, reportedly losing his animals to hog cholera and his assets to the Panic of 1873. Benner gained credibility first as a stargazing "local oracle" who instructed his neighbors in the time-honored tradition of "moon farming," planting and harvesting according to lunar phases.[11] But he found a wider audience for his *Prophecies*, which appeared in over fifteen editions until 1908 with the aim of democratizing market knowledge to give not only the farmer but also the "manufacturer, and legitimate trader, as well as the speculator" a tool for risk management.[12] Forecasting was necessary in all sectors of the economy, according to Benner, who announced in 1876 that "all business

9. Samuel Benner, *Benner's Prophecies of Future Ups and Downs in Prices* (Cincinnati: Samuel Benner, 1876), 17.

10. Benner, *Prophecies* (1876), 19.

11. "Benner's Prophecies Implicitly Believed," *Boston Journal*, January 1, 1904.

12. Benner, *Prophecies* (1876), 28. On Benner's life, see "An Old Resident," *McArthur (OH) Democrat-Enquirer*, January 8, 1913; "Samuel Benner," *McArthur (OH) Republican Tribune*, January 9, 1913; Edward R. Dewey and Og Mandino, *Cycles: The Mysterious Forces That Trigger Events* (New York: Hawthorn Books, 1971), 94–95; Eisenstadt, "Origins and Development of Technical Market Analysis," 338–39. Eisenstadt's 1997 article called attention to the relationships between capitalism and nineteenth-century weather and market forecasting, as well as to Benner's influence on economic forecaster Roger Babson and the longer history of technical analysis. Friedman,

SAMUEL BENNER.

FIGURE 4.1. Samuel Benner. Reprinted from "Benner's Prediction," *Atlanta Constitution*, January 17, 1899.

operations for profit and future contracts are attended with a great deal of risk, and the leading branches of trade demand information on the subject, and that the uncertainties of the future be lessened."[13] For one dollar, Benner promised, in newspaper advertisements from New Hampshire to New Orleans, "fortunes foretold in the ups and downs of prices for the next twenty years; the future judged by the past."[14]

Benner earned national acclaim for his commodity price forecasts for pig iron, hogs, corn, and provisions, which appeared across the country in newspapers that frequently published excerpts from his books. As a West Virginia newspaper proclaimed in 1888, Benner's "predictions . . . attracted more at-

Fortune Tellers, the fullest account of professional economic forecasting in the early twentieth century, traces Benner's influence into the 1920s and 1930s in more detail.

13. Benner, *Prophecies* (1876), vii.

14. *Lowell (MA) Daily Citizen and News*, January 24, 1876, 1. The same advertisement appeared in the *New-Hampshire Patriot*, the *New Orleans Times*, and the *Farmers' Cabinet (NH)*.

tention than the judgment of any modern writer."[15] The "panic-prophet," "financial prophet," and "self styled prophet" commanded considerable attention on Wall Street, with the *New York Times* wryly reporting of "the nimble room trader, with his eye on quotations, [who] says, 'Prophesy unto us, O, Benner.' "[16] Benner's book was known as "Cammack's Bible," after the famed Wall Street trader Addison Cammack, who professed an unswerving faith in Benner (after he reportedly made four million dollars using Benner's forecasts in the early 1890s), and Benner commonly appeared in lists of authorities on the steel and iron industries, including the *Chicago Tribune*'s list of "Seven Wise Men" writing on the economy in 1884.[17] Benner, who claimed never to have owned stock, characterized himself as generally bullish, in keeping with his answer to the money question: he demonized the gold standard and a low tariff as "twin monsters" that were to blame, along with the demonetization of silver, for the Panic of 1893.[18]

Denying otherworldly powers, Benner explained his forecasting method as a combination of twenty years' personal observation and "experienced facts" that yielded precise calculations of recurrent price cycles, which he claimed could be as accurately predicted as a solar or lunar eclipse. Benner compiled historical price data that became the basis for his theory of recurring price

15. "Lines for Laborers," *Wheeling Register*, February 19, 1888.

16. "Literature," *Chicago Daily Tribune*, April 12, 1884; "Benner's Prophecies," *Wall Street Daily News*, January 14, 1889; "Benner Has Got the Blues," *Wall Street Daily News*, October 18, 1893; "The Financial World," *New York Times* (hereafter *NYT*), January 14, 1894.

17. Quotation in "Benner's Prophecies," *Atlanta Constitution*, March 25, 1888. On Addison Cammack's faith in Benner's forecasts, see also Henry Clews, *Twenty-Eight Years in Wall Street* (New York: J. S. Ogilvie, 1887), 671; "Ship 19,000,000 Tons of Iron," *Duluth (MN) News-Tribune*, November 1, 1904; "Ohio Farmer a Good Prophet," *Salt Lake (UT) Telegram*, October 24, 1904. On Cammack's reported four-million-dollar profit, see "Farmer Foretold Trade Conditions," *St. Louis Republic*, October 16, 1904. On Benner's favorable reputation on Wall Street, see "A Hard Question," *Wall Street Daily News*, July 2, 1888; Wallace P. Reed, "Looking Forward," *Atlanta Constitution*, February 7, 1892; "The Financial Markets," *NYT*, January 3, 1898; "Benner's Prophecies," *Wall Street Journal* (hereafter *WSJ*), January 4, 1899. On Benner as an authority on iron and steel, see "Steel and Iron Trade," *WSJ*, October 22, 1904; "Improving Iron Market," *Wall Street Daily News*, October 22, 1904; "Literature," *Chicago Daily Tribune*, April 12, 1884. On Benner's supporters and skeptics, see Friedman, *Fortune Tellers*, 26.

18. *Philadelphia Inquirer*, September 3, 1888, 6. On Benner's claim to have never owned stock, see *WSJ*, September 19, 1933. On the "twin monsters" of the gold standard and low tariff rates, see *Wall Street Daily News*, October 18, 1893. On Benner's analyses of the depression following the Panic of 1893, see "Benner's Prophecies," *Ohio Farmer*, January 25, 1894; *Atlanta Constitution*, May 17, 1896.

cycles—every eleven years for corn and hogs, twenty-seven years for pig iron, and fifty-four years for general trade.[19] Panics were analogous to comets and could be "predicted with much certainty," according to Benner, who forecast their return in sixteen-, eighteen-, and twenty-year intervals, with "widespread 'financial catastrophe'" scheduled to arrive in 1891 (two years early).[20] Known as "the one-horse prophet," Benner gauged the overall state of the economy according to pig iron (a logical choice given his time in his father's ironworks), which he termed the "north star" and "barometer of trade," and he envisioned that foreknowledge of iron prices would eliminate labor unrest. Benner's commodity price forecasts, depicted in a perfectly symmetrical zigzag chart, hinged on his "Cast Iron Rule" in which "*one extreme invariably follows another*," as reliable as magnetic force and as fixed as ancient law.[21] (See figure 4.2.)

Benner espoused a faith in predictive certainty predicated on historical price averages, a deterministic universe, and divine Providence. "God is in prices," Benner proclaimed in his *Prophecies*, and prices are "systematic, and occur in an established providential succession as certain and regular as the magnetic needle points unerringly to the pole." Benner's epistemology allowed no space for chance in the economy, and he similarly denied the workings of chance in nature, pointing to the "unchangeable regularity" of the seasons and the "scientific principles" underlying "the various convulsions of nature." "None of these things happen by chance," Benner insisted.[22] Yet he maintained that agriculture, governed by both seasonal rhythms and unpredictable weather, defied predictability. He pronounced accurate crop forecasting an impossibility on the grounds that government agricultural statistics were retrospective tabulations issued belatedly, commercial and producers' estimates were self-interested, and statistics themselves actually obscured supply and demand. Ironically, given the subsequent influence of his famed chart on technical market analysis, he indicted statistics in general for their inscrutability and inaccuracies: "Statistics are generally large columns of figures, of which no one knows all the channels from whence they came, all the clerical errors in their compilation, and parties interested in their manipulation." In Benner's mind, crop

19. Benner, *Prophecies* (1876), 15–16, 122–23.

20. Ibid., 96–97, 123, 106.

21. "Benner Makes Another Guess," *Wall Street Daily News*, January 12, 1891; Benner, *Prophecies* (1876), 109, 48, 27–28 (emphasis in original).

22. Benner, *Prophecies* (1876), 16–17, 21.

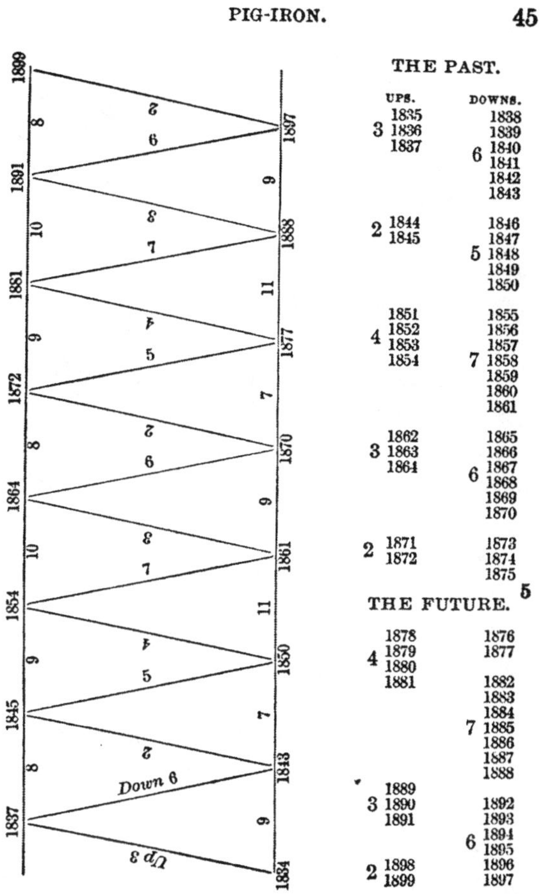

FIGURE 4.2. Samuel Benner's chart combined the ups and downs in pig-iron prices for the past thirty years with his forecast for the next twenty-five years. The ups recur in a cycle of eight, nine, and ten years, while the downs recur in a cycle of nine, seven, and eleven years. The perfectly symmetrical zigzag depicted an underlying order and regularity in the economy. Reprinted from Samuel Benner, *Benner's Prophecies of Future Ups and Downs in Prices* (Cincinnati, 1876).

yields belonged to an unknowable future governed by natural forces. "The seasons make large or small crops, and not the farmers," he wrote.[23]

Benner's skepticism of agricultural statistics followed from his conviction that the short-term temporality of markets was incompatible with the long-range temporalities of weather and harvest. For Benner, the only accurate forecasts

23. Ibid., 22–24.

were long range, since daily market volatility produced short-term fluctuations inscrutable to his systematic study of price cycles. The short term was the domain of the speculator, whose power would be diminished by public knowledge of long-term patterns in prices. The long range was the domain of the forecaster, of both weather and prices. Benner's theory of price cycles hinged on what Peter Eisenstadt has termed the nineteenth-century "meteorologizing" of the business cycle—the use of atmospheric metaphors to conceptualize the economy.[24] Although Benner, believing that "all original causes are invisible," accounted for only effects in his theory of price cycles, he turned to popular long-range weather forecaster John H. Tice of St. Louis to explain the underlying physical laws of commodity markets.[25] Tice, like so many other long-rangers, from W. T. Foster to Willis Moore, claimed to have discovered a physical basis for meteorology in which electrical atmospheric disturbances determined changes in weather. Benner no doubt also embraced Tice's aspirations to predictive certainty: "In the distant future we will attain to sufficient knowledge to understand and comprehend things now problematical and totally incomprehensible to us. Then Meteorology, now classed as one of the exact sciences, will have become so in fact, and we will know the exact time when storms will come, and the localities they will hit on their route over the continent."[26]

Benner emphasized the secular rationality of his theory, declaring that "it does not require a belief in the fabulous to have faith in their [cycles'] periodical appearance," and ranked forecasting alongside scientific discovery and technical innovation as central to late nineteenth-century American progress. Yet Benner's "modern science" of forecasting was not entirely secular but rather a new "science of *price cycles*" that would reveal God's will.[27] Benner's model of predictive reasoning accommodated both scientific and religious epistemologies by combining systematic statistical analysis with belief in the certainties of Providence.

24. Eisenstadt, "Origins and Development of Technical Market Analysis," 337. For an excellent discussion of the metaphors underlying economic forecasting, see Friedman, *Fortune Tellers*, 119–27.

25. Benner, *Prophecies* (1876), 20; Eisenstadt, "Origins and Development of Technical Market Analysis," 338.

26. John H. Tice, *Professor Tice's Meteorological Almanac for the Year 1877* (St. Louis, MO: John H. Tice, 1876), 19–20. I am grateful to the Missouri Historical Society for making this source available. On Benner's appropriation of Tice, see Eisenstadt, "Origins and Development of Technical Market Analysis," 338; Friedman, *Fortune Tellers*, 25.

27. Benner, *Prophecies* (1876), 112, 129–31 (emphasis in original).

Like all well-known forecasters, Benner had supporters and critics who conducted public verifications of his work. Newspapers frequently printed testimonials to his "remarkably correct" and "surprisingly accurate" forecasts.[28] The *Wall Street Journal* compared average prices of twelve stocks with Benner's general predictions on the state of the economy between 1887 and 1895 and found that his forecasts consistently verified, except for his premature prediction of a panic in 1891.[29] Yet verifications did not always convince. James M. Swank, secretary of the American Iron and Steel Association, begrudgingly admitted Benner's accurate forecasts of iron prices but insisted that his method was no better than Henry Vennor's famous almanac: "nothing more than guesswork."[30] In 1894, the *Ohio Farmer* complained that "Mr. Benner knows no more about the future than we do," and a year later, the *Wall Street Daily News* noted that "once in a great while, he hits it, but as a rule he misses it."[31]

After the founding of US Steel in 1901 and its acquisition of the Tennessee Coal, Iron, and Railroad Corporation in 1907, Benner stopped issuing his forecasts in 1908 out of his conviction that the "steel trust . . . would so control the markets that it would prevent the natural trend of affairs."[32] With billion-dollar US Steel the nation's biggest enterprise, the *New York Times* announced that iron was no longer a reliable "barometer" of trade and Benner no longer a reliable prophet of prices:

> So far as the general prosperity of the country is concerned, there is nothing in the condition of the iron and steel industries to warrant setting storm signals. The pessimist of course has as much warrant for gloomy predictions as the weather prophet has for foretelling a storm in clear weather, but that is cheap prophecy and contributes nothing of value to what everybody knows.[33]

28. "Benner's Prophecies," *Wall Street Daily News*, June 25, 1888; *Columbus (GA) Daily Enquirer*, January 24, 1889, 2. For testimonials to Benner's accuracy, see "Cheerful Predictions," *Knoxville (TN) Journal*, January 26, 1890; "Benner's Prophecies for 1895," *WSJ*, January 14, 1895; "Benner's Prophecies," *Chicago Daily Tribune*, February 2, 1895; "Remarkable Fulfillment of Traditions," *Atlanta Constitution*, May 17, 1896; *Minneapolis Journal*, February 2, 1897, 4.

29. "Benner's Prophecies," *WSJ*, January 15, 1895.

30. "An Ohio Man's Dismal Prophecy," *Atlanta Constitution*, December 19, 1882.

31. "Benner's Prophecies," *Ohio Farmer*, January 25, 1894; "Benner's Prophecy," *Wall Street Daily News*, January 15, 1895.

32. "Iron and Steel for 1903," *NYT*, January 1, 1903; quotation in "An Old Resident," *McArthur (OH) Democrat-Enquirer*, January 8, 1913.

33. "Iron and Steel for 1903," *NYT*, January 1, 1903.

But before and after the emergence of US Steel, Benner's forecasts did indeed provide his readers with something of epistemic value: a positivist framework in which future commodity prices were precisely quantifiable economic facts. The *Wall Street Daily News* cited Benner's "philosophic course of reasoning which cannot fail to impress all and convince many," and a Cincinnati newspaper praised his "fixed facts" and his method of "present[ing] his systematic calculations without the usual attempts to hedge or qualify."[34] Benner's public reputation hinged perhaps more on this predictive certainty than on the accuracy of his predictions for any given year. Ohio banker Howard I. Shepard said as much in a 1914 article on business depressions that featured a verification chart comparing Benner's zigzag pig-iron prices with yearly averages beginning in 1834, the starting point for Benner's cycles, and extending to 1914, a year after Benner's death. According to Shepard, Benner "was successful to a greater degree than might be expected, but not sufficiently so to prove his 'Cast Iron Rule.'"[35] This ambivalent assessment would be Benner's legacy, as Wall Street remained interested in his forecasts but unconvinced by his methods. In addition to his direct influence on Roger Babson, one of the first professional business forecasters of the twentieth century, Benner also oriented market watchers toward the chart that has remained the focus of technical analysis ever since.[36] But the most significant aspect of Benner's *Prophecies* was the very idea that commodity prices and the economy overall were not governed by chance but rather within the capacity of humans to predict with certainty. The focus on the distant future to smooth out short-term economic volatility, which was the basis of Benner's *Prophecies*, found its literary expression in the writing of the nineteenth century's most widely read forecaster.

"THE GREAT AMERICAN PROPHET"

Edward Bellamy (1850–1898; see figure 4.3) was born in Chicopee Falls, Massachusetts, where he bore witness to industrialization, growing up near bustling cotton mills and bleak factory housing as well as the homes of wealthy

34. "Benner's Prophecies," *Wall Street Daily News*, January 14, 1889; *Cincinnati Commercial Tribune*, March 20, 1884, 2.

35. Howard I. Shepard, "Business Depressions," *Rotarian* 4, no. 5 (1914): 38–39, quotation on 39.

36. On Benner's influence on Babson, see Friedman, *Fortune Tellers*, 23–24. On the influence of Benner's chart, see Eisenstadt, "Origins and Development of Technical Market Analysis," 339.

FIGURE 4.3. Edward Bellamy, bust portrait, c. 1889, photographic print, Prints and Photographs Division, Library of Congress, LC-USZ62-42652.

manufacturers. With a Baptist minister father and a rigidly Puritanical mother, Bellamy's religious upbringing was a "Calvinistic indoctrination" that he would renounce in his early twenties. Bellamy embarked upon a series of ventures as a young adult, including following his brother to Union College for a year after failing the West Point physical exam, spending a year in Germany with a cousin, and then launching a law career that he abandoned after his first case.[37] Bellamy began his literary career as an editorialist and reviewer for the *Springfield Union* (1872–1877) and a writer of imaginative fiction before becoming cofounder of the *Springfield Daily News* (1880–1884) with his

37. Sylvia E. Bowman, *Edward Bellamy* (Boston: Twayne Publishers, 1986), 1–7; Sylvia E. Bowman, *The Year 2000: A Critical Biography of Edward Bellamy* (New York: Bookman Associates, 1958), 66, 24. Quotation in Bowman, *Edward Bellamy*, 3.

brother.[38] He published five novels and twenty-three short stories that might best be described as sentimental science fiction, but none approached the success of *Looking Backward* (1888), which he "intended, in all seriousness, as a forecast, in accordance with the principles of evolution, of the next stage in the industrial and social development of humanity."[39] Bellamy's novel became the most widely read forecast of the nineteenth century, and his celebrated account of a socialist future articulated a new politics of foresight, a vision of social and economic certainty that originated not only in his critique of late nineteenth-century political economy but also in his lifelong fascination with the promise and peril of looking forward.

The psychological and social costs of anticipating an unknowable future was a central theme of Bellamy's early writing, and in a series of notebooks he scribbled plots for unwritten stories about predictability, including "Might-Have-Been-Land" and one about a "broken down fellow who lives and dreams in the future."[40] Other plots featured a "prophet [who] is to estimate the value of real estate" and a "time-buying business" in which a time speculator buys unwanted bits of time cheaply (e.g., hours until dinner, hours until a prisoner is released) then "sells to lovers and happy people who have as much interest in having time go slow as the prisoner in having it go fast."[41] The logic of this imaginary time market had darker echoes in Bellamy's plans for a story about the marriage market, a futuristic eugenicist fantasy in which a state-controlled system of marriage and reproduction ensured that young people would not be

38. Joseph Schiffman, introduction in Edward Bellamy, *Selected Writings on Religion and Society*, ed. Joseph Schiffman (Westport, CT: Greenwood Press, 1974), xiv–xv.

39. Edward Bellamy, *Looking Backward, 2000–1887* (1888; repr., New York: Signet Classic, 1960), 220. Quotation in Bellamy's postscript to the second edition. Bellamy scholarship, which includes literary histories of futurism, utopian studies, and intellectual histories, has focused overwhelmingly on *Looking Backward*. John L. Thomas's analysis of Bellamyism in *Alternative America: Henry George, Edward Bellamy, and Henry Demarest Lloyd and the Adversary Tradition* (Cambridge, MA: Harvard University Press, 1983) remains indispensable. For an important recent study, see Matthew Beaumont, *The Spectre of Utopia: Utopian and Science Fiction at the Fin de Siècle* (New York: Peter Lang, 2011).

40. "Might-Have-Been-Land" reprinted in Franklin Rosemont, ed., Edward Bellamy, *Apparitions of Things to Come: Tales of Mystery and Imagination* (Chicago: Charles H. Kerr, 1990), 173; Edward Bellamy, Note-book "Plots for Stories #2," p. 8, Edward Bellamy Compositions, 1860–1939 (hereafter Bellamy Compositions) (MS Am 1181.6), Houghton Library, Harvard University, also quoted in Bowman, *Year 2000*, 110.

41. Bellamy, Note-book "Plots for Stories #2," Bellamy Compositions (MS Am 1181.6), Houghton Library; "Time-Buying Business" reprinted in Rosemont, *Apparitions of Things to Come*, 175.

subject to the fickleness of courtship but would enter into marriage certain of their compatibility. Men and women would undergo a psychological examination that classified and color-coded them by temperament, thereby ensuring the "improvement of the species."[42] Bellamy, like many of his contemporaries, subscribed to an evolutionary framework for progress. He considered Darwinian evolution "the most original and noteworthy philosophic idea of the nineteenth century," and he wrote approvingly of selective breeding of humans and animals in *Springfield Union* editorials in the early 1870s.[43]

Despite Bellamy's intellectual affinity for evolution, he remained personally unsettled by the prospect of constant change. "I have a deep seated aversion to change," he wrote in his notebook, where he also longed for "that heavenly quality of constancy . . . because it is so oddly ill-suited in this world where there is so much change in circumstances and characters."[44] Bellamy's short stories include various literary experiments with temporality that sought elusive predictability in a world of change. A teenaged Bellamy wondered whether young people had the capacity to imagine themselves in old age, remarking on "what a strange notion, as we sit listening to our friends' talks, to project ourselves into the future and to a time when we shall vainly strive to recall the tender familiar tones that will then so long time have been silent but yet we listen carelessly, as if no such time would ever come."[45] This observation became the premise of his story "The Old Folks' Party" (1876), in which six bored teenagers dress up as seventy-year-old versions of themselves and reminisce about their childhoods at "a sort of ghost party . . . ghosts of the future, instead of ghosts of the past," until the illusion becomes frighteningly real and the friends are reawakened to an appreciation of their youthful vitality. Bellamy's first thematic exploration of predictability demonstrated foresight's liberating potential—its salutary effect on the present—as well as its emotional risks.[46]

42. Edward Bellamy, "Plots for Stories #1," reprinted in Toby Widdicombe and Herman S. Preiser, eds., *Revisiting the Legacy of Edward Bellamy (1850–1898), American Author and Social Reformer: Uncollected and Unpublished Writings, Scholarly Perspectives for a New Millennium* (Lewiston, NY: Edwin Mellen Press, 2002), 234.

43. Quotation in Bowman, *Year 2000*, 25; "Who Should Not Marry," *Springfield (MA) Union*, September 20, 1873; "Stirpiculture," *Springfield (MA) Union*, October 2, 1875, as cited in Bowman, *Edward Bellamy*, 103.

44. Edward Bellamy, Note-book #2, p. 29; Bellamy, Note-book C, p. 129, both in Bellamy Compositions (MS Am 1181.6), Houghton Library.

45. Bellamy, Note-book #4, p. 2, Bellamy Compositions (MS Am 1181.6), Houghton Library.

46. Edward Bellamy, "The Old Folks' Party," in Edward Bellamy, *The Blindman's World and Other Stories* (Boston and New York: Houghton Mifflin, 1898), 61.

Bellamy's focus on the possibilities and limitations of prospection continued through the mid-1880s, when he shifted away from the individual predictive imagination toward a collective predictive imagination that became the groundwork for his utopian prophecy. In 1886, Bellamy published "The Blindman's World," a short story about an Earth plagued by uncertainty and, although he offhandedly termed it "a fanciful sort of screed," it was his most sustained examination of the theme of prescience.[47] An astronomy and mathematics professor is transported in a dream to a utopian Martian colony where everyone possesses foresight, unlike the unfortunate earthlings who inhabit a "blindman's world" with no clear view of the future.[48] A debilitating lack of foreknowledge is the central problem of the story: a Martian pronounces the uncertainty of earthly life a "disability so crushing" and earthlings "poor blindfolded creatures . . . destitute of the faculty of foresight." Bellamy stated the problem in even bleaker terms in his notes, lamenting that "ninety-nine hundredths of human endeavor is frustrated through ignorance of the future."[49]

Yet the Martians marvel at the capacity of earthlings "to endure the constant buffetings of the unexpected" as they face unforeseen financial and emotional losses, sudden illnesses and deaths, and constant "uncertainty whether you can safely count on the passing day." Martian life was free of apprehension or regret, and, as Bellamy wrote in his notes, disappointment does not exist on Mars. (Not all readers accepted the premise that foreknowledge guaranteed happiness, like the *Literary World* reviewer who insisted that "such psychology seems to contradict human experience.")[50]

Devoid of uncertainty, the Martian economy foreshadowed *Looking Backward*'s socialist utopia, as its inhabitants, cognizant of their future prosperity, have no motive for competition. Bellamy described the Earth's economy as

47. Edward Bellamy to the editor of the *Atlantic Monthly*, July 8, 1886, Edward Bellamy Correspondence, 1850–1898 (hereafter Bellamy Correspondence) (MS Am 1181), Houghton Library, Harvard University.

48. On the popular genre of "superior Martians" observing Earth in this period, see K. Maria D. Lane, *Geographies of Mars: Seeing and Knowing the Red Planet* (Chicago: University of Chicago Press, 2011), 193–97.

49. Edward Bellamy, "The Blindman's World," in *The Blindman's World and Other Stories* (Boston and New York: Houghton Mifflin, 1898), 12, 14–15; Bellamy, "Plots for Stories #1," 242.

50. Bellamy, "Blindman's World," 20, 14; Bellamy, "Plots for Stories #1," 246; "Current Fiction," *Literary World*, December 10, 1898, 434. "The Blindman's World" was published in the *Atlantic Monthly* in November 1886 and reprinted in a posthumous volume of Bellamy's short stories, *The Blindman's World and Other Stories*, published in 1898 with an admiring preface by William Dean Howells.

characterized by the "loss, [through] duplication and competition, such as make man's life below a spectacle of such madness to Martian astronomers." Of course Earth, not Mars, was Bellamy's real focus, and in changing his original title "A Romance of Mars" to "The Blindman's World," he recast his romance as social critique, which the professor articulates in his bleak reassessment of life on Earth: "The lack of foresight among the human faculties, a lack I had scarcely thought of before, now impresses me, ever more deeply, as a fact out of harmony with the rest of our nature, belying its promise,—a moral mutilation, a deprivation arbitrary and unaccountable." Bellamy concluded the story with the professor's utopian yearnings for a wholly different kind of world in which foresight has banished the psychological costs of unpredictability.[51]

Yet the stark contrast between Martian foresight and earthly blindness belies Bellamy's more nuanced conception of temporality in which prospection offered not only the chance to break free of the shackles of Calvinist guilt over a sinful past but also a clearer vision of the present.[52] The question of "The Blindman's World" was not whether foreknowledge was literally possible but rather how to reconceptualize the relationship between present and future. Bellamy's purpose in writing the story was to "describe how anticipation on the earth and on Mars is quite different. On earth it exceeds reality and prepares us to be disappointed with fulfillment, but on Mars it is never excessive, but exactly proportioned to the result."[53] The ideal of Bellamy's Martian utopia was a different kind of predictive imagination that earthlings would do well to emulate.

51. Bellamy, "Plots for Stories #1," 241, 239; Bellamy, "Blindman's World," 28–29.

52. Paul F. Boller, *Freedom and Fate in American Thought: From Edwards to Dewey* (Dallas: SMU Press, 1978), 141. Literary historian Thomas M. Allen has argued that narratives of nationhood in nineteenth-century America hinged upon a "heterogeneous temporal geography" in which time was not a singular fixed abstraction but rather a constellation of different practices. Thomas M. Allen, *A Republic in Time: Temporality and Social Imagination in Nineteenth-Century America* (Chapel Hill: University of North Carolina Press, 2008), 13, 16, quotation on 13. Recent scholarship has emphasized heterogeneous temporalities rather than a singular nineteenth-century conception of increasingly standardized and rationalized time. See Alexis McCrossen, *Marking Modern Times: A History of Clocks, Watches, and Other Timekeepers in American Life* (Chicago: University of Chicago Press, 2013); Lloyd Pratt, *Archives of American Time: Literature and Modernity in the Nineteenth Century* (Philadelphia: University of Pennsylvania Press, 2009); Dana Luciano, *Arranging Grief: Sacred Time and the Body in Nineteenth-Century America* (New York: NYU Press, 2007).

53. Bellamy, "Plots for Stories #1," 245.

But it was the birth of his children, more than anything else, that inspired Bellamy to reimagine futurity and his place in it. As his wife, Emma Sanderson Bellamy, recalled, "the experience of parenthood moved Mr. Bellamy to the depths," and upon becoming a father, to son Paul in 1884 and daughter Marion in 1886, he became preoccupied with the uncertain futures of all children.[54] As William Dean Howells remarked after his friend succumbed to tuberculosis in 1898, Bellamy "told me that he had come to think of our hopeless conditions suddenly, one day, in looking at his own children, and reflecting that he could not place them beyond the chance of want by any industry or forecast or providence; and that the status meant the same impossibility for others which it meant for him."[55] As Bellamy explained in an essay titled "How I Wrote 'Looking Backward,'" parenthood forced him to reckon with the socioeconomic conditions facing "the future of [the] race" and to grapple with the anxiety all parents share, namely, "how to provide for and safeguard their [children's] future" after their own death. Bellamy's personal crisis of certainty led him to begin imagining the year 2000.[56]

"A ROMANCE OF THE FUTURE"

Bellamy wrote *Looking Backward* amid the labor uprisings of the mid-1880s and published it two months after the November 1887 hangings of four anarchists hastily and wrongly convicted of conspiracy to murder in Chicago's Haymarket Square the previous year. But Haymarket did not suddenly transform Bellamy into a radical, reformer, or utopian. Bellamy began his literary career as a social critic, writing editorials on child labor, women's rights, and compulsory education in the 1870s, and through the 1880s, Bellamy's journals reveal a flurry of notes, plot outlines, and titles for a planned utopian text.[57]

54. Emma Sanderson Bellamy, "Edward Bellamy as I Knew Him," undated manuscript, p. 4, folder 2, drawer 1, series 5, Edward Bellamy Memorial Archives, Edward Bellamy House, Chicopee, MA. On parenthood's propelling Bellamy into political economy, see Bowman, *Edward Bellamy*, 15.

55. William Dean Howells, "Biographical Sketch," in Bellamy, *Blindman's World*, xi–xii.

56. Edward Bellamy, "How I Wrote *Looking Backward*," *Ladies' Home Journal*, April 1894, reprinted in Edward Bellamy, *Edward Bellamy Speaks Again!* (Kansas City, MO: Peerage Press, 1937), 222.

57. Franklin Rosemont, "Bellamy's Radicalism Reclaimed," in *Looking Backward, 1988–1888: Essays on Edward Bellamy*, ed. Daphne Patai (Amherst: University of Massachusetts Press, 1988), 155; Bowman, *Year 2000*, 106–11.

Looking Backward is Bellamy's "history of the future" as revealed through the experiences of Julian West, a wealthy Bostonian who undergoes mesmeric treatment for insomnia in 1887 and wakes up to a "moral and material transformation" in the year 2000.[58] Bellamy's utopia is a nationalized industrial democracy based on the principles of equality and Christian solidarity, with none of the economic inequality and volatility of Gilded Age capitalism. Guided through the year 2000 by Bellamy's spokesman, Dr. Leete, and his daughter, Edith, Julian is surprised by the absence of political parties and politicians, merchants and bankers, servants and domestic laborers, and corruption and crime. And he is awed by an array of social and technical advances: an industrial army of equally compensated workers content to work at their chosen vocation from the ages of twenty-one until retirement at forty-five; a centralized system of purchasing and delivering standardized consumer goods; gender equality; and a flourishing artistic and literary culture.

What Bellamy characterized as "a romance of the ideal nation" was also a romance of certainty.[59] As Dr. Leete observes, any venture of a "private capitalist was always a doubtful experiment" since economic crises were no more predictable or preventable than the weather, and he speaks for Edward Bellamy the father in declaring that "the future was so uncertain that to assume parental responsibilities must have often seemed like a criminal risk" in the late nineteenth century.[60] The novel's climactic scene returns Julian to 1887 in a dream, where he is horrified to witness the "specter of Uncertainty" haunting unsuspecting Bostonians. As Julian wanders the streets in his dream, a passerby hands him a card advertising life insurance, which he realizes was only "a partial protection from uncertainty" in the late nineteenth century. "True life insurance" exists only in the year 2000, when policies are underwritten not by individual insurance firms but rather by "one hundred million fellow countrymen."[61] The year 2000 is devoid of inefficient production, unemployment and underconsumption, private wealth accumulation, and market speculation, rendering economic uncertainty an artifact of the late nineteenth century.[62] As Dr. Leete emphasizes, future uncertainties no longer haunt the

58. Daphne Patai, introduction to *Looking Backward, 1988–1888: Essays on Edward Bellamy*, 10; Bellamy, *Looking Backward*, xxi.

59. Edward Bellamy, "How I Came to Write *Looking Backward*," *Nationalist* 1, no. 1 (1889): 2.

60. Bellamy, *Looking Backward*, 157, 159–60, 175.

61. Bellamy, *Looking Backward*, 212.

62. Bellamy, *Looking Backward*, 212. Bowman, *Edward Bellamy*, 31, 38, also characterizes Bellamy's year 2000 as one of economic certainty.

present: "No man any more has any care for the morrow, either for himself or his children, for the nation guarantees the nurture, education, and comfortable maintenance of every citizen from the cradle to the grave."[63] Even unpredictable weather has been rendered inconsequential in the year 2000 by sidewalks' retractable coverings, making umbrellas and boots obsolete.[64]

Although his socialist utopia banished the "specter of Uncertainty," Bellamy distanced himself from socialism and social reform. He claimed that his only knowledge of socialist literature and organizations came from newspapers, insisting that he had read Lawrence Gronlund's *Cooperative Commonwealth* only after the publication of *Looking Backward*, at Gronlund's urging.[65] Bellamy admitted to his friend William Dean Howells that "in the radicalness of the opinions I have expressed I may seem to out-socialize the socialists, yet the word socialist is one I could never well stomach," complaining that this "foreign word . . . smells to the average American of petroleum, suggests the red flag, with all manner of sexual novelties, and an abusive tone about God and religion."[66] Believing the word would doom any political party in the United States, Bellamy opted for the more patriotic-sounding *Nationalist*.[67] Also reluctant to label himself a reformer, Bellamy disavowed any prior association "with any class or sect of industrial or social reformers" and declared that the "self-styled 'reformer' of this day . . . [is] everywhere recognized as a politician who relies upon slander and hypocrisy as his sole weapon."[68] Bellamy was no labor unionist, either: despite his support for the Homestead workers in 1892, he

63. Bellamy, *Looking Backward*, 73.

64. Bellamy, *Looking Backward*, 110. On Bellamy's concern with eliminating weather-related destruction and inconvenience in *Looking Backward* and *Equality*, see William B. Meyer, "Edward Bellamy and the Weather of Utopia," *Geographical Review* 94, no. 1 (2004): 43–54.

65. Howells's *Harper's Monthly* review of *Looking Backward* asserted that Bellamy's and Gronlund's socialist states were virtually identical. But Bellamy, who imagined a singular Nationalist entity, was skeptical of the economic diversity of Gronlund's trade-union socialism. William Dean Howells, "Editor's Study," *Harper's Monthly*, June 1888, 154; John L. Thomas, introduction to Edward Bellamy, *Looking Backward, 2000–1887* (1888; repr., Cambridge, MA: Belknap Press, 1967), 78.

66. Edward Bellamy to William Dean Howells, June 17, 1888, quoted in Joseph Schiffman, "Mutual Indebtedness: Unpublished Letters of Edward Bellamy to William Dean Howells," *Harvard Library Bulletin* 12, no. 3 (1958): 370.

67. Bellamy to Howells, June 17, 1888, quoted in Schiffman, "Mutual Indebtedness," 371.

68. Bellamy, "How I Came to Write *Looking Backward*," 1; Edward Bellamy, *Equality* (New York: D. Appleton, 1897), quoted in Thomas, *Alternative America*, 91.

considered the strike a "blundering instrument" in the long term that exacted more social costs than it did redress individual grievances.[69]

Neither laborite nor reformer nor radical, Bellamy set out in late 1886 to write a work of imaginative fiction that nonetheless electrified a spectrum of political and social reformers.[70] Although Bellamy characterized his political prophecy of evolutionary socialism as conservative, he also acknowledged the appeal of his novel to late nineteenth-century readers who embraced "the necessity and possibility of radical social reform."[71] Bellamy's "sugar-coated" dream tale, as Howells termed it, delivered "a dose of undiluted socialism" that even some of the most fervent antiradicals swallowed without question.[72] As Bellamy confessed shortly after the book's publication, his initial purpose was not social reform but rather "a mere literary fantasy" that imagined, "hanging in mid-air, far out of reach of the sordid and material world of the present, a cloud-palace for an ideal humanity." But by the release of the first edition in January 1888, Bellamy had transformed his novel from "a mere fairy tale of social perfection" into "the vehicle of a definite scheme of industrial reorganization" in which the military was not a metaphor but the model for his nationalized industrial army.[73]

Bellamy's revision paid off, as *Looking Backward* quickly became a best seller with a diverse readership fascinated by what reviewers called a "romance of the future."[74] The novel was first issued by Boston's Ticknor publishing house in January 1888, and after the release of a cheaper edition six months

69. Edward Bellamy, "The Ethics of the Strike," *Springfield (MA) Union*, April 15, 1875, quoted in Milton Cantor, "The Backward Look of Bellamy's Socialism," in Patai, *Looking Backward, 1988–1888: Essays on Edward Bellamy*, 29.

70. On the modernizing reform tradition of Bellamy, Henry George, and Henry Demarest Lloyd, see Thomas, *Alternative America*.

71. Edward Bellamy, "'Looking Backward' Again," *North American Review* 150 (March 1890): 352. Bellamy told Nationalists gathering in Boston in 1890, "We are the true conservative party because we are devoted to the maintenance of republication institutions against the revolution now being effected by the money power. . . . We are not revolutionists, but counter-revolutionists." Quoted in Thomas, introduction to *Looking Backward*, 77.

72. Howells, "Editor's Study," 154.

73. Bellamy, "How I Came to Write *Looking Backward*," 1, 3; Bellamy, "'Looking Backward' Again," 353–54. On the Civil War as "the organizational principle of [Bellamy's] utopia," see Thomas, introduction to *Looking Backward*, 2.

74. "A Look Ahead," *Literary World*, March 17, 1888, 85; Nicholas P. Gilman, "'Nationalism' in the United States," *Quarterly Journal of Economics* 4, no. 1 (1889): 67.

later, sales took off, eclipsing one hundred thousand copies in the novel's second year and four hundred thousand by 1897. Editions were subsequently published in England, France, Germany, Russia, Australia, New Zealand, and Japan.[75] The novel's future orientation fascinated readers, including an enthusiastic Georgia Populist who celebrated Bellamy's "entrancing vision of the twentieth century" and a New Hampshire resident who professed in an admiring sonnet for Bellamy, "Thou hast wrought wonders with thy magic pen, / And set me thinking never as before / About the future, what it has in store / For all who live and love their fellow men."[76]

Agrarian radicals were particularly inspired, as demand for *Looking Backward* spiked in the American West, where a magazine publisher cited "wide and deep interest" in the book, and the paperback edition sold especially well in Midwestern and Plains farming regions.[77] In 1892, a farmer reportedly declared at the St. Louis People's Party convention that "west of the Mississippi we are all Nationalists."[78] Bellamy's book circulated through the countryside thanks to organizations like the Grange and the Farmers' Alliance. At a Grangers' picnic in central Pennsylvania in 1890, five hundred copies of *Looking Backward* were given away, and the Farmers' Alliance distributed several hundred thousand copies of *Looking Backward* to rural readers. In 1889, the Farmers' Alliance marketed the book in conjunction with a subscription to its newspaper (as did other periodicals, like *Ladies' Home Journal*, *Coming Nation*, and the *Indianapolis Leader*).[79]

Along with the Grangers and the Farmers' Alliance, labor unions, socialists, the Women's Christian Temperance Union, the National Council of Women, and religious dissenters like the Theosophists and Christian socialists were

75. Bowman, *Year 2000*, 121; Thomas, introduction to *Looking Backward*, 1; Guarneri, "An American Utopia and Its Global Audiences," 147.

76. James F. King to Edward Bellamy, September 18, 1892; Moses Gage Shirley, "To the Author of 'Looking Backward,'" April 1890, manuscript, both in Bellamy Correspondence (MS Am 1181), Houghton Library.

77. John Walker Brisben to Edward Bellamy, August 7, 1889, Bellamy Correspondence (MS Am 1181), Houghton Library.

78. Edward Bellamy, "Progress of Nationalism in the United States," *North American Review* 154, no. 427 (1892): 750.

79. James J. Kopp, "Looking Backward at Edward Bellamy's Influence in Oregon, 1888–1936," *Oregon Historical Quarterly* 104, no. 1 (2003): 70; Bowman, *Year 2000*, 119, 121; Thomas, introduction to *Looking Backward*, 1. On the role of the rural press and the Farmers' Alliance in spreading Bellamyism, see Christine McHugh, "Midwestern Populist Leadership and Edward Bellamy: 'Looking Backward' into the Future," *American Studies* 19, no. 2 (1978): 58–64.

all enthusiastic readers, each seizing upon different facets of the book's reform vision as congruent with its own.[80] Although Bellamy's detractors faulted the lack of a concrete blueprint to guide them from 1888 to the equality of a Nationalist future, his emphasis on prophecy over prescription was precisely what garnered him such a diverse following.[81] As John L. Thomas has observed, "for a vast number of American readers *Looking Backward* was a moral restorative . . . [that] promised something for everybody."[82]

Bellamy was immediately aware of the political potency of his prophecy. He wrote to publisher Benjamin Ticknor in 1888, "If you will kindly sell 50,000 copies of Looking Backward for me, I will engage to give the voters of 1892 a platform worth voting for, and furnish the voters."[83] Indeed, Bellamy's book—like no other in American history—directly inspired a groundswell of political activism through the establishment of educational clubs that Bellamyites created to discuss and disseminate the principles of Nationalism. Bellamy conceptualized Nationalism not as a discontinuity or revolution but rather an inevitable extension of certain existing conditions, declaring in 1890 that "Nationalism is evolution."[84] The Nationalist clubs that sprung up in response to *Looking Backward* translated Bellamy's long-range forecast of the evolution of Gilded Age capitalism into a short-term political program.

"NO ANXIETY ABOUT THE FUTURE"

Boston was home to the first Nationalist Club, founded in December 1888 by journalists and Theosophists Sylvester Baxter and Cyrus Willard, among others, and subsequently incorporated as the Nationalist Education Association. The largely middle- and upper-class members of the Boston Nationalist Club—among them abolitionist Thomas Wentworth Higginson, Unitarian minister Edward Everett Hale, temperance reformer Frances Willard, and writer William Dean Howells—participated in discussions, lectures, and publication of pamphlets and the *Nationalist*, which was issued monthly from May 1889 to April 1891, when it folded for financial reasons. Hundreds of clubs

80. Bowman, *Year 2000*, 119.

81. On contemporary critiques of *Looking Backward*, see Bowman, *Year 2000*, 138–39; Edward K. Spann, *Brotherly Tomorrows: Movements for a Cooperative Society in America, 1820–1920* (New York: Columbia University Press, 1989), 203–5.

82. Thomas, introduction to *Looking Backward*, 87.

83. Edward Bellamy to Benjamin Ticknor, June 15, 1888, Bellamy Papers, quoted in Thomas, introduction to *Looking Backward*, 69.

84. Bellamy, " 'Looking Backward' Again," 354.

followed nationwide by 1890, at first concentrated in major northeastern cities and California and then in rural areas.[85] The clubs were by nature future oriented, comprising "an organization which looks forward to an entire reconstruction of the political and industrial order of the present day."[86] Some Bellamyites attempted local versions of such reconstruction by forming communitarian utopian colonies in California, Washington, Oregon, Tennessee, and Alabama, as well as in Britain and Paraguay.[87] Others attended turn-of-the-century "Bellamy dances" at Westchester country clubs and Kansas City children's parties, where partygoers put their clothes on backward and wore masks on the backs of their heads as part of what was judged "society's latest fad" (but also "the most freakish dance of the century.")[88] Bellamyite colonies, however short-lived, demonstrated readers' political commitment to Nationalism, and Bellamy dances, however faddish, signaled the cultural influence of the novel, which one critic pronounced "*the* book of the hour in the United States."[89]

Nationalist clubs amplified the program Bellamy set out in *Looking Backward*, advocating government ownership of first the railroad industry, then the integration of telegraph and telephone industries and the express delivery business into the Post Office, then the nationalization of coal mines, and finally the municipal administration of lighting, heating, and transportation services. Other Nationalist principles included civil service reform, referendum and recall, and compulsory education. Nationalism's ultimate goal was the foundational principle of *Looking Backward*: economic equality as the guarantor of political liberty, equal opportunity, and a universal Christian brotherhood.[90] As one reader proclaimed in his sonnet "Looking Backward,"

85. On Nationalism in the context of earlier nineteenth-century cooperative movements like Fourierism, see Spann, *Brotherly Tomorrows*, chap. 12.

86. Gilman, "'Nationalism' in the United States," 64.

87. Kopp, "Looking Backward at Edward Bellamy's Influence in Oregon," 84–87; Guarneri, "An American Utopia and Its Global Audiences," 175.

88. Quotations in "Society's Latest Fad," *Atchison (KS) Daily Globe*, February 3, 1899; "Ellen Osborn's Fashion Letter," *(Grand Rapids, MI) Sunday Herald*, February 12, 1899. On Bellamy dances, see also "Westchester County," *New-York Tribune*, April 13, 1899; "A Bellamy Dance at Troost," *Kansas City (MO) Star*, July 26, 1901; "Bellamy 'Looking Backward' Dances the Latest Fad," *(Denver) Weekly Rocky Mountain News*, February 2, 1899.

89. Gilman, "'Nationalism' in the United States," 53.

90. Edward Bellamy, "Nationalism—Principles, Purposes," address delivered in Boston, December 19, 1889, reprinted in *Edward Bellamy Speaks Again!*, 62–66; Bowman, *Year 2000*, 132–33; Sylvester Baxter, "What Is Nationalism?" *Nationalist* 1 (May 1889): 8–12.

Bellamy's prophecy heralded a "golden age, when Labor holds the throne, / Beside crowned Capital, who now is king!"[91] The Nationalists sought to make this golden age a political reality.

The Nationalist vision of political economy was above all a quest for economic certainty. In 1891, the *Nationalist* published an article by Charles Buell titled "How to Counteract Chance."[92] Chance, according to Buell, was "the tyrant that sits enthroned in a system of competitive industries" and could only be defeated by mutual insurance that would protect factory owners from financial ruin in the event of fire, prevent workers from becoming impoverished during periods of unemployment, and provide housing for old age.[93] As one scholar has noted, "Bellamy makes no room for chance in his theory of evolution; for him, evolution means gradual and peaceful development—nonrandom change toward stability."[94] Bellamy articulated this repudiation of chance in his response to a negative review of *Looking Backward* by Massachusetts Institute of Technology president Francis Amasa Walker, who criticized Bellamy's industrial army as both overly militaristic and fantastic:

> I would ask General Walker whether this "wild, weak dream" of a state in which we should have enough and to spare of necessaries and reasonable luxuries, with agreeable conditions of labor and no anxiety about the future, is not precisely the ideal which all of us spend our days and nights in trying to realize for the benefit of ourselves, our families, our children, and our relatives.[95]

The personal was political for Bellamy, who longed for "no anxiety about the future" of his children or the millions of workers whose labor powered the new industrial society.

91. Harry C. Hochstadter, "'Looking Backward.' A Sonnet," *The Advocate*, undated clipping in Moses Gage Shirley, "To the Author of 'Looking Backward,'" April 1890, manuscript, Bellamy Correspondence (MS Am 1181), Houghton Library.

92. On Buell's essay in the context of social insurance in the United States and Britain, see Jackson Lears, *Something for Nothing: Luck in America* (2003; repr., New York: Penguin Books, 2004), 200.

93. Charles E. Buell, "How to Counteract Chance," *Nationalist* 3, no. 6 (1891): 376–77, quotation on 376.

94. Matthew Hartman, "Utopian Evolution: The Sentimental Critique of Social Darwinism in Bellamy and Peirce," *Utopian Studies* 10, no. 1 (1999): 38.

95. Bellamy, "'Looking Backward' Again," 358. On Walker's critique and Bellamy's response, see Thomas, introduction to *Looking Backward*, 74–75.

Although, as Bellamy's wife recalled, he was "a recluse by nature," he reluctantly accepted the role of political prophet that his readers conferred upon him.[96] The early days of the Nationalist movement, characterized by the didacticism of the *Nationalist*, yielded to a more politically activist era in which Bellamy became more visible in the movement itself—in no small part to address his critics—and in the publication of the weekly magazine *The New Nation*, which he launched in January 1891.[97] In that same year, the Nationalist movement ran its first ticket in a state election in Rhode Island, officially signaling the entry of Bellamyite principles into electoral politics.[98]

William Dean Howells exaggerated when he declared that Bellamy "virtually founded the Populist Party," but some Populist detractors concurred, blaming Bellamy's novel for instigating an agrarian uprising.[99] Nationalism and Populism were ideologically congruent, and Bellamyite aims, such as the nationalization of railroad, telegraph, and telephone industries, migrated into Populist politics.[100] As Georgia Populist James F. King wrote to Bellamy, "If the author of Looking Backward had lived in the forest country of Georgia among the men who guide the plow he would have been a leader in the Farmer's Alliance."[101] For his part, Bellamy, who attended and spoke at People's Party events in the early 1890s, "considered Populism a practical testing of his ideas" and a present incarnation of his future vision.[102]

96. Emma Sanderson Bellamy, "Edward Bellamy as I Knew Him," undated manuscript, p. 2, folder 2, drawer 1, series 5, Edward Bellamy Memorial Archives.

97. The *New Nation* folded in February 1894, a casualty of the depression of 1893 as well as Bellamy's failing health and finances. Bowman, *Year 2000*, 136.

98. Individual Nationalist candidates had run for California and Michigan congressional seats in 1890, as well as for seats in state legislatures. Bowman, *Year 2000*, 133.

99. William Dean Howells, *Literature and Life: Studies* (New York: Harper & Brothers, 1902), 294, quoted in Elizabeth Sadler, "One Book's Influence: Edward Bellamy's 'Looking Backward,'" *New England Quarterly* 17, no. 4 (1944): 538; McHugh, "Midwestern Populist Leadership and Edward Bellamy," 60.

100. On Bellamy's influence on Populism, see McHugh, "Midwestern Populist Leadership and Edward Bellamy," 59–60; Charles Postel, *The Populist Vision* (New York: Oxford University Press, 2007), 233–36.

101. James F. King to Edward Bellamy, September 18, 1892, Bellamy Correspondence (MS Am 1181), Houghton Library.

102. Bowman, *Year 2000*, 135; quotation in Thomas, introduction to *Looking Backward*, 82. Although the Nationalist and Populist movements essentially converged by the election of 1896, their congruence was not complete. Some of the most fervent Nationalists were disgruntled when Bellamy made overtures toward the People's Party in 1892, and Bellamy was wary

Although the Nationalist movement—its fortunes tied to the People's Party—did not survive the folding of Bellamy's publication *The New Nation* in 1893 or the Populist defeat in 1896, *Looking Backward* was widely read into the twentieth century. Bellamy's account of an Americanized socialism had an enormous influence on American socialists, including Daniel De Leon, a Nationalist who would become the Socialist Labor Party leader in 1892, and five-time Socialist presidential candidate Eugene V. Debs, who credited Bellamy with steering him and thousands of others toward socialism. Bellamy's vision of a peaceful transition to socialism devoid of the class struggle of a Marxist revolution also appealed to some Progressive Era reformers, but its biggest resurgence came in the 1930s, both in the United States and in Europe, when *Looking Backward* appeared second to *Das Kapital* on three different lists of "the twenty-five most influential books published since 1885."[103] The establishment of the Utopian Society of America in Los Angeles, the rebirth of Bellamy clubs, the emergence of the Technocracy movement, and a 1934 gubernatorial run in California by socialist Upton Sinclair all contributed to a Depression-era Bellamy revival.[104] The All States Loyal Bellamy Associates, a Los Angeles organization that coordinated membership across twenty-four states, hailed Bellamy as "the greatest Statesman America has ever produced . . . our greatest economist . . . our greatest Moralist," and the Bellamy Club No. 1 of Maine announced "at least 2,000,000 Socialists, Bellamyites, and Technocrats (Technocrats are Bellamyites with a new name)" ready to unite forces in the launching of a national political party.[105] In 1934, journalist Ida Tarbell hailed Bellamy, along with Henry George, as one of the "New Dealers" of his day, and Communist Party leader Earl Browder wrote to Emma Bellamy in

---

of single-plank silverism. On the relationship between Nationalism and Populism, see Spann, *Brotherly Tomorrows*, 198–207.

103. Sadler, "One Book's Influence," 553. On Bellamy's influence on American socialism, see Rosemont, "Bellamy's Radicalism Reclaimed," 162–71; Arthur Lipow, *Authoritarian Socialism in America: Edward Bellamy and the Nationalist Movement* (Berkeley: University of California Press, 1982).

104. Thomas, introduction to *Looking Backward*, 85; Sadler, "One Book's Influence," 544–50; "The Edward Bellamy Group of Brooklyn," undated pamphlet, folder 1, drawer 1, series 6, Bellamy Memorial Archives. On Bellamyite politics in the 1930s, see James J. Kopp, "Edward Bellamy and the New Deal: The Revival of Bellamyism in the 1930s," *Utopian Studies* 4 (1991): 10–16.

105. All States Loyal Bellamy Associates, "Proves He Looked Forward," n.d., folder 20, drawer 1, series 6, Bellamy Memorial Archives; Bellamy Club No. 1 of Maine, "A Definite Plan of Action," n.d., folder 1, drawer 1, series 6, Bellamy Memorial Archives.

1934 that the party's manifesto at the Eighth National Convention represented the "modern development of the ideas of your dead husband, who made so rich a contribution to the revolutionary movement."[106] The enthusiasm for Bellamyite politics in the 1930s waned quickly, however, as Nationalism took on an eerie resemblance to totalitarianism.[107]

Bellamyism ended with the coming of the Second World War, but Bellamy's legacy as a technological forecaster and political prophet endured. The All States Loyal Bellamy Association ran the headline "Proves He Looked Forward" above Bellamy's portrait in the 1930s, and many Bellamyites, futurists, and historians have emphasized the accuracy of Bellamy's predictions, in *Looking Backward* and its 1897 sequel *Equality*, of technical innovations, including radio broadcasting, credit cards, television, and airships.[108] But even more significant was Bellamy's relationship to futurity more broadly. John Dewey would later call him a "great American prophet," and upon his death, the reformist literary magazine *The Arena* asked in a headline, "Is the Prophet Dead?," which historian and editor John Clark Ridpath answered in the negative, observing that all prophets "are always in the present tense."[109] Yet Bellamy belonged to the future tense, for Ridpath and countless late nineteenth-century readers, because of the evolutionary framework for his critique of political economy and his optimistic vision of an alternative to capitalism. As Ridpath wrote,

> He perceived clearly that there is no finality in the human evolution, but only an ongoing and new development for ever and ever. He therefore *looked ahead* and anticipated somewhat the possible state of society to come. He looked beyond the corporation and the trust, beyond the prodigious development of mod-

106. Ida M. Tarbell, "New Dealers of the 'Seventies: Henry George and Edward Bellamy," *Forum and Century* 92 (September 1934): 133–39; Earl Browder to Emma Bellamy, May 8, 1934, folder 36, drawer 2, series 8, Bellamy Memorial Archives.

107. Thomas, introduction to *Looking Backward*, 85. On Bellamy's legacy in the early twentieth century, see Thomas, *Alternative America*, chap. 15.

108. All States Loyal Bellamy Associates, "Proves He Looked Forward," n.d., folder 20, drawer 1, series 6; Alexander Scott, "He Predicted the Coming of Radio, Aeroplanes, Smooth Roads, and Even Television," reprint of radio address (Wellington, New Zealand: Edward Bellamy Society, Inc., 1936), folder 16, drawer 1, series 6; William T. Gay, "Edward Bellamy, Nineteenth Century Futurist," *Futurist* (June 1967): 40–41, folder 19, drawer 1, series 5, all in Bellamy Memorial Archives.

109. John Dewey, "A Great American Prophet," *Common Sense* 3, no. 4 (1934): 6–7; John Clark Ridpath, "Is the Prophet Dead?," *Arena* 20, no. 105 (1898): 284.

ern commercialism and municipality, and saw something higher and grander than these rising in the distance.[110]

Bellamy oriented a generation of readers and writers to the far-off future, and the commercial success of his novel inspired nearly two hundred fictional responses, imitations, and critiques between 1888 and 1900, an outpouring of literary utopianism unequaled before or since. Bellamy's politics of the future looked forward to a society perfected by economic certainty in which, ironically, anticipation would be unnecessary. The elimination of chance at the core of Bellamy's evolutionary socialism would also animate the career of another Massachusetts forecaster who sought to discover recurring long-term patterns in weather and business.

"HUB'S 'LONG DISTANCE' WEATHER PROPHET"

Henry Helm Clayton (1861–1946; see figure 4.4), a rising star in late nineteenth-century observational meteorology and local weather forecasting, was hailed toward the end of his career as a "valiant pioneer" who "persevered against scant understanding and scantier appreciation."[111] A fellow of the American Academy of Arts and Sciences and president of the Boston Scientific Society, Clayton was remembered for his "life of exceptional activity and eminence in meteorology, public service, and business."[112] Clayton's extensive publication record included twelve papers in *Nature*, twenty-eight papers in *Science*, thirty-nine articles in the *American Meteorological Journal*, twenty-two papers in the *Annals of the Observatory of Harvard College*, and forty-two publications on climate and weather cycles.[113] Although he made the *North American Review*'s 1930 list of the ten researchers "who are greatest in science," Clayton has been largely overlooked by historians, perhaps because he worked on the boundaries of government weather science, in private and public institutions, before becoming an independent consultant in the business world, or perhaps because his commitment to cycles and sunspots has relegated him to the margins of

110. Ridpath, "Is the Prophet Dead?," 287 (emphasis in original).

111. "Henry Helm Clayton and Old Sol Foretell Boston Weather," *Boston Evening Transcript*, February 2, 1929.

112. Sterling P. Fergusson and Charles F. Brooks, "Henry Helm Clayton, 1861–1946," *Science*, n.s., 105, no. 2723 (March 7, 1947): 247.

113. H. Helm Clayton, folder "Bibliography of Henry Helm Clayton," box 12, H. Helm Clayton Papers, Smithsonian Institution Archives.

FIGURE 4.4. Henry Helm Clayton, meteorologist at Blue Hill Observatory in Milton, Massachusetts, from 1886 to 1909. Reprinted from Raymond L. Bridgman, "Weather Studies at Blue Hill," *New England Magazine*, March 1895, Collection of Newspaper Clippings, Offprints, Photographs, etc., Relating to the Blue Hill Meteorological Observatory, Abbot Lawrence Rotch Papers, Houghton Library, Harvard University.

scientific meteorology.[114] But Clayton was a highly regarded figure in the meteorological profession from the late nineteenth to the mid-twentieth century, when he worked tirelessly to establish a correlation between meteorological and economic cycles and to popularize his theories of periodicity. He came to believe, but could never prove, that changes in the weather governed markets, and he remained convinced that a science of cycles would reveal predictable patterns in both.

114. E. E. Free, "Who Are Greatest in Science?," *North American Review* (January 1930): 17, reprint in folder "Free, E. E. 1928–1936," box 4, Clayton Papers.

Born and raised in Tennessee, Clayton became fascinated with meteorology as a teenager, when, recovering from an illness, he stayed in Missouri with an aunt who swore by lunar influences on the weather. Duly inspired, Clayton returned home and collected three years' worth of observations that sought (unsuccessfully) to demonstrate a correlation between the moon and the weather. But his efforts, which he described in the pages of the *American Meteorological Journal*, earned him a job as an assistant in the University of Michigan Observatory, after which he moved east in 1886 to become an observer at the new Blue Hill Observatory in Massachusetts, where he remained until 1909.[115] During the first three decades of his career, Clayton fell in and out of favor with government institutions: he advocated in the late 1880s for the transfer of the weather service from military to civilian oversight and served as a Weather Bureau forecast official in Boston from 1891 until 1893, when, as the previous chapter revealed, he was drummed out of the bureau for inconsequential errors on his weather maps. However, he subsequently drew the attention of Weather Bureau officials curious about his surprisingly successful weekly forecasts.[116]

Clayton's work would take him literally above and beyond Boston. His work in aerology included participating in the first experimental use of kites to carry self-recording instruments into the atmosphere at Blue Hill in 1894, joining Blue Hill Observatory founder Abbot Lawrence Rotch and French meteorologist Teisserenc de Bort to launch kites and balloons from the deck of a steamer in the Atlantic to collect data on trade winds in 1905, and contributing to Blue Hill's development of the radiosonde in 1935.[117] He contributed to publications of global significance, including the US Hydrographic Office's *Atlas of*

115. "H. Helm Clayton and Old Sol Foretell Boston Weather," *Boston Evening Transcript*, February 2, 1929; "Henry Helm Clayton, 85, Expert on Sun Spots," *NYT*, October 28, 1946; Fergusson and Brooks, "Henry Helm Clayton," 247.

116. On Clayton's call for a civilian national weather service, see H. Helm Clayton, "A Plea for Civilian Control of the U.S. Weather Bureau," *Science* 9, no. 209 (February 4, 1887): 113–14. On the Weather Bureau's renewed interest in Clayton's work in the early twentieth century, see Frank Waldo, "Noted Meteorologist Will Help 'Make the Weather,'" *NYT*, April 15, 1906.

117. On the use of kites to carry self-recording meteorological instruments, see H. Helm Clayton, "Scientific Kite-Flying at Blue Hill," pamphlet reprinted from the *Boston Commonwealth*, May 9, 1896, Collection of Newspaper Clippings, Offprints, Photographs, etc., Relating to the Blue Hill Meteorological Observatory, Abbot Lawrence Rotch Papers, Houghton Library, Harvard University. On the Rotch–de Bort expedition, see "Blue Hill Meteorological Observatory, 1885–1910," reprint from *Technology Review* 12, no. 2 (1910): 2, Collection of Newspaper Clippings, Offprints, Photographs, etc., Relating to the Blue Hill Meteorological Observatory, Rotch Papers, Houghton Library.

Clouds (1897), and he edited the first four volumes of the Smithsonian's *World Weather Records* (1927, 1934, 1944, 1947), which collected monthly data on average temperature, average barometric pressure, and precipitation from nearly four hundred land stations around the globe and "became the *de facto* standard data source for large-scale climatology" until the 1980s.[118] In the 1910s and 1920s, he headed Argentina's forecasting service in Buenos Aires, and upon his return to the United States in 1925 he launched a private weather service that would count numerous businesses among its subscribers. In 1943, the Foundation for the Study of Cycles awarded him a gold medal, and indeed, periodicity was the unifying theory for his research and private forecasting work. As he explained to a Massachusetts clergyman in 1929, "the entire progress of science depends upon discovering and relying upon an orderly sequence in nature. Even in that variable thing we call the weather the scientist is finding an orderly sequence of cause and effect."[119]

The Blue Hill Meteorological Observatory in Milton, Massachusetts, where Clayton spent over twenty years, played a crucial role in his early investigations into periodicity. The "half-monastery, half-castle" was founded in 1885 by Abbot Lawrence Rotch, a wealthy young Bostonian who had graduated from the Institute of Technology (as MIT was then called) the year before.[120] Rotch purchased land near his family's summer estate southwest of Boston, financed the construction and operation of the observatory, and oversaw a handful of men in his employ.[121] Great Blue Hill, which "possessed many of the characteristics of a mountain," was an ideal site given its status as the highest point

118. Robert G. Quayle, "Philanthropy and Global Climate Monitoring," *Earth Sciences History* 20 (2001): 193–95; Paul N. Edwards, *A Vast Machine: Computer Models, Climate Data, and the Politics of Global Warming* (2010; repr., Cambridge, MA: MIT Press, 2013), 58–59, 309–11, quotation on 309.

119. H. H. Clayton to Rev. Edward H. Cotton, July 12, 1929, folder "C, general," box 3, Clayton Papers.

120. John Collier, "Blue Hill Observatory," *Boston Daily Globe*, May 8, 1887.

121. "Blue Hill Meteorological Observatory," reprint from *Journal of Education*, 1903, folder "Blue Hill Observatory—Miscellaneous Clippings and Illustrations," Collection of Newspaper Clippings, Offprints, Photographs, etc., Relating to the Blue Hill Meteorological Observatory, Rotch Papers, Houghton Library; "Blue Hill Meteorological Observatory, 1885–1910," reprint from *Technology Review* 12, no. 2 (1910), Collection of Newspaper Clippings, Offprints, Photographs, etc., Relating to the Blue Hill Meteorological Observatory, Rotch Papers, Houghton Library; "Blue Hills Observatory Has a Long History," *Milton (MA) Times*, April 1, 1999, folder "Blue Hill Observatory—Newspaper Articles," box 7A, Blue Hills Observatory Papers, Milton Historical Society.

(635 feet above sea level) on the east coast within ten miles of the Atlantic Ocean, which afforded lower-altitude observations to complement the Signal Service stations at Mount Washington (six thousand feet) and Pikes Peak (fourteen thousand feet) as well as clear views of distant cloud formations.[122] Blue Hill conducted pioneering cloud observations (including hourly observations of movement, changes in form, and changes in altitude) from 1886 to 1890 that were the only data of that kind for the western hemisphere.[123] As one journalist remarked, Clayton "takes a professional look into the clouds hourly," and indeed, Blue Hill is known in the present day as "the oldest continuously operating weather observatory in the United States and a benchmark for world climatology."[124] Renowned as a research site and a local forecasting center, Blue Hill, with its impressive library and instrumentation, was considered "among the best observatories of the world." Blue Hill would become affiliated with Harvard University after Rotch's death in 1912, but until then, its identity as a private facility apart from the federal government and other scientific institutions cultivated its "thoroughly independent attitude," which in many ways characterized Clayton as well.[125]

Clayton's belief in periodicity and his labors of data collection converged at Blue Hill, which focused both on operational weather forecasting and climatology during his tenure. In 1885, a year before his arrival at Blue Hill, Clayton found an apparent cycle of twenty-five months in barometric pressure at distant stations. Although he could not account for an underlying cause, he assumed one existed. That periodicities could "all be the work of chance seems to me incredible," he wrote.[126] Barometric periodicity was visible only in the long term, according to Clayton, who also posited a seven-day period in barometric pressure and mean daily temperature in Chicago as well as a thirty-day weather period.[127] Situating his work on periodicity within a contemporary

122. Collier, "Blue Hill Observatory"; Frank Waldo, "The Blue Hill Meteorological Observatory," reprint from *Popular Science Monthly*, July 1901, p. 292, Collection of Newspaper Clippings, Offprints, Photographs, etc., Relating to the Blue Hill Meteorological Observatory, Rotch Papers, Houghton Library.

123. Fergusson and Brooks, "Henry Helm Clayton," 247.

124. Collier, "Blue Hill Observatory"; "The Blue Hill Observatory," http://www.bluehill.org/history.html.

125. Waldo, "Blue Hill Meteorological Observatory," 293, 304, 302–3.

126. H. Helm Clayton, "A Lately Discovered Meteorological Cycle," *American Meteorological Journal* 1, no. 12 (1885): 534.

127. H. Helm Clayton, "Weather Changes of Long Period," *American Meteorological Journal* 2, no. 3 (1885): 133.

tradition of meteorological observation of cycles, he acknowledged the epistemic contests over periodicity and correlation, citing "that almost universal popular belief in a connection of the changes of the weather with the changes of the moon . . . against which modern science has brought her battering rams of argument and proof in vain, for it seems to retain a well nigh invincible hold on the popular mind."[128]

In 1886, Clayton put his theories of periodicity into practice when he used his data on cycles to make monthly, ten-day, and one- to two-day weather forecasts.[129] Long-range forecasting was, Clayton maintained, the only kind worth pursuing: "Predictions become of value only when they are extended so far in advance, that to the average man they become a better guide to the coming weather, than the appearance of the sky and the conditions surrounding him."[130] Although his verification percentages were not high enough to establish the predictive value of cycles, Clayton still rejected what he deemed "an unreasonable faith in the goddess of chance."[131] In the 1890s, determined to turn his theoretical investigations into a forecasting method, Clayton focused on the somewhat paradoxical problem of irregular periodicities, which he defined as "apparent periodicities which vary much in intensity from time to time and whose causes are not yet known."[132] In 1893, Clayton calculated the probability of seven- and six-day weather periods being accidental in an attempt to "preclude the theory of chance." He compared irregular periodici-

128. Clayton cited historical observations of storm cycles made by Smithsonian Secretary Joseph Henry in 1858, Connecticut politician Thomas Belden Butler in 1870, and famed almanac writer Henry Vennor in 1883. H. Helm Clayton, "The Seven Day Weather Period," *American Meteorological Journal* 2, no. 4 (1885): 167–68.

129. In the mid-1880s and 1890s, Blue Hill Observatory was well known for its pioneering work in local forecasting, which combined the Signal Service's synoptic charts of data from approximately 150 observation stations with observations from the Boston area. On the competition between Blue Hill and Signal Service forecasts, see Bergman, "Knowing Their Place: The Blue Hill Observatory and the Value of Local Knowledge." On Clayton's call for a partially decentralized government weather service that would combine Washington's general forecasts of high- and low-pressure areas with more specific local forecasts based on volunteer observations "from a net-work of secondary stations in their locality," see H. Helm Clayton, Letter to the editor, *American Meteorological Journal* 4, no. 9 (1888): 417.

130. H. Helm Clayton, letter to the editor, *American Meteorological Journal* 4, no. 10 (1888): 484.

131. H. Helm Clayton, "An Experiment in Long Range Prediction," *American Meteorological Journal* 2, no. 10 (1886): 463.

132. H. Helm Clayton, "Long Range Weather Predictions," *American Meteorological Journal* 8, no. 2 (1891): 68.

ties to "a series of waves of various lengths moving over the surface of a bowl of water and rebounding back and forth [that] might throw the surface into irregular agitation without any visible trace of periodicity."[133] Unlocking their secret would be, Clayton believed, the key to long-range forecasting. As he remarked in 1895, "to most persons nothing seems more irregular and fitful than the weather. Yet I think there is a good reason to believe that through all this seeming irregularity runs a web of harmony and rhythm."[134]

Clayton found a theoretical explanation for irregular periodicities in what he called "weather rhythms," similar to harmonic sound waves or waves produced from throwing multiple pebbles into water, which made it impossible to find a pattern in overlapping and complex waves. In 1895, he proposed periods of 5 days plus 10.8 hours and 4 days plus 15 hours to go along with his previously discovered periods of 7 days plus 6 hours and 6 days plus 4 hours, and he illustrated these periods and their harmonic intervals with temperature data from summer months at the Blue Hill Observatory. Clayton acknowledged that the complexity of these "weather harmonics" posed three major obstacles to forecasting: "the multiplicity of the cycles and their independent variations in amplitude according to some unknown law; the oscillations of the cycles . . . simultaneously in different phases in different parts of the world; the sudden inversion of the phase of the cycle from time to time at any one point on the earth's surface."[135] Nonetheless, he faulted skeptics as akin to those who would expect to find a neatly concentric path of rubble in a tornado's aftermath. Although Clayton acknowledged weather harmonics as "one of the complexities which has helped to obscure weather periodicity," he maintained his faith that "the[se] difficulties will in time be solved, and that forecasting by means of weather cycles will supplant largely, if not entirely, all other forms of weather forecasting."[136] But the problem of complex and irregular periodicities would haunt Clayton throughout his career, as he publicly acknowledged at the Symposium on Cycles in 1937; he privately admitted that "this is a very complicated question and I sometimes get discouraged with the results."[137]

133. H. Helm Clayton, "Six and Seven Day Weather Periodicities," *American Meteorological Journal* 10, no. 1 (May 1893): 43, 64.

134. Henry Helm Clayton, "Rhythm in the Weather," *American Meteorological Journal* 11, no. 10 (1895): 376.

135. Clayton, "Rhythm in the Weather," 377, 244–45.

136. H. Helm Clayton, "Weather Harmonics," *Science*, n.s., 7, no. 164 (February 18, 1898): 243, 245.

137. H. Helm Clayton, "Moving Waves of Weather: Remarks by H. H. Clayton at Symposium on Cycles, Washington, D.C., April 25, 1937," manuscript, folder "Manuscripts and Papers

At the turn of the century, Clayton broadened the scope of his investigations into the economic and political implications of weather and climate. He ventured a causal link between the weather and the market in a 1901 *Popular Science Monthly* essay titled "The Influence of Rainfall on Commerce and Politics." He sought to connect "links in the chain of causation" between drought, famine, financial panics, and their political repercussions, correlating the panics of 1837, 1857, 1873, and 1893 with periods of below-average rainfall and characterizing the prevailing explanations for panics as "akin to superstitions, because they attribute to human action what is due largely to natural causes beyond the control of man." Clayton urged further scientific research into the correlation between changes in the weather and changes in the market, with the grand hope that, with an independent private research institution founded by "some wise benefactor" (on the model of Rotch at Blue Hill), an equivalent to Darwinian theory would emerge to account for causation in the world of political economy.[138]

Clayton's efforts to find rainfall in prices were well received, with climatologist Robert DeCourcy Ward pronouncing his essay "extremely suggestive" and indicating its value to "those who take pleasure, not only in studying the correlations of meteorological conditions and politics in the past, but who also wish to try their luck at forecasting the political changes of the future." Yet the *Boston Globe* denounced Clayton's argument as naïvely apolitical, editorializing against "a violent reversal of a common practice which Mr. Clayton invites us to make. Instead of holding the administration responsible for the weather he would have us hold the weather responsible for the administration." The *Globe* sneered at the notion of Coxey's Army "pray[ing] for rain instead of marching on Washington" or politicians campaigning for precipitation, concluding that "the full dinner pail cannot be so easily transformed from a political to a meteorological emblem."[139] Clayton's vision of the economic future hinged on a meteorological determinism that did not sit well with the *Globe*'s editors, who pointed to the political power of the electorate in Beacon Hill and beyond.

---

of Henry Helm Clayton, 1877–1943, and undated," box 12, Clayton Papers; H. H. Clayton to W. B. Schostakowitsch, August 20, 1932, folder 3, box 7, Clayton Papers.

138. H. Helm Clayton, "The Influence of Rainfall on Commerce and Politics," *Popular Science Monthly*, December 1901, 163, 164, 165.

139. Robert DeC. Ward, "Notes on Climatology," *Bulletin of the American Geographical Society* 34, no. 1 (1902): 45; Robert DeC. Ward, "Current Notes on Meteorology," *Science*, n.s., 25, no. 368 (January 17, 1902): 110; "Politics and the Weather!," *Boston Daily Globe*, December 8, 1901.

Clayton ultimately found the most support for his work on cycles in the business world when he launched a private forecasting firm in 1925 after spending two years conducting sunspot research in conjunction with the Smithsonian, which was funded by John A. Roebling II, grandson of the famous Brooklyn Bridge engineer.[140] The Clayton Weather Service operated out of Clayton's home in Canton, Massachusetts, employing three people: a computor, a radio operator, and his daughter Frances Clayton, who was responsible for secretarial work, accounting, and marketing.[141] Clayton steadily expanded his business despite "limited resources" but was earning upward of $2,500 a year (about $35,000 today) before being "hard hit" by the Great Depression.[142] For the North Atlantic and Great Lakes regions, the Clayton Weather Service issued twice-weekly bulletins with a general weekly forecast and a detailed five-day forecast, along with what one newspaper called "a prophecy for the season ahead" and long-range forecasts six to ten months in advance. Clayton's list of between forty and one hundred clients in the 1930s (which included the city of Boston, department stores Filene's and Lord & Taylor, food manufacturers like Canada Dry Ginger Ale and General Seafoods, as well as coal, gas, electric, and transportation companies) paid one hundred dollars per year for access to all forecasts and fifty dollars per year for a partial subscription.[143]

Clayton's forecasts combined the local and the global, incorporating data from the Smithsonian's mountain solar observatories in California, New Mexico, and Chile with upper-air cloud observations at his Massachusetts home.[144] Clayton based his temperature and precipitation forecasts (which required

140. H. H. Clayton to Harold Manchester, September 17, 1940, folder 13, box 5, Clayton Papers.

141. H. H. Clayton to Harold Manchester, September 17, 1940, folder 13, box 5, Clayton Papers; H. H. Clayton to Howard A. Poillon, December 6, 1930, folder 11, box 6, Clayton Papers.

142. H. H. Clayton to C. G. Abbot, September 9, 1929, folder "Abbot, Charles G., 1929," box 1; H. H. Clayton to Clients of the Clayton Weather Service, July 28, 1942, folder 10, box 6; H. H. Clayton to Howard A. Poillon, December 6, 1930, folder 11, box 6; H. H. Clayton to Melvin L. Morse, December 19, 1931, folder 19, box 5, all in Clayton Papers.

143. H. H. Clayton to H. W. Clough, July 2, 1929, folder 8, box 3, Clayton Papers; C. G. Abbot to Stephen T. Lockwood, May 10, 1939, folder 9, box 5, Clayton Papers; "Henry Helm Clayton and Old Sol Foretell Boston Weather," *Boston Evening Transcript*, February 2, 1929; H. H. Clayton to Howard A. Poillon, December 6, 1930, folder 11, box 6, Clayton Papers; Frances Clayton to Leon W. Stetson, July 30, 1930, folder 14, box 6, Clayton Papers.

144. C. G. Abbot, radio address, April 1939, folder "Abbot, Charles G., 1939," box 1, Clayton Papers; H. H. Clayton to F. W. Reichelderfer, November 20, 1942, folder 10, box 6, Clayton Papers. On the relationship between local and global in the history of meteorology, see James Rodger Fleming, Vladimir Jankovic, and Deborah R. Coen, eds., *Intimate Universality: Local*

"smoothing" of irregular periodicities) on an established eleven-year sunspot period during the "vogue" of solar radiation research led by Charles Greely Abbot, secretary of the Smithsonian from 1928 to 1944, who undertook controversial astrophysical research into the solar constant's presumed effect on weather.[145] Clayton's method of forecasting, which he tried unsuccessfully to patent in 1931, involved analyzing climatological data to find periodicities, correlating those atmospheric cycles with observed sunspot cycles, mathematically aligning the amplitudes of solar and terrestrial waves, and temporally extending the waves to produce a forecast for individual locations in a network of observation stations.[146] Clayton published an article on his long-range forecasting method in the *Monthly Weather Review* in 1937, but government meteorologists privately remarked that "there is some question as to whether these periods have any reality at all, or are only imposed by Clayton's smoothing method," noting "many cases in which this extrapolation becomes pretty much a matter of imagination."[147]

Clayton characterized his forecasts as "qualitative" and "reasonable estimate[s] of certain coming changes" rather than predictions of quantifiable temperature differences. The Clayton Weather Service bulletin reminded its readers that "the purpose of the forecast is to indicate the TRENDS in weather over a given period of time in the future. Definite dates for changes are given but actual changes may be earlier or later than the time forecasted owing to

---

and Global Themes in the History of Weather and Climate (Sagamore Beach, MA: Science History Publications/USA, 2006).

145. Clayton Weather Service, "Long Period Weather Changes Forecasted," December 30, 1935, folder "Manuscripts and Papers of Henry Helm Clayton," box 12, Clayton Papers; David H. DeVorkin, "Defending a Dream: Charles Greely Abbot's Years at the Smithsonian," *Journal for the History of Astronomy* 21, no. 1 (1990): 121, 129, quotation on 129.

146. Thomas E. Robertson to Byrnes, Townsend & Potter, March 8, 1932; Byrnes, Townsend & Potter to H. H. Clayton, March 24, 1931, both in folder 15, box 2, Clayton Papers. Other scientists pursuing studies of cycles and correlation at the same time included oceanographer George F. McEwen on Pacific Ocean and air temperatures; astronomer A. E. Douglass on sunspots, climate, and tree growth; and astronomer Dinsmore Alter on sunspots and rainfall. S. P. Fergusson, "Recent Discussion of Long-Period Forecasting," memorandum to Chief of Weather Bureau, February 23, 1928, folder "Fergusson, S. P., 1911, 1922–1929," box 3, Clayton Papers.

147. H. H. Clayton, "Long Range Weather Changes and Methods of Forecasting," *Monthly Weather Review* 64, no. 11 (1936): 359–76; Hurd C. Willett, "Review of H. H. Clayton on Long-Range Weather Changes and Methods of Forecasting," c. 1937, pp. 5, 6, folder 16, box 7, Clayton Papers.

irregularities in the movement of storm areas."[148] A bulletin from August 25, 1932, for example, included the following monthly forecast:

> In the NORTH ATLANTIC STATES warmer periods are expected about Sept. 1st to 5th and 15th to 19th; cooler periods about Sept. 7th to 13th and 21st to 25th. Frosts are likely in Northern New York and Northern New England during this latter period. The days when rainfall is most probable are the 3rd, 7th, 12th, 16th, 20th, 25th, and 29th.[149]

The range of dates in this forecast, not unlike the "storm periods" of late nineteenth-century long-rangers, was qualitatively probabilistic and did not make claims to predictive certainty. Frances Clayton explained to a potential client that "we do not try to make people think our service is 100% accurate," and indeed, printed at the bottom of a Clayton Weather Service bulletin was a disclaimer that "our reports, forecasts, and opinions contain the possibility of error and are not guaranteed." But the source of this uncertainty, Clayton maintained, was the "not yet precise" understanding of sunspot periodicity. With a breakthrough in solar radiation research, Clayton predicted, "we will be able to tell you the precise day to day weather, not a few days in advance, but months, and perhaps years ahead."[150]

In the late 1920s, Clayton boasted average yearly verification rates of approximately 70 percent (according to the secretary of the Retail Trade Board and the Yale professor Ellsworth Huntington), and in 1929, a Detroit Edison Company fuel agent pronounced "the Clayton Weather Service Reports to be the best gauge upon which to depend and their weekly reports to be equally as accurate as the 48-hour Government Report."[151] The Weather Bureau disagreed, finding only a 52 percent verification of Clayton's temperature and precipitation forecasts in the mid-1930s.[152] Nonetheless, requests for forecasts

148. H. H. Clayton to Edward M. Burd, January 9, 1933, folder 6, box 2; Science Service, "Rainfall for Drought Area Forecast in Next Few Weeks," March 6, 1931, folder 4, box 7; Clayton Weather Service Bulletin (blank), n.d., folder 17, box 8, all in Clayton Papers.

149. Clayton Weather Service Bulletin, August 25, 1932, folder 6, box 2, Clayton Papers.

150. Frances Clayton to Leon W. Stetson, July 30, 1930, folder 14, box 6; Clayton Weather Service Bulletin (blank), n.d., folder 17, box 8; "Scientist Hopes to Tell Weather Far in Advance," *Boston Traveler*, April 7, 1934, folder 5, box 6, all in Clayton Papers.

151. H. H. Clayton to Harold Manchester, September 17, 1940, folder 13, box 5; quotation in Harold B. Tyree to H. B. Gates, May 10, 1929, folder 12, box 3, both in Clayton Papers.

152. Hurd C. Willett, "Review of H. H. Clayton on Long-Range Weather Changes and Methods of Forecasting," c. 1937, p. 7, folder 16, box 7, Clayton Papers.

came in from across the country, including inquiries from a small farmer in California, a New York fire and crop insurance company, and the Petroleum Economics Division of the US Bureau of Mines.[153] Opposed by Weather Bureau chief Charles F. Marvin upon its founding, the Clayton Weather Service survived the Great Depression but was shut down by the federal government in 1942 as part of a wartime ban on commercial forecasts (for fear that they could be exploited by enemy forces), but Clayton himself was invited to join a new government research program on long-range forecasting, which he did. Indeed, from the turn of the twentieth century until the end of Clayton's life in 1946, the US Weather Bureau continued to be both interested in and skeptical of his work in long-range forecasting and his search for the elusive secret of weather cycles, and Clayton's correspondence reveals a network of connections to not only Weather Bureau officials but also their targets, such as long-rangers W. F. Carothers and Herbert Janvrin Browne.[154]

According to Weather Bureau meteorologist S. P. Fergusson, the long-range predictions of the Clayton Weather Service in the 1920s were similar to those of economic forecasting firms like Babson's and Brookmire's.[155] Indeed, Clayton posited correlations between stock prices and rainfall in the interior of the United States, observing that rainfall "had an irregular period of about three years corresponding closely with the shorter cycles in the prices of stock" that a market analyst had charted. And the correlation was potentially predictive: Clayton noted that the rainfall maxima generally preceded the stock price maxima by a few months.[156] (See figure 4.5.)

153. Clarence E. Hawkins to H. H. Clayton, October 31, 1931, folder 11, box 4; J. M. Wennstrom to H. H. Clayton, January 31, 1929, folder 11, box 4; E. B. Swanson to H. H. Clayton, October 19, 1931, folder 14, box 6, all in Clayton Papers.

154. H. H. Clayton to H. W. Clough, July 2, 1929, folder 8, box 3; H. H. Clayton to Clients of the Clayton Weather Service, July 28, 1942, folder 10, box 6; H. H. Clayton to Albert Leman, June 1944, folder 9, box 5, all in Clayton Papers. On the Weather Bureau's skepticism of Clayton's weather periods, see Larry Page to H. H. Clayton, October 12, 1936, folder 13, box 7, Clayton Papers. For Clayton's connections to other long-rangers, see H. H. Clayton to Herbert Janvrin Browne, December 25, 1932, folder 13, box 2, Clayton Papers.

155. S. P. Fergusson, "Recent Discussion of Long Period Forecasting," memorandum to Chief of Weather Bureau, February 23, 1928, folder "Fergusson, S. P., 1911, 1922–1929," box 3, Clayton Papers. On Babson and Brookmire, see Friedman, *Fortune Tellers*, 12–50, 126.

156. H. H. Clayton, "Rainfall and the Price of Stocks," undated manuscript, folder "Manuscripts and Papers of Henry Helm Clayton, 1877–1943, and undated," box 12, Clayton Papers, quotation on p. 1.

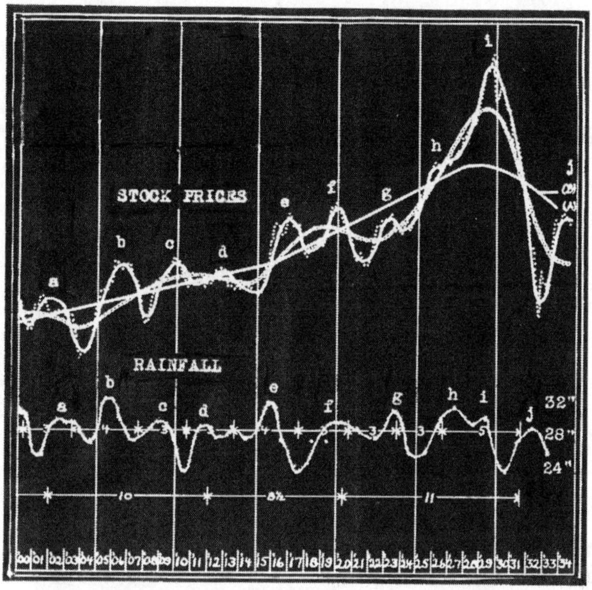

FIGURE 4.5. Henry Helm Clayton's attempt to correlate climatological and economic cycles in the 1930s yielded a "stock prices" curve (based on C. H. Welden Jr., "Concurrent Long and Short Cycles in Industrial Stock Prices, 1872–1934," *Annalist*, November 9, 1934, p. 647) and a "rainfall" curve (based on US Weather Bureau data on rainfall between the Appalachians and the Rockies). The lowercase letters indicate the "striking similarity" of maxima in both curves, with the rainfall maxima preceding the stock price maxima by a few months. H. H. Clayton, "Rainfall and the Price of Stocks," undated manuscript, Smithsonian Institution Archives, H. Helm Clayton Papers, image SIA2016-012207.

Toward the end of his career, sometime in the late 1930s or early 1940s, Clayton typed up the seven commonsensical steps that comprised his personal investment policy, which included buying stocks in "well established growing industries," diversifying investments, and keeping cash reserves on hand during a depression. The seventh part of Clayton's investment policy invoked the predictive value of historical data: "Changes in management and increases or decreases in orders have an important bearing on the future earnings of a corporation but they are not matters easily discovered by an individual investor. He has to depend on reports of earnings and past performance in order to estimate the future of any industry."[157] It is not clear whether this particular

157. H. H. Clayton, "My Investment Policy," undated manuscript, folder "Manuscripts and Papers of Henry Helm Clayton, 1877–1943, and undated," box 12, Clayton Papers.

strategy—which echoed Benner's model of commodity price forecasting—paid off for Clayton, but he did find reward in the form of material and moral support from leading businessmen in the 1920s and beyond who funded the study of cycles. He wrote,

> Men with new ideas need freedom to develop such ideas and need encouragement. Under any sort of government control original ideas are likely to remain undeveloped for lack of encouragement. Their promotion by universities is better, but even there the leaders are likely to consider university training an essential for research, while the history of scientific advance shows that many of the leaders in research . . . had no college training. Businessmen are accustomed to judging the capacity of men by accomplishments rather than by formal training and are often the best administrators in such matters.[158]

Undeterred by periodicity's elusive causation, Clayton pursued his investigation of cycles until the end of his life. At the time of his death in 1946, he was working on a manuscript titled "Terrestrial Cycles," which his daughter, Frances Clayton, submitted to *Popular Astronomy*, where it was published in 1948. The article represented Clayton's attempt to correlate the established eleven-year sunspot cycle with apparent ten-year terrestrial cycles revealed by recent research on cycles of animal populations, tree growth, commodity and stock prices, and death rates of infectious diseases. Clayton observed, "Thus once in ten years or a little less, something seems to happen which causes an increase and then a decrease in the vital activities of both animals and plants all over North America, from the borders of Alaska to the Maritime Province and Northern United States, and also in adjacent seas."[159] Clayton's article included graphs depicting cycles of hare and lynx populations, the tent caterpillar population in New Jersey, cycles of redwood tree growth, stock prices, and mortality rates associated with whooping cough and measles. In synthesizing these accounts of terrestrial cycles, Clayton signaled his debt not only to the newly formed Foundation for the Study of Cycles but also to the business leaders who had encouraged the systematic study of cycles.[160] Clayton acknowl-

158. H. Helm Clayton, "Terrestrial Cycles," undated manuscript, p. 34, folder "Manuscripts and Papers of Henry Helm Clayton," box 12, Clayton Papers.

159. H. Helm Clayton, "Terrestrial Cycles," *Popular Astronomy* 56 (1948): 361.

160. Edward R. Dewey, formerly an economic analyst in President Hoover's Commerce Department tasked with determining the causes of the Great Depression, established the Foundation for the Study of Cycles in 1941 and became the leading figure in the field. See Edward R.

edged that the ten-year terrestrial cycle would be "of varying length and amplitude, and not a cycle constant in length or a combination of such cycles"—thus nearly impossible for others to find—and that, like other cycles, it fell short of an empirical account of causation.[161] Clayton began his career by investigating meteorological cycles, then sought to apply them to the economy, and finally, in this essay, gestured toward a broader theory of periodicity encompassing atmospheric, terrestrial, and market conditions. Yet his search for a unified theory of predictability would remain incomplete.

*

Samuel Benner, Edward Bellamy, and Henry Helm Clayton were all long-range forecasters of future economies who shared a conviction that short-term forecasts were of relatively little value. All three based their forecasts on theories—evolution for Bellamy and periodicity for Benner and Clayton—that depended on a long-range time frame. All three constructed epistemic frameworks—banishing the "specter of Uncertainty" and finding God and rainfall in prices—that promised predictability and rejected the workings of chance. Benner and Clayton, whose careers illustrate the migration of theories of predictability from weather to the economy, sought to explain correlated effects rather than isolated causes, but their ultimately futile quests for predictive certainty did not diminish their public reputations. Histories of weather and business forecasting focus on scientific, government, and academic institutions as the locus of knowledge production, but Benner and Clayton serve as two notable examples of the late nineteenth century's self-taught forecasters who operated outside of or on the margins of institutions to produce expert knowledge. And although Bellamy's politics of the future did not put American capitalism on the path to socialism, his utopia of economic certainty had hundreds of thousands of readers from the late 1880s through the 1930s and inspired a Bellamyite revival at the height of the Great Depression. Bellamyism would disappear with the rise of totalitarianism in Europe, but theories of periodicity had lasting appeal beyond Benner and Clayton and continue to have adherents in the present day.

Benner and Clayton would reappear in studies of cycles and publications on financial astrology in the mid to late twentieth century, thereby revealing an

---

Dewey and Edwin F. Dakin, *Cycles: The Science of Prediction* (New York: Henry Holt, 1947); Dewey and Mandino, *Cycles: The Mysterious Forces That Trigger Events*.

161. H. Helm Clayton, "Terrestrial Cycles," *Popular Astronomy* 56 (1948): 362.

enduring connection between the worlds of weather and market forecasting and the world of fortune-telling. Evangeline Adams, the "world's most famous astrologer," counted New York's social and financial elite among her patrons in the first three decades of the twentieth century, noting that "Wall Street men often come to me for such readings of the stars."[162] Adams was the most prosperous of the countless fortune-tellers in the late nineteenth and early twentieth centuries who looked into the future and sold promises of love and money to a clientele well accustomed to consulting forecasts on a daily basis. Yet the success of a booming industry in occult forecasting came at a cost, as Adams and many of her colleagues found themselves prosecuted under antidivination statutes and embroiled in legal disputes over whether fortune-tellers were violating the law by predicting with certainty.

162. Evangeline Adams, "Keep Your Eye Upon Your Step in 1922," *Seattle Star*, December 31, 1921; Marguerite Mooers Marshall, "Astrology More Certain in Its Diagnosis Than Medicine or Law, Says Miss Adams," *New York Evening World*, December 14, 1914.

CHAPTER 5

Promises of Love and Money

A young Edward Bellamy believed, writing in the 1870s, that the new industrial society would render fortune-telling obsolete. In Europe and the United States, Bellamy declared, "the progress of education, and the systematization of industry . . . leaves no function for the Gypsey to perform," and in the United States especially, the forces of education and capitalism were "destroying the market for love-philter, amulets and charms." He cited "Webster's Spelling-book" as a threat "more formidable than rack or gibbet, to the principles of Gipseydom."[1] Bellamy's prediction of disenchantment and the decline of fortune-telling was, however, as inaccurate as it was confident. The new organizational society of professions and corporations did not rationalize occult forecasting out of existence, but rather, fortune-telling became a form of knowledge production considered systematic, scientific, and modern by many of its practitioners and patrons.

The ancient practice was transformed in late nineteenth-century American cities, where "purveyor[s] of futurity" proliferated.[2] Fortune-tellers—often poor and working-class immigrant and African American women, and occasionally

1. Edward Bellamy, "The Gypsies," manuscript, 187[?], Edward Bellamy Compositions, 1860–1939 (MS Am 1181.4), Houghton Library, Harvard University.

2. "Swindling Fortune-Tellers," *New York Times* (hereafter *NYT*), July 24, 1873. On fortune-tellers in the late eighteenth and early nineteenth centuries, see Jackson Lears, *Something for Nothing: Luck in America* (2003; repr., New York: Penguin Books, 2004), 55–68. On dream books as "vernacular divination" in the nineteenth century, see Ann Fabian, *Card Sharps and Bucket Shops: Gambling in Nineteenth-Century America* (1990; repr., New York: Routledge, 1999), 142–50, quotation on 144.

white middle-class women and men—emerged in increasing numbers to protect Americans from the "specter of Uncertainty" by selling a glimpse (however fragmentary) and a sense of control (however illusory) of one's future. Fortune-tellers' predictions of money gained or lost countered the uncertainties of booms and busts, and their "romantic forecasting" offered a way to navigate the unfamiliar and potentially threatening social terrain of anonymous urban life.[3]

Bellamy was correct, however, in implying that fortune-tellers both faced and posed a social threat. Although instruments of torture were no longer the primary tools for disciplining occult practitioners in the late nineteenth century, fortune-tellers regularly encountered discursive and legal mechanisms designed to control their labor. In response to police raids and consumer demand, many urban fortune-tellers refashioned themselves as modern scientific practitioners, distancing themselves from the time-honored clairvoyant arts of palmistry, card reading, and tea leaf reading.[4]

As fortune-tellers reimagined their practices, so too did state courts, which reshaped the regulation of fortune-telling through a series of early twentieth-century decisions that centered not on the empirical question of whether one was reading palms, cards, or tea leaves but on the epistemological question of whether one was genuinely attempting to foretell the future or merely pretending. The concept of "pretending to tell fortunes," a legal construct from Elizabethan antidivination statutes, proved complicated to interpret as fortune-tellers and the courts debated the definition and legitimacy of fortune-telling. From the late nineteenth to the early twentieth century, legal discourses on fortune-telling shifted their focus from the nature of the forecaster to the nature of the forecast, focusing in particular on the question of whether fortune-tellers were pretending to predict the future with certainty. The ambiguity that fortune-telling critics and moral reformers considered fraudulent in the

3. Quotation in Lears, *Something for Nothing*, 111. On the social threat of imposture in nineteenth-century American cities, see Karen Halttunen, *Confidence Men and Painted Women: A Study of Middle-Class Culture in America, 1830–1870* (New Haven, CT: Yale University Press, 1982); John F. Kasson, *Rudeness and Civility: Manners in Nineteenth-Century Urban America* (New York: Hill and Wang, 1990).

4. The term *fortune-telling* as it was used in the nineteenth century was a broad category encompassing diverse practices, including astrology, palmistry, medical and business clairvoyance, card reading, tea leaf reading, channeling spirits, and conjuring. Periodicals and fortune-tellers' own advertisements sometimes used *fortune-teller* interchangeably with *seer* and *clairvoyant* but other times distinguished between fortune-telling and astrology, Spiritualism, and other practices. In this chapter, I opt for the more widely used *fortune-teller*, making clear instances in which historical actors had specific motives for choosing one particular term over another.

nineteenth century became legally acceptable in the early twentieth century as courts acknowledged the uncertainties of occult prediction as a feature of economic life and consumer culture. State courts ruling on antidivination prosecutions by the 1910s disallowed predictions that made claims to certainty but permitted those that acknowledged their own indeterminacy, thereby constructing a new legal framework for fortune-telling that accepted uncertainty embedded in claims of occult foreknowledge.

Fortune-tellers most commonly engaged in "futurework" as a means of economic survival.[5] As one folklorist observed in 1895, "the well-to-do fortune-tellers are, however, few in number; the majority earn precarious livelihoods." Contemporary accounts depicted impoverished fortune-tellers operating in the economic and cultural shadows, like the 1892 "sketch of gypsy life" in the *Brooklyn Daily Eagle* that pronounced "the gypsy . . . the true twin of the shadow, the clairvoyant of half lights," who "came originally out of the shadows of the dark continent."[6] Descriptions like this have led to the classification of fortune-telling as part of the "market activity that historians have by and large dismissed as 'underground,' 'marginal,' and 'informal.'"[7] But fortune-telling, advertised and practiced widely, was squarely within the mainstream of late nineteenth-century American capitalism, as fortune-tellers joined those "ordinary people, [who] through their varied struggles to survive and succeed, created capitalism as much as captains of commerce and industry did."[8]

The history of the policing and prosecution of fortune-tellers makes visible how contests over the production of future knowledge hinged on ideologies

5. Gary Alan Fine, *Authors of the Storm: Meteorologists and the Culture of Prediction* (Chicago: University of Chicago Press, 2007), chap. 3, considers the nature of "futurework" in the professional culture of National Weather Service forecasters in the early twentieth century. I use the term here to underscore the centrality of labor and commercial exchange to the history of fortune-telling.

6. Henry Carrington Bolton, "Fortune-Telling in America To-Day: A Study of Advertisements," *Journal of American Folklore* 8, no. 31 (1895): 300; "Tent of La Zingara," *Brooklyn Daily Eagle*, August 14, 1892.

7. Brian P. Luskey and Wendy A. Woloson, introduction, in *Capitalism by Gaslight: Illuminating the Economy of Nineteenth-Century America*, ed. Brian P. Luskey and Wendy A. Woloson (Philadelphia: University of Pennsylvania Press, 2015), 3. For a recent history of African American spirit mediums in the "informal economy" of early twentieth-century Harlem, see LaShawn D. Harris, "Dream Books, Crystal Balls, and 'Lucky Numbers': African American Female Spiritual Mediums in Harlem, 1900–1945," *Journal of Afro-Americans in New York Life and History* 35, no. 1 (2011): 74–110, quotation on 77.

8. Luskey and Woloson, introduction, *Capitalism by Gaslight*, 2.

of gender, race, and class—as well as their subversion.[9] As Shane White has recently demonstrated, "fortune telling and conjure, and eventually policy, served as a means of mediating race relations in New York during the long transformation from slavery to freedom in the first half of the nineteenth century, contributing to the halting process by which blacks and whites came to new ways of living together in this rapidly transforming city."[10] In the second half of the century, racial ideologies, gender norms, and class distinctions were central to the arrests as well as the appeal of fortune-tellers, and the appropriation of fortune-tellers' racial identities by white middle-class consumers and practitioners led to the assimilation of fortune-telling into twentieth-century mass consumer culture.[11] Like the *Eagle*'s "sketch of gypsy life," tales of "gypsy

9. Fortune-telling has not received extensive scholarly attention, but this chapter is informed by a handful of important studies. I draw on Tammy Stone-Gordon's 1998 dissertation for its analysis of novels, films, and nearly 1,500 fortune-telling advertisements from New York, San Francisco, Boston, St. Louis, and Chicago newspapers between 1850 and 1930 in an account of the self-fashioning of "occult workers" and the middle-class consumption of their labors. Stone-Gordon finds that the efforts of fortune-tellers to define their labor as professional intersected with cultural debates over middle-class identity as well as the "social purity and nativism" that engendered antidivination discourses. Tammy Stone-Gordon, "'Fifty-Cent Sybils': Occult Workers and the Symbolic Marketplace in the Urban U.S., 1850–1930" (PhD diss., Michigan State University, 1998), quotation on 213. David Allen Harvey has attributed the declining prosecution of French fortune-tellers in the twentieth century to the emergence of an increasingly liberal modern French culture that repudiated their nineteenth-century legal punishment as "a form of archaic intolerance." This chapter shares Harvey's methodological approach of using the published appeals of antidivination prosecutions to trace shifting legal and cultural frameworks for fortune-telling. Quotation in David Allen Harvey, "Fortune-Tellers in the French Courts: Antidivination Prosecutions in the Nineteenth and Twentieth Centuries," *French Historical Studies* 28, no. 1 (2005): 155. Shane White's recent article on African American fortune-tellers in antebellum New York City pieces together newspaper and legal accounts of fortune-telling into a history of community formation and cultural practice after the end of slavery in New York in 1827. Although the percentage of African Americans in the population of New York City declined as European immigration increased after the mid-nineteenth century, White's characterization of fortune-tellers as "cultural brokers" who helped patrons navigate urban life can be applied more broadly throughout the late nineteenth century, as this chapter reveals. Shane White, "The Gold Diggers of 1833: African American Dreams, Fortune-Telling, Treasure-Seeking, and Policy in Antebellum New York City," *Journal of Social History* 47, no. 3 (2014): 673–95, quotation on 679.

10. White, "Gold Diggers," 675–76.

11. On professionalization as the impetus for the assimilation of occult workers into twentieth-century American consumer culture, see Stone-Gordon, "Fifty-Cent Sybils," 251.

fortune-tellers" and "negro fortune-tellers" constituted a narrative genre of their own in nineteenth-century newspapers and periodicals. Essays and book-length treatments offered sensationalized literary accounts or polemics in order to sell copies, justify moral reform and nativism, or delineate the boundaries of middle-class respectability—and sometimes all three—revealing more about their white middle-class authorship than about fortune-tellers themselves. Because these published narratives and fortune-tellers' advertisements constitute a voluminous yet fragmentary record, this chapter relies on the voices embedded in newspaper accounts and legal proceedings to reconstruct the history of fortune-telling as it was experienced by its practitioners and patrons.

"FORTUNE-TELLING FAD"

Fortune-tellers worked their magic on American cities in the late nineteenth century and transformed their practice into a profession.[12] The sensational *National Police Gazette* remarked in 1880 that "the manners of fortune-telling have changed with the materialistic tendencies of the age" and identified the emergence of a "modern" professional fortune-teller with an "office . . . very little different from that of a real estate firm, and who go[es] about the business in a cold-blooded manner." In 1855, the *New York Times* called attention to a "nest" of fortune-telling parlors that included "some two hundred astrologers, clairvoyants, and fortunetellers," and by 1909, the *Times* counted one thousand fortune-tellers, who reportedly took in ten thousand dollars each day.[13] By 1890, the *Brooklyn Daily Eagle* announced a "fortune-telling fad," declaring that "there never was a time when fortune-telling in drawing rooms was so rife or apparently so much believed in. . . . The gypsies are avenged for ages of ridicule and penalties and palmistry is a rage," and by the turn of the century, a

12. On fortune-telling advertisements and professionalization, see Stone-Gordon, "Fifty-Cent Sybils," chap. 4; Alana Piper, "Women's Work: The Professionalisation and Policing of Fortune-Telling in Australia," *Labour History* 108 (May 2015): 37–52.

13. "Flirting with Dame Fortun," *National Police Gazette*, December 18, 1880; Paul Prowler, "Glimpses of Gotham," *National Police Gazette*, January 10, 1880; "Fortune-Tellers and Fools," *NYT*, November 23, 1855; "One Thousand Fortune-tellers," *NYT*, December 12, 1909. The US Census also counted fortune-tellers, although they were surely underreported due to the increasing criminalization of fortune-telling in the late nineteenth century. As Stone-Gordon notes, the 1860 Census listed eight astrologers, but fortune-tellers of any kind did not appear again until the 1910 Census, which counted 1,600 "fortune-tellers, hypnotists, spiritualists, etc." Stone-Gordon, " 'Fifty-Cent Sybils,' " 223–24.

Utah newspaper observed that the "morbid desire to get a peep into the future has become almost a craze."[14]

This craze had its roots in the early 1870s, when fortune-tellers' advertisements became a regular feature in urban dailies nationwide.[15] The *Eagle* regularly listed clairvoyants for hire through the turn of the century, featuring, among many others:

> Mme. Barras . . . reads your life with great power; business changes, lawsuits, wills, divorces, health; tells whom and when to marry; brings separated together; causes speedy marriages; settles lovers' quarrels, etc.[16]

> Noor Mahal, the famous Indian fortune teller, tells your life from the cradle to the grave, gives lucky numbers, cures rheumatism.[17]

> Madame Latrove, reliable medical and business clairvoyant born with second sight, tells without questioning names, dates, events; information on winning horses, stocks, mines, medical examination by hair; magnetic healing.[18]

> Countess Habeba. The only distinguished Persian gypsy; scientific palm reader; remarkable, most successful clairvoyant; reveals your past, present and future; gives indispensable advice on courtship, business speculations, health, marriage; removes all troubles; gives luck; satisfaction guaranteed; has unfailable herb remedy for all disorders of the blood, nerves, and stomach.[19]

As these advertisements reveal, fortune-tellers constructed a range of professional personas, some purely fictional. Countess Habeba was Mary Kroeger, who lived in Brooklyn with her husband, a traveling salesman in the woolen

14. "The Fortune-Telling Fad," *Brooklyn Daily Eagle*, September 28, 1890; "Fortune-Telling," *Deseret (UT) News*, February 12, 1898. On a "renewed interest in the therapeutic value of superstition" in the late nineteenth-century United States, see Lears, *Something for Nothing*, 215. On the Victorian intellectual allure of the occult, see Alex Owen, *The Place of Enchantment: British Occultism and the Culture of the Modern* (Chicago: University of Chicago Press, 2004).

15. Stone-Gordon, "'Fifty-Cent Sybils,'" 120.

16. "Clairvoyants," *Brooklyn Daily Eagle*, June 16, 1887.

17. "Clairvoyants," *Brooklyn Daily Eagle*, September 14, 1881.

18. "Clairvoyants," *Brooklyn Daily Eagle*, June 20, 1883.

19. *Brooklyn Daily Eagle*, April 5, 1896, 27.

business. The Kroegers earned a comfortable living, judging by the $1,500 in diamond and gold jewelry that was stolen from their home in 1902 while Countess Habeba was traveling in Europe.[20] Newspaper advertisements commonly invoked "gypsy" along with Eastern and Western Europe, a mystical Orient, India, and Africa.[21] (See figure 5.1.) As Tammy Stone-Gordon has demonstrated, an exoticized racial other figured prominently in fortune-telling advertisements: "Occult workers . . . utilized the classified advertisement to play freely with and often invert dominant notions of gender, race, class, age, ableness, and nation. . . . Immigrant status meant exotic knowledge rather than foreign threat."[22] Fortune-tellers' titles signaled different personas, from "professors" who promised scientific expertise and masculine authority to "madames" and "countesses" who offered the wisdom and camaraderie of female confidantes and the refinement of Old World aristocracy. Their methods equally varied; fortune-tellers read horoscopes, palms (including thumbs and fingernails), cards, tea leaves, and character (using phrenological techniques).[23] Fortune-telling practices varied regionally, from the African American conjure tradition in the South to the "calculating" tradition of Chinese fortune-tellers

20. "Legal Notes," *New York State Journal of Medicine* 4, no. 5 (1904): 173; "Chloroform and Pistol Found on This Burglar," *Brooklyn Daily Eagle*, December 19, 1902; "Zeno Had Stories of Plunder," *Brooklyn Daily Eagle*, December 20, 1902; "Burglars Come and Go; Big Dog Sleeps Sweetly," *Brooklyn Daily Eagle*, November 24, 1902.

21. Stone-Gordon, "'Fifty-Cent Sybils,'" 124, 141. The scholarship in Romani studies has been largely dominated by folklorists, anthropologists, and sociologists. For classic studies in the field, see Rena C. Gropper, *Gypsies in the City: Culture Patterns and Survival* (Princeton, NJ: Darwin Press, 1975); Anne Sutherland, *Gypsies: The Hidden Americans* (New York: Free Press, 1975). For more recent work, see the academic journal *Romani Studies*; Ethel C. Brooks, "The Possibilities of Romani Feminism," *Signs* 38, no. 1 (2012): 1–11; Carol Silverman, "Education, Agency, and Power among Macedonian Muslim Romani Women in New York City," *Signs* 38, no. 1 (2012): 30–36; Michael L. Chohaney, "Hidden in Plain Sight: Mixed Methods and the Marble Orchards of the Vlach Rom (Gypsies) in Toledo, Ohio," *Journal of Cultural Geography* 31, no. 1 (2014): 57–80.

22. Stone-Gordon, "Fifty-Cent Sybils," 118, 141, quotation on 118.

23. "Palmistry," *Phrenological Journal and Science of Health* 51, no. 6 (1870): 436; "Character Read by the Thumbs," *Current Literature* 20, no. 6 (1896): 515; "Fortune-Telling by Finger Nails," *Godey's Lady's Book and Magazine* 123, no. 735 (1891): 227; C. I. Bailey, "The Magic Cards," *The Aldine, the Art Journal of America* 9, no. 3 (1898): 93–94; "Fortunes in Teacups," *Ladies' Home Journal* 13, no. 2 (1896): 16; "Fortune-Telling by Phrenology," *Phrenological Journal and Science of Health* 57, no. 1 (1873): 37–38. For an overview of fortune-telling practices, see Stone-Gordon, "'Fifty-Cent Sybils,'" 14–15.

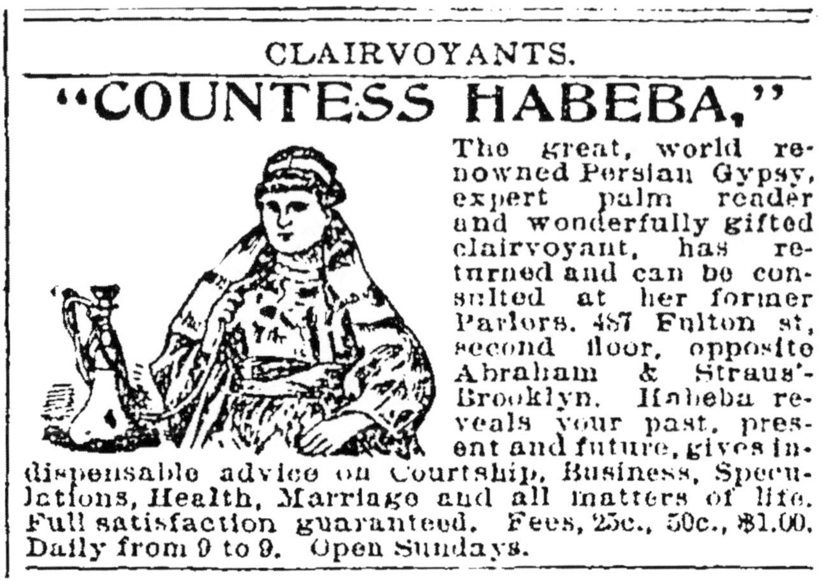

FIGURE 5.1. Mary Kroeger, who lived in Washington, DC, and then New York City, frequently ran illustrated newspaper advertisements as Countess Habeba. She saw clients in her Brooklyn parlor as well as on the Coney Island boardwalk, where she hired assistants to work as her business expanded. Reprinted from the *Brooklyn Daily Eagle*, October 17, 1897, 23.

in the West who based their predictions on tortoise shells, coins, dominoes, or characters printed in a book.[24]

Fortune-tellers were predominantly but not exclusively female practitioners, and the majority of their clients were women. In 1885, a Brooklyn police officer ventured that 90 percent of fortune-tellers and 99 percent of their clients were women, and ten years later, folklorist Henry Carrington Bolton confirmed that "the advertisements of the 'Madames' far outnumber those of the

24. On conjuring in African American culture, see Yvonne Chireau, *Black Magic: Religion and the African American Conjuring Tradition* (Berkeley and Los Angeles: University of California Press, 2003); Jeffrey E. Anderson, *Conjure in African American Society* (Baton Rouge: Louisiana State University Press, 2005); Carolyn Morrow Long, *Spiritual Merchants: Religion, Magic, and Commerce* (Knoxville: University of Tennessee Press, 2001), chap. 4. Quotation in "Chinese Fortune-Tellers," *Harper's Bazaar*, September 20, 1873, 604–5. On Chinese fortune-telling, see also W. C. Benton, "The Chinese at Home," *Phrenological Journal and Science of Health* 68, no. 4 (1879): 167–70; Stewart Culin, "Divination and Fortune-Telling among the Chinese in America," *Overland Monthly and Out West Magazine* 25, no. 146 (1895): 165–72.

'Professors.'"[25] Fortune-tellers marketed their services according to gender, with "a common phrase being, 'ladies, 50 cts.; gents, $1.'"[26] Fortune-tellers who declared "no gents" were reluctant to invite male clients into the intimate space of the home because they were potentially dangerous or potentially incredulous. As one fortune-teller commented, "We don't like men to call on us, for it is seldom that we can get them to believe what we say." Yet women too were skeptics, like the young Brooklyn woman who insisted, "Of course, we don't believe anything they tell us."[27] For this woman and countless others, the appeal was not the relative accuracy of the fortune but rather the experience of the telling, a form of self-reflexive epistemic play that transcended fixed categories of authenticity and quackery such that clients understood the entertainment value of "artful deceptions."[28]

Fortune-telling became a fad in late nineteenth-century American cities because it offered, in addition to promises of love and money, a sense of control over one's present and future during a time when economic life was increasingly governed by action at a distance, when "ordinary people's livelihoods increasingly depended on decisions made in distant cities, on circumstances beyond the individual's control."[29] Patrons often asked fortune-tellers to look deep into the present, whether to locate lost jewelry, confirm a spouse's fidelity, or find a missing relative, and fortune-tellers provided reassurance with predictions of "a long life, and the three great desideratums thereto, health, wealth and happiness." Like long-range weather forecasters, fortune-tellers couched their predictions in such general terms that skeptics could easily

25. "Fortune-Tellers," *Brooklyn Daily Eagle*, October 25, 1885; Bolton, "Fortune-Telling in America," 299.

26. Fees also varied according to the nature of the fortune being told; for instance, an English medical and business clairvoyant charged one dollar for medical predictions and two dollars for business predictions in 1884. *Brooklyn Daily Eagle*, July 15, 1884; quotation in Bolton, "Fortune-Telling in America," 300.

27. "Fortune-Telling. The Black Art as Practiced in This City," *NYT*, December 12, 1869; "Danger to Girls. The Alarming Number of Young Ladies Who Visit Clairvoyants," *Brooklyn Daily Eagle*, May 14, 1882.

28. James W. Cook, *The Arts of Deception: Playing with Fraud in the Age of Barnum* (Cambridge, MA: Harvard University Press, 2001), 28. On cultural practices, including mesmerism, Spiritualism, and stage magic, whose explanatory power endured in an age commonly characterized by the rise of a secular, rational, bureaucratic order, see Michael Saler, "Modernity and Enchantment: A Historiographic Review," *American Historical Review* 111, no. 3 (2006): 692–716.

29. Lears, *Something for Nothing*, 151.

point to ambiguity as the trick of the fortune-telling trade.[30] The vaguer the prediction, the greater the chance that it might come to pass, as one observer noted in 1873: "Persons believe in the predictions of fortune-tellers for the seemingly excellent reason that such predictions are repeatedly fulfilled. They do not notice that . . . there would be as many fulfillments if every prediction had been precisely reversed."[31]

The appeal of fortune-telling also hinged on the intimate space of the private fortune-telling parlor as well as the spatial organization of urban leisure during the heyday of dance halls, amusement parks, and slumming.[32] A fortune-teller's parlor, often located in a crowded tenement, offered middle-class clients escapism tinged with exoticism during a period characterized by what Chad Heap describes as "both the cultural ascendancy of a vibrant and sizable urban middle class and the increasing geographic segregation of U.S. cities along class lines." Slumming, according to Heap, was "a practice that required the advent of the slum—as an identifiable urban space and as an ideological concept that carefully separated the lives and homes of well-to-do white women and men from the deprivation and degradation associated with urban poverty."[33] Fortune-telling critics reinforced the same boundary through their condemnation of the squalor and disorder of clairvoyants' domestic lives, depicting "old women, dirty, foul and unpleasant" living "in dirty tenement houses, where, in the midst of offensive odors, swarms of unkempt children, piles of dirt and oceans of filth, they commune with the stars through the congenial *media* of bad gin and worse rum."[34]

Yet by the end of the century, fortune-tellers were commonly welcomed into the homes of wealthier patrons. In 1890, a *Pittsburg Dispatch* headline

30. "Seeking Seers," *Brooklyn Daily Eagle*, March 5, 1891; "Missing Emily," *Brooklyn Daily Eagle*, July 17, 1889; quotation in "Fortune-Telling. The Black Art as Practiced in This City," *NYT*, December 12, 1869; "Fortune-Tellers and Fools," *NYT*, November 23, 1855; "Soothsayers in Trouble," *NYT*, April 23, 1875; "Fortune-Tellers," *Brooklyn Daily Eagle*, October 25, 1885.

31. "Superstitions," *NYT*, January 5, 1873.

32. Kathy Lee Peiss, *Cheap Amusements: Working Women and Leisure in Turn-of-the-Century New York* (Philadelphia: Temple University Press, 1986); John F. Kasson, *Amusing the Million: Coney Island at the Turn of the Century* (New York: Hill and Wang, 1978); Chad Heap, *Slumming: Sexual and Racial Encounters in American Nightlife, 1885–1940* (Chicago: University of Chicago Press, 2009).

33. Heap, *Slumming*, 19. On the Victorian context, see Seth Koven, *Slumming: Sexual and Social Politics in Victorian London* (Princeton, NJ: Princeton University Press, 2004).

34. "A Growing and Dangerous Class," *NYT*, February 4, 1867.

declared that the "Fortune-telling Fad" was "Making Its Way into Fashionable Drawing Rooms," and the *Ladies' Home Journal* announced in 1895 that "during the past year fortune-telling has become a prominent social fad."[35] Turn-of-the-century commentaries drew sharp racial and class distinctions in accounting for fortune-telling's appeal, emphasizing a divide between skeptics ("the thinking public") and enthusiasts ("the other part of the public").[36] Folklorist Henry Carrington Bolton, explaining "how can such charlatans flourish," classified fortune-telling patrons into two mutually exclusive groups: "(1.) The superstitious who ignorantly believe that mankind has power over the supernatural. . . . Probably a large proportion of this credulous class are of foreign birth. (2.) The curious non-believers in the pretensions of the fortune-tellers, who visit them 'just for the fun of the thing.'" But Bolton reserved his harshest judgment for fortune-tellers themselves: "Financially and socially, these people who live by preying on credulity born of ignorance have no standing in this world, and in the next they are consigned by Dante to one of the lowest divisions of the Inferno, 'Malebolge;' the poet represents them as having their heads turned around on their shoulders."[37]

"LYING SORCERERS"

As Bolton reminded his readers, Dante had relegated fortune-tellers and other frauds to the eighth circle of Hell, and more than five centuries later, fortune-tellers would receive similar treatment in well-known exposés, including P. T. Barnum's *The Humbugs of the World* (1866) and Anthony Comstock's *Frauds Exposed* (1880). Barnum's catalogue of humbugs included "Alchemy . . . Magic . . . Astrology, and above all, Fortune-telling."[38] Comstock denounced "a nineteenth century scamp" named Calvin who sold a "Chart of Destiny" for three dollars and a business chart for five dollars that promised "YOUR FUTURE DESTINY CAN BE CLEARLY REVEALED with as much certainty as the solving of

35. "The Fortune-Telling Fad," *Pittsburg Dispatch*, reprinted in *Brooklyn Daily Eagle*, September 28, 1890; Amelia E. Barr, "The Fad of Fortune-Telling," *Ladies' Home Journal*, December 1895, 14. On the 1890s' fortune-telling vogue, see also "A Truthful Fortune-Teller," *Christian Advocate*, June 6, 1895, 358; "Parlor Fortune-Telling," *Harper's Bazaar*, May 16, 1891, 376; "Piercing the Veil of Futurity," *Current Literature* 20, no. 1 (July–December 1896): 66.

36. "A Fortune Teller's Ill Fortune," *Brooklyn Daily Eagle*, December 12, 1896.

37. Bolton, "Fortune-Telling in America," 306–7.

38. P. T. Barnum, *The Humbugs of the World: An Account of Humbugs, Delusions, Impositions, Quackeries, Deceits and Deceivers Generally, in All Ages* (New York: Carleton, 1866), 16.

the mathematical problem. It can also foretell with perfect certainty the result of all commercial and business transactions and speculations."[39]

Fortune-tellers also appeared regularly in newspaper articles condemning their practice as a foolish, fraudulent, and dissolute feature of urban life. Mid-nineteenth-century newspapers denounced a "dishonorable trade" and an "immoral and abominable business," "quiet and unobtrusive robbery," and "a vulgar exhibition of pretended sorcery, which should not deceive even a child." The *New York Times* branded fortune-tellers "vile and unscrupulous hags," "imposters," "lying sorcerers," and "pests of society," who, "having constructed their webs, like the spider, wait for the unsuspecting fly to enter." In these accounts, fortune-tellers preyed upon a naïve public with simple assurances and false promises, and customers who dropped anywhere from five cents to five dollars into fortune-tellers' palms were not, critics argued, getting anything in return. Stories abounded of unwitting upper-class patrons swindled out of thousands of dollars by unscrupulous seers. But detractors feared more serious consequences, pronouncing it "impossible to calculate the amount of ruin which clairvoyantes and fortune-tellers work on the simple and weak-minded."[40] Personal misfortune and "terrible domestic misery" could easily result, one critic warned: "Sick people are prevented from consulting respectable medical practitioners, by a blind belief in these swindling quacks. Simple country people lose their time, their money and sometimes their reason in fruitless searches after lost or concealed treasures, aided and abetted by those self-instituted magicians."[41]

Sometimes the ruin wrought by fortune-tellers lasted a lifetime, as in the case of a forty-two-year-old Pennsylvania woman who confessed that she had fallen in love twenty years earlier but remained alone and unhappy after fortune-tellers deemed her feelings unrequited.[42] In the worst case, putting too much stock in a fortune-teller's prediction could be deadly. When forty-eight-year-old Mrs. George Thompson was struck and killed by a Brooklyn

39. Anthony Comstock, *Frauds Exposed; or, How the People Are Deceived and Robbed, and Youth Corrupted* (New York: J. Howard Brown, 1880), 270, 272, 275.

40. "Letters to the Editor," *NYT*, September 18, 1876; "Fortune-Tellers," *NYT*, August 6, 1852; "Fortune-Tellers and Fools," *NYT*, November 23, 1855; "A Growing and Dangerous Class," *NYT*, February 4, 1867; "Swindling Fortune-Tellers," *NYT*, July 24, 1873; "Soothsayers in Trouble," *NYT*, April 23, 1875; "Fortune-Telling. The Black Art as Practiced in This City," *NYT*, December 12, 1869.

41. "Fortune-Tellers and Fools," *NYT*, November 23, 1855.

42. "Down on the Fortune Tellers," *Brooklyn Daily Eagle*, March 2, 1861.

train in 1889, newspaper coverage blamed the fortune that was found wrapped in blue paper in her purse, which landed fifteen yards away. Mrs. Thompson had carried with her three predictions, the second of which was that she would live to the age of seventy-one. Witnesses believed she could have moved to safety in time, but, the *Brooklyn Daily Eagle* wondered, perhaps the prophecy in her pocketbook "made her disbelieve in the necessity of taking ordinary precautions." These accounts blamed fortune-tellers for emboldening Mrs. Thompson to take a fatal risk, but fortune-telling could also have the opposite effect, leading one to end one's life to avoid risk and ruin. In 1900, a wealthy retired Brooklynite named George Beller reportedly committed suicide after a fortune-teller informed him that he would soon suffer a significant financial setback.[43]

Moralizing and sensational tales of ruination and tragedy far outnumbered accounts of fortune-telling's salutary effects, but in some cases fortune-tellers saved lives. The *New York Age* reported in 1906 on Aunt Nancy, an African American fortune-teller whose pronouncement in the aftermath of a violent assault in Atlanta reportedly prevented one instance of racial violence during an era when lynchings claimed thousands of lives across the South. After the white victim, Mrs. Dupree, identified her assailant as black, "the papers printed extra editions and the newsboys on the streets were yelling about the 'Negro brute that cut the white woman's throat,'" and a headline proclaimed "a Lynching Bee Was in Readiness." Soon inconsistencies in Dupree's story surfaced, and amid whispers that her injuries might have been self-inflicted, she allowed that the attacker might have been her estranged husband. Dupree's mother then consulted Aunt Nancy, "a local Afro-American fortune-teller, who for years has been regarded as a sort of oracle." Aunt Nancy declared that Dupree's husband had "blacked his face to conceal his identity, when he sprang upon his wife and cut her." Mrs. Dupree and her mother accepted Aunt Nancy's conclusion, and the police began a search for Mr. Dupree, who was nowhere to be found.[44]

Life-and-death stories notwithstanding, some maintained that fortune-telling posed a greater threat to society than to any individual. Critics feared that fortune-telling would turn laborers away from the morality of a Protestant work ethic by engendering "dissatisfaction with honest work" and inspiring "the foolish to endeavor to pry into the secrets of the future instead of going

43. "Not Fulfilled," *Brooklyn Daily Eagle*, August 13, 1889; "Day of Death in New York," *NYT*, June 11, 1900.

44. "When It Pays to Be Black," *New York Age*, February 15, 1906.

to work to make a future for themselves."[45] An 1867 editorial titled "A Growing and Dangerous Class" sounded this alarm: "We recognize the fact that the evil is growing and . . . we detect an element which, when fostered and encouraged, will breed a race of future gamblers, a nation of enthusiasts and adventurers."[46] Less than a decade later, an indignant *New York Times* reader described fortune-telling as

> far more nefarious than [lotteries and gambling], for in lotteries or gambling there is at least a chance of winning, and a man buys a ticket with his eyes open. But the fortune-teller, (although she assures her dupes that she can unfold the future to them,) yet will laugh behind their backs at their credulity, and makes money most shamelessly by false pretenses.[47]

And some critics insisted that fortune-telling would promote not only idleness and gambling but also a turn to "insanity and crime," exposing the "real business of many of the women, whose fortune-telling is merely a screen to intercept the public gaze from it."[48]

Fortune-tellers were often accused of operating illicit businesses like counterfeiting and prostitution and providing abortion procedures in the pages of chronicles of urban vice like *The Spider and the Fly; or, Tricks, Traps, and Pitfalls of City Life*, a publication attributed posthumously to well-known novelist and journalist Henry William Herbert.[49] *The Spider and the Fly* railed against the "humbug sorceresses" responsible for "vile assignation-houses" and "infant-murder" in making its central argument that "*there is a straight path between the fortune-teller's den and the brothel.*"[50] Of course these allegations—and indeed the exact words themselves—were taken directly from humorist Mortimer Thomson's 1858 exposé *The Witches of New York*, which chronicled his trips to fortune-tellers in an attempt to bring about "the day . . . when

45. "Swindling Fortune-Tellers," *NYT*, July 24, 1873; "Fortune Telling. The Black Art as Practiced in This City," *NYT*, December 12, 1869. On the "secularization" of the Protestant work ethic in industrial capitalism, see Daniel T. Rodgers, *The Work Ethic in Industrial America, 1850–1920* (1974; repr., Chicago: University of Chicago Press, 1978), 9.

46. "A Growing and Dangerous Class," *NYT*, February 4, 1867.

47. "Letters to the Editor," *NYT*, September 18, 1876.

48. "Swindling Fortune-Tellers," *NYT*, July 24, 1873; Henry William Herbert, *The Spider and the Fly; or, Tricks, Traps, and Pitfalls of City Life* (New York: C. Miller, 1875), 60.

49. For another example of this genre, see Anthony Comstock, *Traps for the Young* (New York: Funk and Wagnalls, 1883).

50. Herbert, *Spider and the Fly*, 60, 61 (emphasis in original).

they will be no longer classed with harmless mountebanks, but with dangerous criminals."[51]

In bustling late nineteenth-century American cities, saloons, brothels, gambling establishments, and obscenity represented a threat to middle- and upper-class social critics and reformers who warned of sullied individual virtue and the moral disintegration of society. Responses to anxieties about urban vice consisted of regulation and reform efforts targeting disorderly conduct, as preventing moral disorder in late-century cities became a major concern of state and local governments, police forces, volunteer charity organizations, and the writers who published "a torrent" of etiquette manuals in the last quarter of the century.[52] Social purity proponents, urban vice societies, and moral reformers all sought to impose Anglo-Saxon standards of order, propriety, and respectability on growing immigrant populations as "morals regulation . . . exploded" in the nineteenth century.[53]

Fortune-telling, as Ann Fabian has observed of gambling, "was neither production nor consumption" but rather a transaction in which patrons exchanged a small amount of money for the hope of more rewarding financial and romantic futures.[54] The site of the exchange—in cities, most often a private residence that clients visited individually or in small groups—was not a visible

51. Mortimer Thomson, *The Witches of New York, as Encountered by Q. K. Philander Doesticks, P. B.* (New York: Rudd & Carleton, 1859), 18.

52. William J. Novak, *The People's Welfare: Law and Regulation in Nineteenth-Century America* (Chapel Hill: University of North Carolina Press, 1996), 149, 151; Paul S. Boyer, *Urban Masses and Moral Order in America, 1820–1920* (1978; repr., Cambridge, MA: Harvard University Press, 1992), 121–219; quotation in Kasson, *Rudeness and Civility*, 5.

53. Novak, *People's Welfare*, 189. The postbellum explosion of morals regulation included the passage of the first federal law against mailing "obscene" material in 1865; the passage of the Comstock Law in 1873, which broadened definitions of "obscene literature" and criminalized the mailing of material pertaining to birth control and abortion; the establishment of the New York Society for the Suppression of Vice and the Women's Christian Temperance Union in 1874; the establishment of the Society for the Prevention of Cruelty to Children in 1875; the passage of "little Comstock laws" in twenty-four states by 1885; and the establishment of the national Anti-Saloon League in 1895. Antialcohol and antiprostitution efforts escalated in the 1890s but had little demonstrable effect. The Progressive Era response to urban vice translated the "moral coercion" approach of the Gilded Age into investigative commissions, social-scientific data gathering and analysis, and a public health imperative. Boyer, *Urban Masses and Moral Order in America*, 121–219; Molly McGarry, "Spectral Sexualities: Nineteenth-Century Spiritualism, Moral Panics, and the Making of U.S. Obscenity Law," *Journal of Women's History* 12, no. 2 (2000): 8–29.

54. Fabian, *Card Sharps*, 10.

public space like a saloon or brothel. The transactional and private nature of fortune-telling rendered it difficult to police and difficult to interpret in courts of law, and by the end of the century, fortune-tellers and the courts had begun to redefine the practice and its legality.

POLICING AND PROSECUTING FORTUNE-TELLERS

Legislation against fortune-telling in the United States had its roots in the antidivination tradition in English law, dating from the vagrancy provisions of the 1597–1601 Poor Laws. An act against "rogues, vagabonds and sturdy beggars" mandated punishment for vagrants, gamblers, physiognomists, palmists, and fortune-tellers, and English witchcraft acts of the eighteenth and nineteenth centuries, as well as the Vagrancy Act of 1824, also contained provisions against fortune-telling.[55] American antidivination laws originated in the early republic, when nearly all of the newly independent states passed legislation against "rogues, vagabonds, common beggars, and other idle, disorderly and lewd persons."[56] The "open-ended enumerations of classes of people deemed vagrants" almost always included fortune-tellers.[57]

State laws against fortune-telling appeared on the books from the turn of the eighteenth century to the early twentieth century. A New Jersey statute classifying "all persons who shall use or pretend to use or have skill in physiognomy, palmistry or like crafty science" as disorderly persons dated from 1799.[58] Massachusetts law in this era included "all rogues, vagabonds and idle persons, . . . or persons using any subtle craft, juggling or unlawful games or plays, or feigning themselves to have knowledge in physiognomy, palmistry, or pretending that they can tell destinies or fortunes."[59] Baltimore's vagrancy legislation, in effect in 1804, also included prostitutes, fortune-tellers, and gamblers. By the mid-1920s, most major American cities had laws that pro-

55. Statute of the Realm 1597–8, Eliz 39, Ch. 4, p. 89. On sixteenth- and seventeenth-century antidivination statutes, see Blewett Lee, "The Fortune-Teller," *Virginia Law Review* 9 (1923): 249, 258. On fortune-telling as a misdemeanor under the Napoleonic Code, see Harvey, "Fortune-Tellers in the French Courts," 133.

56. Quoted in Novak, *People's Welfare*, 167.

57. Novak, *People's Welfare*, 167.

58. State v. Kenilworth, 69 N.J. 144, 54A 244 (1903).

59. "An Act for suppressing and punishing Rogues, Vagabonds, common Beggars, and other idle, disorderly and lewd persons," in *The General Laws of Massachusetts, 1780–1822*, ed. Asahel Stearns and Lemuel Shaw, 2 vols. (Boston, 1823), I (1787), c. 54, 322, quoted in Novak, *People's Welfare*, 319n108.

hibited fortune-telling outright or required exorbitant licensing fees.[60] Yet fortune-telling continued to be a booming business in late nineteenth- and early twentieth-century cities amid "an atmosphere of tolerated illegality."[61]

Antidivination laws were selectively enforced in public and private spaces by urban police forces who conducted raids with increasing frequency throughout the late nineteenth century. Newspapers regularly covered the arrests of fortune-tellers, who were fined, jailed if they could not post bond, or released on the condition that they cease telling fortunes or leave town.[62] In 1858, when New York City police rounded up eleven astrologers, clairvoyants, and Spiritualists, the *New York Times* published their names and addresses, providing free advertising as well as censure.[63] Pittsburgh police raided fortune-telling parlors in 1859, ordering the famous Madame Morrow out of town, fining her twenty-five dollars for being a "common cheat."[64] Baltimore police conducted raids in 1875 that resulted in the prosecution of so many fortune-tellers that the *New York Times* feared a migration to New York or other nearby cities. Those arrested had little recourse other than to deny the charges against them and, as Brooklyn astrologist Madame Phebe did in 1884, claim that they did not tell fortunes but merely provided advice.[65] But at the turn of the century, fortune-tellers began to appeal their disorderly persons convictions on the grounds that they were engaged in something other than fortune-telling.

In August 1897, Coney Island underwent a "purifying" campaign initiated by Kings County district attorney Foster L. Backus, who, after visiting with his wife and friends, announced that "all immorality must be suppressed" in a

60. Novak, *People's Welfare*, 319n108; House Committee on the District of Columbia, *Fortune Telling Hearings on H.R. 8989 Before the Subcommittee on Judiciary*, 69th Cong., 1st sess. (1926), 24.

61. Harvey, "Fortune-Tellers in the French Courts," 139.

62. For newspaper coverage of the arrests of fortune-tellers, see, for example, "Fortune-Tellers in Trouble," *NYT*, July 25, 1873; "Soothsayers in Trouble," *NYT*, April 23, 1875; "The Law on Fortune-Telling," *Brooklyn Daily Eagle*, November 2, 1882; "Fortune-Tellers Come to Grief," *Brooklyn Daily Eagle*, September 4, 1885; "The Stars in Their Courses Could Not Save the Astrologer from the Hands of a Process Server," *Brooklyn Daily Eagle*, March 18, 1895; "Fortune Teller Arrested," *Brooklyn Daily Eagle*, June 24, 1897; "Fortune Teller Fined," *Brooklyn Daily Eagle*, December 24, 1897; "Fortune-Telling Charged," *Brooklyn Daily Eagle*, May 2, 1898.

63. "Unfortunate Fortune-Tellers—Eleven of Them Before the Mayor," *NYT*, October 23, 1858.

64. "Fortune-Tellers in Trouble at Pittsburgh," *NYT*, June 21, 1859.

65. "Soothsayers in Trouble," *NYT*, April 23, 1875; "Unlucky Star," *Brooklyn Daily Eagle*, June 23, 1884.

public space frequented by so many women and children.[66] Known as "Sodom by the Sea" since the 1880s, Coney Island was transformed in the mid-1890s with the construction of enclosed amusement parks, including Sea Lion Park and Steeplechase Park. But the "New Coney Island" still offered a carnival of amusement and escapism that flouted Victorian codes of social propriety and exasperated moralists like Backus, who targeted fortune-tellers, palm readers, and astrologists, as well as belly dancers, "immoral" motion picture houses, and games of chance.[67] But ridding Coney Island of its fortune-tellers proved more difficult than shutting down the "couchee couchee" dancers and scantily costumed performers in Backus's "questionable theaters and improper shows," due to the fact that local government and law enforcement officials could not agree on what constituted fortune-telling.[68]

Fortune-telling was both illegal and highly visible at Coney Island before Backus's visit. In 1895, fortune-tellers congregated less than a quarter of a mile from the police station and prominently displayed signs advertising their services. Backus tolerated those who professed to tell the events of the past but arrested as disorderly persons any who "pretend[ed] to foretell the future," under Section 899 of New York's Code of Criminal Procedure. In mid-August 1897, seven fortune-tellers and palm readers were arrested and held on three hundred dollars bail.[69] The mayor and police commissioner rejected Backus's literal interpretation of New York's antidivination statute, arguing instead that palm readers were practicing the science of palmistry, not pretending to

66. " 'Fake' Shows Must Close," *Brooklyn Daily Eagle*, August 4, 1897; "Reform on Coney Island," *NYT*, August 4, 1897. A Republican elected Kings County district attorney in 1895, Backus had previous careers in education and law and also served as assistant district attorney from 1877 to 1884. "Foster L. Backus Dead," *NYT*, March 11, 1907; "Foster L. Backus," undated clipping, The Miriam and Ira D. Wallach Division of Art, Prints and Photographs: Print Collection, New York Public Library Digital Collections, http://digitalcollections.nypl.org/items/510d47dc-7f5f-a3d9-e040-e00a18064a99.

67. Kasson, *Amusing the Million*, 57–58, 36; " 'Fake' Shows Must Close," *Brooklyn Daily Eagle*, August 4, 1897; "Reform on Coney Island," *NYT*, August 4, 1897.

68. "Coney's Touch of Reform," *Brooklyn Daily Eagle*, August 23, 1897.

69. "Wilted at Coney Island," *NYT*, August 12, 1895; Blewett Lee, "Spiritualism and Crime," *Columbia Law Review* 22 (1922): 439, 446; "Reforming Coney Island," *NYT*, August 19, 1897. Section 899 enumerated the categories of disorderly persons, including "Persons pretending to tell fortunes, or where lost or stolen goods may be found." John T. Fitzpatrick, *Penal Law and the Code of Criminal Procedure of the State of New York*, 10th ed. (Albany, NY: Matthew Bender, 1918), 381.

foretell the future. Many fortune-tellers exploited this disjuncture between the language of the statute and its enforcement by reinventing themselves as perfectly legal "character readers," thwarting Police Captain Thomas H. Collins's attempts to evict them. Captain Collins, for his part, acknowledged that character reading was "fake" but not illegal.[70]

When Backus returned to Coney Island in late August, he was satisfied that "fakirs and costumed entertainers" had ceased operation, but he found fortune-tellers still very much in business. In a paradoxical bit of self-promotion, they hung signs advertising that they could not tell fortunes, offering instead to show photographs of future spouses for a ten-cent fee. Backus visited two fortune-tellers, both of whom pressed a piece of purportedly photographic paper against his palm and then revealed the picture that had "developed." Backus reportedly became indignant when the second fortune-teller, nervous about the presence of the district attorney, inadvertently showed him a picture of a young African American woman as his bride-to-be.[71]

This strategic fortune showing did not allow fortune-tellers to escape punishment for long, however, as Backus ordered the arrest of a Madame Dora and numerous others for advertising such a service.[72] During a late-August trial in Coney Island police court, Madame Dora's attorney argued that she merely read her clients' palms and did not pretend to tell fortunes.[73] When Backus cross-examined her, she admitted to selling pictures of future spouses that she bought in bulk from a New Jersey supplier. Madame Dora went on to describe the principles of palmistry that revealed how many times a client would marry, whether the spouse would be "light or dark complexioned," and how many children they would have. Unpersuaded, Police Justice J. Lott Nostrand ordered Madame Dora placed under bonds. The other fortune-tellers who appeared

70. "Backus Is Not Satisfied," *Brooklyn Daily Eagle*, August 16, 1897; "Gamblers Hastily Depart," *NYT*, August 9, 1897.

71. "Coney's Touch of Reform," *Brooklyn Daily Eagle*, August 23, 1897.

72. Ibid.

73. In police-station courtrooms of late nineteenth-century cities, justices of the peace (court employees without formal legal training) presided over civil cases and lesser criminal cases in a casual atmosphere that was nonetheless extremely influential in the lives of the workers, debtors, tenants, spouses, and parents who appeared before the court. As New York mayor Abram Hewitt declared in 1888, the police justice "in nearly all cases affect[s] the life, liberty, and property of the citizens." Michael Willrich, *City of Courts: Socializing Justice in Progressive Era Chicago* (New York: Cambridge University Press, 2003), 3–6, quotation on 5.

before Nostrand that morning were found guilty but released with suspended sentences contingent on their promise to no longer ply their trade on Coney Island.[74]

Backus's prosecution of fortune-tellers was not always successful, however. The following week, he was unable to convict self-described "scientific palmist" Mlle. Agnes Charcot of Bergen Beach when a jury accepted her defense that "palmistry was a recognized science" and took only two minutes to acquit. The verdict probably did not surprise onlookers, since, according to the *Brooklyn Daily Eagle*, "it was a well known fact" that Coney Island juries were reluctant to convict those in the business of local entertainment.[75]

Still, by the end of August 1897, Backus judged his efforts a success and believed he had fulfilled his mission of "protect[ing] the fools who insist upon spending their money with the fortune-tellers."[76] Backus may have temporarily banished fortune-tellers from Coney Island, but they soon returned with new professional personas. In 1901, the *Eagle* identified a shift in the industry:

> Instead of the old gypsy fortune-tellers, who were satisfied to prophesy a glorious future with great wealth and any number of husbands or wives . . . for the modest sum of 25 cents, there was formed during the summer a "trust" which included four of the best known "palmists" on the Island. They hired a large place on the Bowery, fitted it up sumptuously, and were soon doing a land office business.[77]

By 1903, business was booming for Coney Island fortune-tellers, whose "shanties [were] crowded at all hours" with men and women eager to have their fortunes told by "philosophers of physiognomy and psychology," like Mary Kroeger, more famously known as Countess Habeba.[78] (See figure 5.2.) As newspaper accounts suggested, the "old gypsy fortune-tellers," premodern purveyors of the occult, had become modern practitioners of a new, scientific brand of divination.

74. "Told Miss Ahngren's Fate," *Brooklyn Daily Eagle*, August 27, 1897.

75. "Mlle. Charcot Acquitted," *Brooklyn Daily Eagle*, September 7, 1897; "Will Make His Own Lists," *Brooklyn Daily Eagle*, August 23, 1897.

76. "Backus Is Satisfied," *Brooklyn Daily Eagle*, August 30, 1897; "Coney's Touch of Reform," *Brooklyn Daily Eagle*, August 23, 1897.

77. "Coney Island in the Throes of Craze for Consolidation," *Brooklyn Daily Eagle*, September 15, 1901.

78. "A Day with the Fortune Tellers of Coney Island," *NYT*, August 2, 1903.

FIGURE 5.2. Countess Habeba drew the attention of passersby on the Coney Island boardwalk with signs advertising her as a "gypsy" practitioner of "scientific palmreading" and "phrenologie." The man in the center of the image, wearing a hat and vest with his head lowered, is in conversation with a woman, either the so-called "Persia Gypsy" herself or one of her employees in what was known as "Countess Habeba's gypsy encampment." Robert L. Bracklow, *Persia Gypsy Countess Habeba Scientific Palmreading Phrenologie*, c. 1900, photograph, Archive Photos, New-York Historical Society/Getty Images.

Backus was neither the first nor the last to attempt to impose law and order on the "laborator[y] of the new mass culture" that was Coney Island, and he was not alone in his moral crusade against its revelry and irreverence.[79] A Presbyterian clergyman who cheered Backus on condemned Coney Island as "a nest for the breeding of moral scorpions."[80] Backus's campaign was also

79. Quotation in Kasson, *Amusing the Million*, 8. In 1894, Mayor Schleren had ordered sweeping reform of Coney Island, targeting dance halls and theaters, banning all costumes and games of chance, and calling for the regulation of saloon licenses. "Old Coney Island No More," *NYT*, May 6, 1894.

80. "Reform at Coney Island," *NYT*, August 5, 1897.

part of a broader "sustained assault on popular superstitions" spearheaded by scientists, liberal Protestants, and secularists in the 1890s.[81]

Backus's moral reform of Coney Island, however temporary, is significant because it reveals the instability of fortune-telling's legal definitions and the extent to which that instability disrupted the policing of fortune-tellers, some of whom navigated loopholes in the language of the law to escape prosecution. As one of the "ill-defined offenses" characteristic of nineteenth-century morals regulation, fortune-telling would be repeatedly redefined by practitioners and jurists in the late nineteenth and early twentieth centuries.[82]

"PRETENDING TO TELL FORTUNES"

In 1896, the Michigan State Supreme Court rejected the appeal of Arthur Elmer, convicted under a disorderly persons statute because "he pretended to tell fortunes." At the heart of the case was Elmer's unsuccessful reformulation of a professional identity that distinguished itself from "fortune-telling." Elmer claimed to be a "magnetic healer" and a "clairvoyant physician" engaged in "prognosticating" and "looking into the future," but not "telling fortunes." His self-promotion, however, suggested otherwise.[83] When Elmer sent advance notice of his arrival in Ionia, Michigan, in 1895, he billed himself a "modern day seer" and a "clairvoyant, trance medium, and healer." In newspaper ads he promised "a peep through the keyhole of the mysterious future," business and legal advice, romantic counsel, the locations of buried treasure and long-lost relatives, character reading, career counseling, and the presumably less-sought-after ability to divine the contents of one's stomach. Witnesses reported that Elmer "pretended to go into a trance . . . to tell the future and the past," advising a Mrs. Webber to leave her husband after discovering that her husband would murder her. But according to Elmer, none of this amounted to "fortune-telling." The court disagreed, ruling it "immaterial what name the respondent applied to his acts" and finding Elmer guilty of "professing to possess a power and pretending to exercise that power."[84]

Arthur Elmer was not alone in his efforts to distance himself from the term *fortune-telling* at the turn of the century, when some fortune-tellers recast their

81. Lears, *Something for Nothing*, 176.

82. Novak, *People's Welfare*, 167.

83. People v. Elmer, 109 Mich. 493, 67 N.W. 550 (1896).

84. Ibid.

practice as scientific, in contrast to the "cheap 25 cent fake fortune teller[s]" who read palms or tea leaves.[85] In 1896, three Coney Island fortune-tellers were arrested and appealed their sentences on the grounds that palmistry and phrenology were legitimate scientific practices.[86] Lady Gonzalez of Brooklyn advertised herself as "the greatest mind, sight and character reader and scientific palmist in the world," declaring, "Do not class me among fortune-tellers," and the *Brooklyn Daily Eagle* ran an ad for a "Mrs. Hutchings, business clairvoyant, not a fortune teller."[87] In 1899, a self-described character reader named Haleel took umbrage with newspapers that labeled him a soothsayer, insisting that he was "no fortune teller, but a scientific reader of character from the voice, face and hands."[88] Haleel and his colleagues transformed fortune-telling from art to science by simply renaming it.

Recognizing changing perceptions of their practice, fortune-tellers traded on the cultural authority of science. In 1901, the *Eagle* advised, "To Produce a New Science Add 'Ology' to Your Name . . . jumble up all your methods into one and your own financial success for a time at least is assured." The *Eagle*'s interview with Mrs. Glyn—Brooklyn's only practitioner of "Glynology"—revealed that consumers were in the market for "something new," something more exotic: "The public . . . has an idea that there's something uncanny, strange and mysterious about an 'ology.' Look at astrology or phrenology. Fortune telling by cards or even palm reading sounds plain compared to it." But however mysterious this new science of the future, when a client asked for "a séance of glynology," he got, Mrs. Glyn cheerfully acknowledged, "just fortune telling," plain and simple. For her part, Mrs. Glyn did her best to "lead them away from the word," offering specific practices like palmistry, horoscope reading, and spiritualist trance, but not the promise of a future foretold. Redefining fortune-telling as a scientific enterprise paid off for Mrs. Glyn, who reported doubling her business to fifty clients a day once she became an "ologist" instead of "a common every day fortune teller."[89] Fortune-tellers also invoked scientific authority for its predictive certainty. Professor Clay, who advertised as a "clairvoyant, palm reader, palmist and medium," in Washington, DC, at

85. "Clairvoyants," *Brooklyn Daily Eagle*, December 7, 1896.

86. "Science in Fortune Telling?," *Brooklyn Daily Eagle*, September 15, 1896.

87. "Clairvoyants," *Brooklyn Daily Eagle*, November 30, 1897; *Brooklyn Daily Eagle*, February 8, 1899.

88. "'Haleel' No Soothsayer," *Brooklyn Daily Eagle*, January 30, 1899.

89. "To Produce a New Science Add 'Ology' to Your Name," *Brooklyn Daily Eagle*, September 29, 1901.

the turn of the century, was known for "foretell[ing] with a certainty all commercial and business transactions," and the advertisements of "Mme. Dr. Thomas, scientific revealer" promised to predict "entire life, past, present, future . . . mistake impossible."[90]

Although Arthur Elmer and Mrs. Glyn crafted professional identities as "prognosticator" and "Glynologist" to disguise the ordinariness of their practices, both were clearly pretending—Elmer pretended he was less than a fortune-teller, and Mrs. Glyn pretended she was more. But the concept of *pretending* proved far more complex in a series of early twentieth-century legal decisions in which fortune-tellers and state appellate courts faced off over the legality of fortune-telling and its very nature, with both fortune-tellers and the courts advancing more circumscribed definitions of the practice. A literal interpretation of the phrase "pretending to tell fortunes," first formulated in the Elizabethan act against "rogues, vagabonds, and sturdy beggars" at the turn of the seventeenth century, considers the act of foretelling improbable, as a New York appellate court did in 1915 in accepting the prosecution's interpretation of "pretending to tell fortunes" as skepticism embedded in the language of the law, "simply signify[ing] the fact that the legislature deemed it an impossible thing to tell fortunes."[91] But the rest of the Elizabethan statute, which also named those "fayning themselves to have knowledge in Physiognomie, Palmestry, or other like crafty Scyence," hinged on a dichotomy between authenticity and fraudulence that figured into numerous early twentieth-century antidivination rulings in which rhetorical appeals to astrology, palmistry, and Spiritualism as occult ways of knowing implied that certain people could legit-

90. "Prof. Clay," *Alexandria (DC) Gazette*, September 3, 1902; Bolton, "Fortune-Telling in America," 302, 303.

91. The Elizabethan statute against rogues, vagabonds, and beggars included "all idle persons going about in any Country eyther begging or using any subtile Crafts or unlawfull Games and Playes, or fayning themselves to have knowledge in Physiognomie, Palmestry, or other like crafty Scyence, or pretending that they can tell Destenyes, Fortunes, or such other like fantasticall Ymagynacõns." Statute of the Realm 1597–8, Eliz 39, Ch. 4, p. 89. For the 1915 New York appellate decision, see People v. Malcolm, 90 Misc. 517, 154 N.Y.S. 919 (1915). The word *pretending* did not appear in the antidivination provision of the Napoleonic Code but did appear in a jurist's comment in 1858 that identified "fortune-tellers, supposed sorcerers, card readers, . . . all those who make a profession of pretending to uncover hidden facts, to predict the future, by the means of superstitious practices." P. Gilbert, ed., *Codes d'instruction criminelle, pénal et forestier*, vol. 3 of *Les codes annotés de Sirey, contenant toute la jurisprudence des arrêts et la doctrine des auteurs* (Paris, 1858), 608–16, quoted in Harvey, "Fortune-Tellers in the French Courts," 134.

imately foretell the future.[92] To determine the legality of a particular practice, courts came to focus on whether or not a fortune-teller's predictions could be deemed genuine and whether or not her intent was to defraud.[93]

In 1903, the "Distinguished Palmist, of London," Zoza Kenilworth (born Walter Winston Kenilworth), had his disorderly persons conviction for practicing palmistry on the Atlantic City boardwalk upheld by the State of New Jersey.[94] The statute under which Kenilworth was convicted and fined encompassed "all persons who shall use or pretend to use or have skill in physiognomy, palmistry or like crafty science," and the ruling on his appeal was determined by the court's defining palmistry as a "crafty science—that is, one by which the simple-minded are apt to be deceived." Kenilworth had read a man's palm and then "foretold the age at which the witness would marry and the duration of his life." The court determined that Kenilworth intended to defraud, noting, "If ever there shall be discovered any rational evidence that palmistry is a real science, its use for honest purposes will pass beyond the range of this statute."[95] But that time had yet to come, so Kenilworth's palmistry was judged not a "real science" but a "crafty science."

In 1904, Cleveland spirit medium Lena Wolf won a dismissal of her conviction in a case that centered on questions of intent, self-representation, and the public meaning of *fortune-teller*. Wolf was charged with "unlawfully

92. Historians have often considered Spiritualism separately from fortune-telling, but the two converge in the history of antidivination prosecutions as spirit mediums found themselves subject to laws against "pretending to tell fortunes." The literature on Spiritualism is extensive. Important works include Robert Laurence Moore, *In Search of White Crows: Spiritualism, Parapsychology, and American Culture* (New York: Oxford University Press, 1977); Ann Braude, *Radical Spirits: Spiritualism and Women's Rights in Nineteenth-Century America* (1989; repr., Bloomington: Indiana University Press, 2001); Alex Owen, *The Darkened Room: Women, Power, and Spiritualism in Late Victorian England* (1989; repr., Chicago: University of Chicago Press, 2004); Robert S. Cox, *Body and Soul: A Sympathetic History of American Spiritualism* (Charlottesville: University of Virginia Press, 2003); Molly McGarry, *Ghosts of Futures Past: Spiritualism and the Cultural Politics of Nineteenth-Century America* (Berkeley: University of California Press, 2008).

93. Gregory G. Sarno, *Annotation: Regulation of Astrology, Clairvoyancy, Fortunetelling, and the Like*, 91 A. L. R. 3d 766, 770–71 (1999). French jurists were animated by similar questions of fortune-tellers' intent in the nineteenth and twentieth centuries. See Harvey, "Fortune-Tellers in the French Courts," 135.

94. *Washington Evening Star*, January 9, 1901, 12; "Crowns and Omens," *St. Paul (MN) Globe*, August 3, 1902.

95. State v. Kenilworth, 69 N.J. 114, 54A 244 (1903).

represent[ing] herself to be a fortune-teller" to undercover police officer James Dolan.[96] She and Dolan had faced each other across a table, upon which Wolf had placed her palms and "pretend[ed] to place herself in a tranced condition . . . and while pretending to be in said tranced condition" took Dolan's right hand and informed him that his current business would change by the new year and that a profitable new venture awaited. Wolf also predicted an uneasy relationship with an intimidating business partner but assured Dolan that he would ultimately be "all right." Her final vision was ambiguous: she saw "a gold field; lots of gold" but warned Dolan not to venture toward it.[97]

In overturning Wolf's conviction, the Ohio appellate court narrowed its definition of fortune-telling. Simply facing a man across a table and "pretending" to enter a trance did not mean that Wolf was presenting herself as a fortune-teller, the court ruled, noting that her prediction of a promising business opportunity was the kind of innocuous comment anyone might make to an acquaintance and did not legally constitute fortune-telling: "It would not be regarded as fortune-telling for any one to say 'I do not think you will continue in the business in which you now are.'" In observing that Wolf's vague prediction regarding a troubled partnership, perhaps in a gold venture, did not foretell a specific future event, the court established prediction as a ubiquitous and thus legally reasonable act. The court distinguished between the capacity to foretell and the act of foretelling, declaring that merely claiming the ability to read palms, for example, did not itself constitute fortune-telling. The ruling also differentiated between fortune-tellers' private and public self-representations, finding that Wolf's individual session with Dolan did not make her a fortune-teller in the minds of the public, who would be foolish to fall for this "ignorant Indian girl" whose predictions were nothing but "trash."[98] In this ruling, "pretending" ultimately meant that Lena Wolf had only adopted a persona, nothing more, and that she was not predicting any more specifically (or successfully) than the average person might.

In the 1908 case of *Fay v. Lambourne*, a less sympathetic New York appellate court adhered to a more literal interpretation of "pretending to tell fortunes" in which such an act was inherently deceptive. Ironically, the court ruled that

96. Early twentieth-century police departments commonly sent undercover officers to elicit a fortune-telling. On undercover New York City police matrons, see Stone-Gordon, "Fifty-Cent Sybils," 232. On undercover police officers in France at the turn of the twentieth century, see Harvey, "Fortune-Tellers in the French Courts," 141.

97. Wolf v. State of Ohio, 24 Ohio C.C. (n.s.) 526 (1904).

98. Ibid.

those who pretended to predict the future for a living were not eligible for legal protection from those who tried to cheat them, given that "equity does not adjust the differences between rogues."[99] Two of the rogues involved were no strangers to the discourse of fraudulence that animated the culture of late nineteenth-century stage Spiritualism and in fact were born into it: John Fay was the son of the "indescribable phenomenon" Anna Eva Fay, and his wife, Eva, stepped into her mother-in-law's stage persona (but only a fraction of her acclaim) as a mind-reading specialist on the vaudeville circuit.[100] Promising to "unfold the mysteries of the mind," the Fays' show, which they claimed to have spent over two hundred thousand dollars promoting, "caus[ed] quite a sensation" and became "the centre of much controversy" in the first decade of the twentieth century.[101] Billed as "a skilled and scientific manipulator of the arts . . . of the East Indian yogi and famed Mahatmas, being a daughter of India," an entranced Eva answered—correctly, a reported 95 percent of the time—questions that her audience had written during the performance.[102] In 1906, John Fay told an interviewer that "Mrs. Fay when she is blindfolded feels as if she were floating about on ether. The questions that people write then flash before her, as she explains it, somewhat like the flickering of a moving picture machine."[103]

A different explanation of Fay's perspicacity came from the affidavits of the two other "rogues" in the lawsuit, Herbert Lambourne and Louis Granat, former Fay employees who toured under the name "The Fays" (sometimes "The

99. Fay v. Lambourne, 124 A.D. 245, 108 N.Y.S. 874 (N.Y. App. Div. 1908).

100. One of the most popular spirit mediums of the American stage, Anna Eva Fay was a fixture on the vaudeville circuit by the 1870s and a household name by 1900. She toured widely in the United States and Europe and was the subject of extensive press coverage as well as the target of numerous exposés that sought to discredit her theatrical performance of spirit manifestations. On Fay's career, see Barry H. Wiley, *The Indescribable Phenomenon: The Life and Mysteries of Anna Eva Fay* (Seattle: Hermetic Press, 2005). For an excellent discussion of *Fay v. Lambourne*, see Erika White Dyson, "Spiritualism and Crime: Negotiating Prophecy and Power at the Turn of the Twentieth Century" (PhD diss., Columbia University, 2010), chap. 5.

101. Quotations in advertisement, Lowell, Massachusetts, 1901, reprinted in Wiley, *Indescribable Phenomenon*, 272; Hereward Carrington, *The Physical Phenomena of Spiritualism* (1907; repr., New York: Dodd, Mead, 1920), 310; "Exposing the Fays Again," *New York Sun*, September 14, 1906. On the Fays' advertising costs, see Amended Complaint of John T. Fay and Eva N. Fay against Chauncey Herbert and Louis Granett, October 19, 1906, New York Supreme Court, filed with *Fay v. Lambourne*, GA 573/1907, Records of the Supreme Court: 1848–1960, Division of Old Records, New York County Clerk's Office, New York City Municipal Archives.

102. "Feats of Thaumaturgy," *Washington Post*, December 10, 1905; "Exposing the Fays Again," *New York Sun*, September 14, 1906.

103. "Wonderful Occult Powers," *Boston Daily Globe*, May 21, 1906.

Phays") and put on a show titled "The Flamaturgists," an obvious swipe at "thaumaturgy," the Fays' term for Eva's onstage clairvoyance. Lambourne and Granat insisted that they were justified in "exposing such fraud and cheat" for the sake of "the public welfare" since the Fays' act hinged on "tricks, devices, and . . . mechanical means."[104] (And Eva Fay was not from India but rather a small town near St. Louis, Granat revealed.) The Fays' "thaumaturgy," Lambourne and Granat charged, was accomplished with the help of charcoal powder applied to pages of the audience's notepads that revealed what had been written on the previous sheet, a cohort of Fay confederates scattered throughout the audience who "developed" these pieces of paper, and a system that the *New York Sun* called "mind reading by phone." Indeed, a former "advance man" for Fay revealed that her team would conduct research on local residents weeks before a performance and then relay the answers to audience members' questions over a telephone connection from beneath the stage, with Fay wearing her long hair over her ears to conceal the earpiece.[105]

John and Eva Fay sought an injunction to prevent their former employees from using their name in advertisements and posters that attracted an audience expecting to see the real Fays.[106] (The audience's misunderstanding was hardly surprising, considering that the posters read "The Flamaturgists—In Fazing THE FAYS," with "THE FAYS" in large print and the rest visible only up close.)[107] The New York State Supreme Court originally ruled in their favor,

104. "Call 'The Fays' Frauds Who Trick Audiences," *NYT*, September 14, 1906; "Exposing the Fays Again," *New York Sun*, September 14, 1906. Quotations in Answer to Amended Complaint of John T. Fay and Eva N. Fay against Chauncey Herbert and Louis Granett, October 19, 1906, New York Supreme Court, filed with *Fay v. Lambourne*, GA 573/1907, Records of the Supreme Court: 1848–1960, Division of Old Records, New York County Clerk's Office, New York City Municipal Archives.

105. "Call 'The Fays' Frauds Who Trick Audiences," *NYT*, September 14, 1906; "Exposing the Fays Again," *New York Sun*, September 14, 1906; Thomas J. Minnock, "Confessions of a Reformed Grafter," *Chicago Daily Tribune*, August 3, 1913.

106. Perhaps an unusually litigious mind reader, Eva Fay threatened to sue well-known journalist and romantic novelist Richard Harding Davis in October 1909 for depicting her and her husband, who had reportedly committed suicide in December 1908, as frauds. Eva Fay vowed to seek twenty thousand dollars in damages from Davis on the grounds that his exposé would ruin her career, but I have found no evidence that she followed through on her threat. "Mrs. Fay to Sue Author," *NYT*, October 16, 1909; "John T. Fay Kills Himself," *NYT*, December 22, 1908. On the publication of Davis's novel *Vera, the Medium* and Fay's reaction, see Stone-Gordon, "'Fifty-Cent Sybils,'" 165.

107. Amended Complaint of John T. Fay and Eva N. Fay against Chauncey Herbert and Louis Granett, October 19, 1906, New York Supreme Court, filed with *Fay v. Lambourne*,

but the judgment was overturned by an appellate court based on an interpretation of "pretending to tell fortunes" in which Mrs. Fay's "ability to answer these questions and to foretell the future is pretended to come from supernatural powers possessed by her." The "pretense of occult powers," the opinion read, "is a species of fortune-telling and is a fraud upon the public." Since the Fays were "engaged in deceiving the public, and the most entertaining part of their performance is in fact fortune-telling," the court ruled, they were not entitled to protection of the signature part of their act that was "intended and calculated to deceive the public."[108] *Fay v. Lambourne* involved two sets of pretenders: John and Eva Fay, who were legally accountable for "pretending to tell fortunes," and Lambourne and Granat, whose advertisements pretended they were the real Fays. Of course the intent of the pretend Fays was to profit from exposing the real Fays as frauds. According to the somewhat dizzying logic of this ruling, "pretending to tell fortunes" was not legally permissible, but pretending to be the ones pretending to tell fortunes was. Eva Fay would continue on the vaudeville stage after her husband's sudden death in 1908, but, as the program for her solo act made clear, "she is not a fortune teller, palmist or pretender of any kind."[109]

These turn-of-the-century rulings—*People v. Elmer* (1896), *State v. Kenilworth* (1903), *Wolf v. State of Ohio* (1904), and *Fay v. Lambourne* (1908)—sought to determine whether a defendant was pretending to tell fortunes and thus focused on the accused fortune-teller's character and intent. A series of subsequent decisions shifted their attention to the nature of the prediction itself. Courts in the 1910s focused less on whether fortune-tellers were pretending or genuinely predicting and more on whether they were predicting with certainty.

"LEARN TO READ PEOPLE YOU MEET"

It should not be surprising that in an age of segregation, fortune-telling raids attempted to maintain a racial order of white supremacy. The history of the Clio School of Mental Sciences, a character-reading training school founded and operated by Adena Minott, reveals how antidivination law was used to enforce segregation in early twentieth-century cities as well as how the professional

---

GA 573/1907, Records of the Supreme Court: 1848–1960, Division of Old Records, New York County Clerk's Office, New York City Municipal Archives.

108. Fay v. Lambourne, 124 A.D. 245, 108 N.Y.S. 874 (N.Y. App. Div. 1908).

109. Quotation in program at B. F. Keith's, Boston, 1909, reproduced in Wiley, *Indescribable Phenomenon*, 290.

authority of a black educational institution challenged racial discrimination. Adena Minott, known variously as Professor, Doctor, and Madame Minott, "considered one of the best readers of phrenology" in New York City and a "popular character analyst and metaphysician," was charged with "pretending to tell fortunes" after a visit by police matron Isabella Goodwin in 1911. But her case, heard in the Jefferson Market Street Court, was dismissed by a judge who "ruled that scientific delineation of character was not fortune telling, but a recognized business" in the absence of any evidence "prov[ing] that Mme. Minott offers to tell the past or the future." For her part, Minott supported fortune-telling raids, "stat[ing] that the crusade being made by the police against the fakers is a good move, which will relieve the present unhealthy condition of pretenders, who now injure the health of mental sciences."[110]

Minott, whom historians have described as a "mental scientist, phrenologist, and child care specialist" and "local businesswoman and antilynching activist," defied easy categorization at the turn of the twentieth century as well.[111] Born in Jamaica, she had grown up in New York as the daughter of an architect, excelling in the Girls' Technical High School and then studying at the MacDonnall College of Phrenology and Psychology in Washington, DC. As the *Chicago Defender* would later report, "because she was a Race woman, she was not permitted to study with the regular classes, but forced to take the courses by private instruction. Despite this disadvantage, Miss Minott completed the studies in one-half the regular time, graduated with honors and received the degree of Bachelor of Philosophy." Minott subsequently studied at the American Institute of Anthropology in New York City and was named a fellow, the first black woman to hold that distinction, and she also studied at New York's Fowler and Wells Institute of Phrenology and Anthropology.[112] (See figure 5.3.)

In 1906, she opened the Clio School of Mental Sciences in New York City near Sixth Avenue and 29th Street, where students could "Learn to Read Peo-

110. "All Graduates Were White," *New York Age*, March 2, 1911; "Former Chicagoan Is Making Good in East," *Chicago Defender*, February 17, 1923; "Mme. Minott Gains Strong Legal Point," *New York Age*, January 19, 1911.

111. Anne Meis Knupfer, *Toward a Tenderer Humanity and a Nobler Womanhood: African American Women's Clubs in Turn-of-the-Century Chicago* (New York: NYU Press, 1996), 152; Judith Weisenfeld, *African American Women and Christian Activism: New York's Black YWCA, 1905–1945* (Cambridge, MA: Harvard University Press, 1997), 87.

112. Quotation in "Doctor of Metaphysics to Adena C. E. Minott," *Chicago Defender*, April 9, 1921; "All Graduates Were White," *New York Age*, March 2, 1911.

MME. ADENA C. E. MINOTT.

FIGURE 5.3. After earning credentials in phrenology, Adena Minott opened the Clio School of Mental Sciences in New York and then Chicago, which offered training in phrenology, physiognomy, psychology, and palmistry as part of a curriculum in character reading that would prepare students to take control over their own futures by anticipating others' behavior. Minott's work received extensive press coverage, and in 1914 she was featured in a *New York Age* article on successful black businesswomen. Reprinted from "Clio School in Third Year," *New York Age*, June 17, 1909, 3.

ple You Meet" and embark on "A Certain Road to Success."[113] Minott's school combined the imperatives of economic and racial betterment into a curriculum that equipped students to take control of their futures, and indeed, educator and civil rights activist Mary Bethune would commend Minott for "stand[ing] for the uplift of our race." As the *New York Age* reported, "Prof. Minott feels that despite the seeming drawbacks which people of color encounter in this

113. On the rise of "occult science and esoteric educational programs and schools" at the turn of the century, see LaShawn Harris, *Sex Workers, Psychics, and Numbers Runners: Black Women in New York City's Underground Economy* (Urbana: University of Illinois Press, 2016), 102.

land, the United States offers better advantages and superior opportunities to every individual than can be found in any other country. The secret is in knowing where these advantages lie and how to take hold of the opportunity." The school offered individual and classroom instruction, as well as correspondence courses, in four branches of mental science: "Phrenology—Head Reading, Physiognomy—Face Reading, Psychology—Mind, and Scientific Palmistry—Hand Reading."[114] "KNOW YOURSELF. Appreciate your abilities, and Prosper. You need to know people you meet to succeed, to avert deceptions, sorrows, losses, pain," the school's advertisement read. In her frequent lectures, Minott emphasized that "phrenology is a science and not mere fortunetelling," and a former student, New York physician Dr. York Russell, remarked that "neither does she countenance advertisements of clairvoyants and other pretenders who injure occult science, but is an esteemed patron of a genuine and scientific psychic curricula."[115]

In November 1910, Minott advertised the opening in Harlem of the "elegant comfortable" Clio School Home at 121 West 136th Street between Lenox and Seventh Avenues, which offered "first-class accommodations only at moderate prices" for those affiliated with the Clio School of Mental Sciences. An opening reception drew one hundred guests for an evening of musical performances and speeches on phrenology, including one by Jessie Fowler, the president of the American Institute of Phrenology.[116]

But as soon as the Clio School Home opened, it became a target of the 121 members of the newly formed Association of Property Owners of West 136th Street, a branch of the Property Owners' Protective Association of Harlem, which had raised twenty thousand dollars in an attempt to, as the *New York Times* reported, "keep the negroes of 'Little Africa,' just east of Lenox Avenue, from further encroaching on the street." The *Times* cited the protective association's vice president, John G. Taylor, who blamed Minott and the address of the Clio School Home for "the first encroachment of negroes upon the exclusive residential section of 136th Street, west of Lenox Avenue." Taylor had complained at the West 125th Street police station "that the house was occu-

114. "Reopening of Clio Studio," *New York Age*, May 30, 1912; "Clio School in Third Year," *New York Age*, June 17, 1909; "Learn to Read People You Meet," *New York Age*, October 29, 1908; "A Course through the Mind," *New York Age*, November 18, 1909.

115. "Learn to Read People You Meet," *New York Age*, October 29, 1908; "Field Notes," *Phrenological Journal and Science of Health* 123–24 (1910): 167; "Learn the Certain Road to Success: Honor, Fortune and Fame," *New York Age*, April 22, 1909.

116. "Clio School Home," *New York Age*, November 17, 1910; "Clio Home Opens," *New York Age*, December 8, 1910.

pied by disorderly persons," but when his complaint was ignored, he appealed to the police commissioner, and within a week Minott's tenants had, according to Minott's lawyer, been "compelled to vacate" the premises. Perhaps not coincidentally, it was within the next month that Minott found herself on the New York City police department's list of fortune-tellers to investigate. The *Times* article cited Taylor's allegations "that there are some property owners who are using the 'negro scare' for financial gain, and contemplate renting their houses to negroes in order to induce owners of adjacent property to buy it at an enhanced price." Taylor vowed that his protective association would spare no expense "in buying up, whenever possible, mortgages on the property."[117]

Minott sued the *New York Times* for libel and won, with the decision upheld by an appellate court in November 1911. The innuendo submitted by Minott's counsel explicated the nature of the libel as suggesting that Minott was "guilty of the crime of keeping a disorderly house," that she "knowingly let to and permitted the said premises to be occupied by disorderly persons," and that she "has let the said premises to disorderly persons and to negroes for the purpose of blackmail and in order to extort from neighboring owners and tenants money as an inducement to cause the removal therefrom of such disorderly persons and tenants, and the said false and libelous article was extensively circulated and read by numerous persons." The appellate court ruled that the *Times* article "was intended by inference to charge plaintiff with acts which are perilously akin to blackmail."[118] But a year later, Minott left the Clio School Home address, a move that the Property Owners' Protective Association took credit for in a February 1912 *New York Times* advertisement celebrating the departure of "the last of the colored tenants to leave the restricted section of 136th Street, west of Lenox Avenue" and their efforts against "the Negro Invasion." Undeterred, by the end of May, Minott held a formal reopening of the Clio School Home in its new location in a four-story brownstone just down the street at 135 West 136th Street, where one of the board members of the Clio School spoke against "the storm of race prejudice" that had driven Minott from her previous address, declaring, "We congratulate her and we rejoice that she has struck this blow to enforced segregation."[119]

117. "$20,000 to Keep Negroes Out," *NYT*, December 8, 1910; Minott v. New York Times Co., 146 App. Div. 857 (N.Y. App. Div. 1911).

118. Minott v. New York Times Co., 146 App. Div. 857 (N.Y. App. Div. 1911).

119. "Restricted from Negroes," *NYT*, February 11, 1912; "Reopening of Clio Studio," *New York Age*, May 30, 1912. On the new location, see "Many Women of Race Large Wage Earners," *New York Age*, October 1, 1914.

Minott later opened a branch of the Clio School in Chicago, where it was in operation by January 1917. She explained, "The real problem of the day is to KNOW, and be PREPARED." Despite the future-oriented nature of the Clio School's curriculum in character reading, which taught students to better anticipate their own career paths and their social interactions, African American newspapers, like Minott herself, explicitly denied any connection to fortune-telling. The *New York Age* insisted that "the school has nothing to do with fortune-telling, neither does it pretend to give the power to predict future events," and the *Chicago Defender* noted that the "Clio School of Mental Sciences must not be confounded with the cheap fortune-telling institutions with which Chicago today is infested to such a disgraceful degree. It is an institution of higher education and scientific research and is the pioneer school of its kind in the world."[120]

Just as Minott established professional authority for character reading in the institutional context of the Clio School of Mental Sciences, twentieth-century astrologers and spirit mediums increasingly established organizations to confer cultural legitimacy on their forecasting practices and thus avoid prosecution.[121] In 1912, the Washington State Supreme Court, upholding a vagrancy conviction for reading horoscopes and telling fortunes for a fee, based its judgment on an astrologist's claim to predictive certainty. An undercover police officer paid a visit to F. F. Neitzel, an ordained minister in the National Astrological Society, and asked to have his fortune told. Neitzel responded that he could not engage in fortune-telling but would attempt to "figure it out," after which he gave the officer a horoscope reading of past and future for one dollar. Rejecting the argument that Neitzel was practicing the science of astrology, the court ruled that the defendant was clearly "engaged in fortune-telling, for he was professing to tell future events in the life of the witness."[122] The

120. "Clio School," *Chicago Defender*, January 27, 1917; "The Clio School of Mental Sciences," *Chicago Defender*, September 7, 1918; "Many Women of Race Large Wage Earners," *New York Age*, October 1, 1914; "Clio School," *Chicago Defender*, January 27, 1917. On Minott's subsequent work as an antilynching activist and a leader in the National Association of Colored Women in the 1920s, followed by her 1941 arrest for grand larceny after promising to double a client's insurance settlement, see Harris, *Sex Workers, Psychics, and Numbers Runners*, 105.

121. Dyson, "Spiritualism and Crime," frames the institutional history of American Spiritualism as a series of efforts to bring the movement within the mainstream of American religious practice from the founding of the National Spiritualists' Association in 1893 through the first half of the twentieth century.

122. State v. Neitzel, 69 Wash. 567, 125 P. 939 (1912).

court thereby denied Neitzel's vague promise to "figure it out" as categorically distinct from fortune-telling and found him guilty of predicting with certainty the police officer's future.

Occasionally the intent and actions of the complaining witness—usually an undercover police officer—determined a ruling. In 1912, undercover police officer Peter Sicokis arranged a visit to Chicago spirit medium Rosa Westergren, whereupon he paid fifty cents for a reading involving the whereabouts of a (fictitious) missing brother. Westergren "closed her eyes and said, 'It seems to me he is sick; I can't find him; he is not dead.'" She was fined fifty dollars for violating a section of the Chicago penal code that stipulated that "any person or persons who shall obtain money or property from another by fraudulent devices and practices in the name of or by means of spirit mediumship, palmistry, card reading, astrology, seership, or like crafty science, or fortune telling of any kind, shall be deemed guilty of a misdemeanor and on conviction thereof shall be punished by a fine." But Westergren's conviction was overturned on appeal when the court ruled that "it is clear from the record that he did not rely upon or believe the said statements and was in no manner deceived or defrauded thereby. It is apparent that under such circumstances the defendant cannot be held guilty of obtaining money by false pretenses."[123]

THE BUSINESS OF ASTROLOGY

The most high-profile antidivination ruling of the time involved Evangeline Adams, who by 1914 was well on her way to becoming "the leading exponent of astrology in this country." Adams hailed from Boston (and implied that she descended from the presidential Adamses) and moved to New York in 1899 in an attempt to jumpstart her fledgling career as an astrologist, which she did in dramatic fashion by allegedly (and retrospectively) predicting the deadly St. Patrick's Day fire at the Windsor Hotel, where she was staying.[124] As Adams would later recall, "I was 'made' by the Windsor fire."[125] In 1909, Adams announced the opening of her studio at Carnegie Hall, where she offered readings in "scientific astrology" for those who visited her office as well as a free

123. City of Chicago v. Westergren, 173 Ill. App. 562 (1912).

124. Allison Gray, "People Who Try to Get Tips from the Stars," *American Magazine*, December 1921, 34. On Adams's life, see her autobiography, Evangeline Adams, *Bowl of Heaven* (1927; repr., New York: Dodd, Mead, 1926); Karen Christino, *Foreseeing the Future: Evangeline Adams and Astrology in America* (Amherst, MA: One Reed Publications, 2002).

125. Gray, "People Who Try to Get Tips from the Stars," 140.

"circular, including General Indications for the current year" by mail.[126] Adams, whom the *New York World* termed the "Society Palmist," regularly drew astrological charts for "her clients [who] are of a higher order of society," operating a lucrative business that netted her up to a thousand dollars per week by the mid-1920s (over thirteen thousand dollars today). Adams's clientele grew steadily through the 1910s, and by the mid-1920s she was overseeing a business and a staff in a much larger office space at Carnegie Hall as well as appearing on her own radio program, "Listening In on the Stars."[127] In her "richly furnished" office at Carnegie Hall, Adams received clients, drew astrological charts, and answered correspondence. An interviewer described her in 1914 as "a businesslike person" who "dressed quietly but modishly, with no mummery of Oriental robes, and her hair . . . parted simply over a wide, intellectual forehead." With rows of clerical workers and filing cabinets just outside her door, Adams's office was more reminiscent of a corporation than a fortune-telling parlor. As Adams noted in the first sentence of her autobiography, "My working day begins, as a business man's does, with the never-ending task of answering letters." (See figure 5.4.) By the time of her death in 1932, Adams had accumulated considerable wealth and renown as well as three arrests for "pretending to tell fortunes."[128]

In 1911, the New York City Police Department deployed undercover police matrons as clients in a fortune-telling raid, which resulted in the arrest of six fortune-tellers, including Adams, whose case was dismissed by the magistrate.[129] But three year later, Adams would be arrested again in a raid, which led to a more complex decision in which she was acquitted based on a ruling that accepted predictive uncertainty in the science of astrology. In May 1914, New York police detective Adele D. Preiss arrived at Carnegie Hall for a five-dollar horoscope reading, after which a warrant for Adams's arrest was issued. The ensuing trial in a packed municipal courtroom sought to answer the questions posed by New York City magistrate John J. Freschi: "What is fortune-telling, and who is a fortune-teller?" After Preiss and Adams presented

126. *New-York Tribune*, September 25, 1909, 10.

127. Christino, *Foreseeing the Future*, 78, 124–25; "Professional Spook Raisers Wax Scornful of Amateurs," *Sun and New York Herald*, February 8, 1920; "By Radio Today," *Washington Evening Star*, August 25, 1922.

128. Marguerite Mooers Marshall, "Astrology More Certain in Its Diagnosis Than Medicine or Law, Says Miss Adams," *New York Evening World*, December 14, 1914; Adams, *Bowl of Heaven*, 1; "Evangeline Adams, Astrologer, Dies, Foretelling Her Own Death," *Palm Beach Post*, November 12, 1932.

129. Christino, *Foreseeing the Future*, 77–78, 82.

FIGURE 5.4. Evangeline Adams's offices at Carnegie Hall in the 1920s featured ornate decor, a staff of clerical workers, and filing cabinets in a professional atmosphere devoted to the business of astrology. *Portrait of Evangeline Adams, Famed Astrologer and Prognosticator, in Her Study*, photograph, Bettmann/Getty Images.

conflicting accounts of their interaction, in which Adams prepared an astrological chart for Preiss and examined her hands, Adams described the epistemic limitations of astrology and her unwillingness to offer clients predictive certainties:

> I explain in the beginning I simply give what is shown by the stars. . . . If I read for a client I always say "This is indicated" and I always explain that no astrologer can conscientiously say that any one thing will happen. Sometimes a lady will say "Will I be married in 1914?" I say, "I don't know whether you will or not. I think you have an opportunity to be but . . . whether you will I can't tell. Astrology does not indicate you are going to be haled to the alter." . . . I simply say, "This is likely to be, guard against it, if it is bad. If it is good, make the most of it."[130]

130. People ex rel. Preiss v. Adams, 32 N.Y. Crim. 326 (1914).

Convinced by Adams's testimony, Freschi deemed her practice both a business and a science: "There is no claim here that the defendant was garbed in special garments or that there was any air of mysticism about the place; it was a simple apartment with library furniture without signs of any kind in or about the studio, except to indicate that it was the office of the defendant," who "raises astrology to the dignity of an exact science." Yet the potential exactitude of astrology did not make claims to future certainties, in Freschi's judgment, and he characterized Adams's statements as "but a guess or conjecture" in which she "was giving no assurance that this or that eventually would take place." He legitimized astrology by differentiating it from fortune-telling when he declared that

> every fortune-teller is a violator of the law; but every astrologer is not a fortune-teller. I believe that there is a line of distinction between the person who pretends to be able to read the future and tell with positiveness what will or shall happen; and the one who merely reads a sign as indicating what ought to happen but is particular to make it plain that he is not attempting to predict future events. The former is a charlatan, an oppressor and an imposter; the latter is surely not a fortune-teller as he is commonly understood.[131]

To be found guilty of "pretending to tell fortunes" (a "peculiar" wording, Freschi noted), one must intend to dissemble and defraud, and he judged Adams to be no fraud but rather "a woman of learning and culture, and one who is very well versed in astronomy and other sciences."[132] Adams would later characterize her strategic choice to have the case tried rather than dismissed as part of her "ambition . . . to legalize Astrology in the state of New York" and proudly claimed that the case "raised my profession to a legal par with law and medicine."[133]

Adams turned Freschi's decision into a ringing endorsement of her business in a nineteen-page promotional pamphlet titled "Astrology and the Law: The Scientific Practice of Astrology Is Upheld by the New York Courts," which combined the epistemic authority of science and law by juxtaposing an explanation of astrology's methods and benefits with excerpts from Freschi's ruling. Characterizing astrology as "the oldest science," "an exact science," and "an empirical science," exactly as Adams's counsel had done at her trial,

131. Ibid.
132. Ibid.
133. Adams, *Bowl of Heaven*, 57, 54.

the pamphlet included three pages of testimonials to the value of Adams's forecasts in equipping individuals to navigate daily uncertainties. A reprint from a 1911 *Paris Herald* article noted that "while predictions forewarn the subject, threatened disaster may be averted if known in time, for life is not a journey in an environment of cast-iron fate, unalterable and relentless, but a progress in which the intelligence, the sense of moral responsibility, the power of will that seeks co-operation always with the Divine will, determines its cause." The writer and editor Lillian Whiting sent Adams an admiring letter that celebrated astrology's conquering of chance in its "enabling [of] one to be MASTER rather than servant—dictator, not the victim of circumstances." Whiting went on to discuss the horoscope Adams had provided for a friend, noting that it had "the same value as that of the well-equipped seaman with chart, rudder and compass, over one who should sail without these and drift blindly along at the mercy of wind and wave."[134]

Given Adams's celebrity, the case became front-page news in New York, where the *Evening World* ran the headline "Astrology More Certain in Its Diagnosis Than Medicine or Law, Says Miss Adams." Yet much of the article, based on an interview with Adams, revealed astrology's predictive uncertainties, emphasized in a heading that read, "Woman Declared by Court to Be No Fortune Teller Asserts That Stars Indicate What a Mortal's Fate MAY Be, and Then It's Up to the Knowledge Seeker to Fend Off Disaster or Invite Happiness." And in boldface type appeared this quotation from Adams:

> I can read what the fate MAY be . . . No astrologer can conscientiously say that any one thing is sure to happen. . . . My role is similar to that of the weather prophet. He doesn't say, "It will pour tomorrow; you will go out, get your feet wet, have an attack of pneumonia and die." He says, "Rain is expected," and if you are wise, you will wear rubbers and take an umbrella. So when I tell a person that the stars show danger on a certain day, he is simply warned to be brave and wary.[135]

Adams's invocation of the weather prophet was more apt than perhaps she knew, given the similar efforts of weather forecasters and astrologers to establish professional authority by constructing an epistemic boundary between

134. Evangeline Adams, "The Law and Astrology" (New York: Schulte Press, n.d.), 3, 16, 18. I am grateful to Karen Christino, Adams's biographer, for providing a copy of this pamphlet.

135. Marguerite Mooers Marshall, "Astrology More Certain in Its Diagnosis Than Medicine or Law, Says Miss Adams," *New York Evening World*, December 14, 1914.

scientific forecasting and the "superstition" of commercial "long-rangers" and fortune-tellers, while simultaneously communicating a measure of predictive uncertainty.[136] The article made another telling analogy in a cartoon—captioned "Astrology is similar to weather reporting"—of a man scratching his head in confusion while holding a paper that reads, "Probable success followed by probable failure."[137] Adams operated according to a kind of qualitative probabilism in which she issued forecasts in broad outline, the particulars of which defied astrological prediction. As she explained in 1921, "I send out regular monthly 'forecasts.' The first sheet is a statement of general conditions, as they are indicated by the astrological charts. In addition there is a personal forecast, or what you might call the 'current horoscope' of the individual receiving it." (See figure 5.5.) Adams resisted calls to make specific short-term market forecasts, cautioning that "it is impossible to predict, through astrology, the daily fluctuations of the market. The relative positions of the planets change gradually. Predictions based on these relative positions are for a longer or shorter period, but not from day to day. Even if general conditions are favorable, the influences ruling one certain individual may be *un*favorable."[138] (Still, in 1927, Adams estimated that a third of her clientele sought stock-market advice, and demand for her financial outlooks would only increase after the 1929 crash.)[139] But Adams's careful delineations of the epistemic limitations of astrology, coupled with her status as a member of elite New York society, legitimized her practice and spared her punishment. As one writer noted in 1920, "As she casts horoscopes what she tells you cannot be condemned as fortune telling," and indeed, Adams was arrested for a third time in 1923, but the charges were summarily dismissed.[140]

In 1915, a New York appellate court followed the precedent of *People ex rel. Preiss v. Adams* in upholding a disorderly persons conviction against Maude Malcolm, a self-described astrologist and palm reader, because her predictions made claims to certainty. In *People v. Malcolm*, the court ruled that palmists

136. On the "superstition" of astrology, see Allene Talmey, "Evangeline Adams and Her Stars," *Outlook and Independent*, February 18, 1931, 258.

137. Marguerite Mooers Marshall, "Astrology More Certain in Its Diagnosis Than Medicine or Law, Says Miss Adams," *New York Evening World*, December 14, 1914.

138. Allison Gray, "People Who Try to Get Tips from the Stars," *American Magazine*, December 1921, 34, 136.

139. Adams, *Bowl of Heaven*, 130; Walter A. Friedman, *Fortune Tellers: The Story of America's First Economic Forecasters* (Princeton, NJ: Princeton University Press, 2014), 1; Christino, *Foreseeing the Future*, chap. 14.

140. "Professional Spook Raisers Wax Scornful of Amateurs," *Sun and New York Herald*, February 8, 1920; Christino, *Foreseeing the Future*, 119.

FEES FOR

ASTROLOGICAL WORK

ORAL

FEE FOR ORAL READING . . . . . $ 5.00
(HALF HOUR INTERVIEW)

FEE FOR COMPLETE ORAL READING . . 10.00

WRITTEN

MANUSCRIPT HOROSCOPE . . . . . $ 5.00

BIRTH INDICATIONS

LIFE REVIEW AND A GENERAL GUIDE, THE MAIN FEATURES OF ABILITY, THE COURSE THE BIRTH PLANETS INDICATE TO AVOID WASTE OF TIME, WHAT TO CONCENTRATE UPON TO SUCCESSFULLY ACHIEVE NATURE'S INTENTION. BUSINESS, FINANCIAL, SOCIAL EVENTS, WHAT TO EXPECT, WHEN TO DO.

MORE EXTENDED CALCULATIONS AND SPECIFIC ADVICE BASED ON PERSONAL CHARACTERISTICS, AND BEARING DIRECTLY ON QUESTIONS OF HEALTH, FINANCE, MARRIAGE, ETC., ETC., CAN BE HAD BY SPECIAL REQUEST, AND FEE WILL BE IN KEEPING WITH AMOUNT OF TIME PUT ON THE WORK.

ASTROLOGICAL FORECASTS . . . $10.00

ABSTRACTS GIVING THE TREND OF COMING EVENTS FOR THE YEAR, MONTHLY CONSIDERED, AS INDICATED BY THE VARIOUS ASPECTS OF THE PLANETS.

ASTROLOGICAL YEARLY INDICATIONS . . $ 5.00

THE YEAR SURVEYED AS A WHOLE ACCORDING TO THE TRANSITS TO THE NATAL SUN.

FIGURE 5.5. Evangeline Adams's fee schedule, undated. On the reverse was a form with blanks for a client's name, address, and date and time of birth, along with a space labeled "READING DESIRED—QUESTIONS—STATE JUST WHAT YOU WISH TO KNOW." Courtesy of Karen Christino.

and astrologists could give "general advice as to the future of persons" or draw inferences from character reading without technically fortune-telling. However, those who referred to specific future events involving "marriage, death and travel" were telling fortunes and therefore considered disorderly persons. The court's ruling asserted that statements about the future in broad outline were permissible, but predictive certainty, which Maude Malcolm promised, was not.[141]

Malcolm's place on West 109th Street had been visited by Detective Isabella Goodwin, a police matron at the time of Evangeline Adams's first arrest

141. People v. Malcolm, 154 N.Y.S. 919 (1915).

in 1911 who had since been promoted to "the first municipal woman detective in the world" for going undercover to track down the perpetrators of a violent robbery who made off with twenty-five thousand dollars. Goodwin, a widowed mother of four, was particularly well suited to undercover work, a *New York Times* journalist decided, on account of her "unassuming" nature and "kind, motherly face" not "spectacularly made up in dime novel fashion." Such a "nondescript appearance was a studied art" for Goodwin, who visited hundreds of fortune-tellers, often with her face hidden behind a veil, and she prided herself on her ability to avoid detection. As she told the *New York Times* after being promoted to detective, "I have found that these fakers are just as gullible as their victims, and it is just as easy to fool them." Evangeline Adams and Maude Malcolm were but two of the many fortune-tellers who unwittingly looked into Detective Goodwin's future and paid the price.[142]

After learning Goodwin's birthday, Malcolm warned her to avoid hurrying (for fear of a foot injury) and eating red meat. After gazing at Goodwin's palm, Malcolm declared that in 1916, she would marry a medium-skinned man interested in machinery, that her aunt Anne would die from indigestion the same year, and that in 1917, Goodwin would take a transatlantic voyage. For this glimpse of her future Goodwin paid two dollars, but at trial she identified numerous inconsistencies in Malcolm's account, chief among them that she did not have an aunt named Anne. Malcolm, who was found guilty and put under $1,500 bond for one year's good behavior, appealed her conviction on two grounds: first, in order to convict those who "pretend" to tell fortunes, the prosecution must demonstrate "an element of deceit or fraud" on the part of the fortune-teller; and second, the "so-called science or system" of astrology and palmistry rendered such practices separate from fortune-telling. The appellate court disagreed on both counts, ruling that the intent of "deceit or fraud" was not feasible to prove, since one would have to wait an indefinite length of time to verify a prediction, and that the basis or method of making predictions was irrelevant.[143]

In Maude Malcolm's case, as well as those involving astrologers F. F. Neitzel, Rosa Westergren, and Evangeline Adams, the courts ruled that making uncertain, ambiguous, or vague predictions was legally permissible, whereas making predictions with specificity and "positiveness," as Magistrate Freschi had put it, constituted "pretending to tell fortunes" and was therefore illegal.

142. "The First Municipal Woman Detective in the World," *NYT*, March 3, 1912.

143. People v. Malcolm, 154 N.Y.S. 919 (1915).

"ALL OF US PREDICT THE FUTURE MORE OR LESS"

Cultural anxieties over the fate of soldiers killed in World War I fueled a Spiritualist revival in the United States and Europe as well as renewed legal and epistemological debates over the classification of the knowledge produced by fortune-tellers.[144] Debunkers of Spiritualism also intensified their efforts. Escape artist extraordinaire and "arch-enemy of spiritism" Harry Houdini lamented that he had "watched this great wave of Spiritualism sweep the world in recent months and realized that it has taken such a hold on persons of a neurotic temperament, especially those suffering from bereavement, that it has become a menace to health and sanity."[145] As spirit mediums in the United States claimed the constitutional protection of religious freedom, antidivination rulings drew more precise distinctions between Spiritualism and fortune-telling, as they had done with respect to astrology in the 1910s.[146] In 1923, for example, the New Jersey Supreme Court overturned the disorderly persons conviction of Spiritualist "message bearer" Mary De Laney, ruling that the constitutional freedom of religious expression permitted her to make predictions in the context of a ceremony at the First Spiritualist Church in Camden.[147] In 1929, the New York state legislature amended Section 899 of its Code of Criminal Procedure to exempt religious organizations from antidivination law. The New York statute classifying "persons pretending to tell fortunes" as disorderly persons was narrowed such that it "shall not be construed to interfere with the belief, practices or usages of an incorporated ecclesiastical governing body or the duly licensed teachers or ministers thereof acting in good faith and without personal fee."[148] The revision of the statute signaled an important shift in

144. Blewett Lee, "The Fortune-Teller," *Virginia Law Review* 9 (1923): 249, 252. On the resurgence of fortune-telling's appeal during World War I, see Harvey, "Fortune-Tellers in the French Courts," 145.

145. Travis Hoke, "The Heyday of the Fortune Tellers," *Harper's Magazine*, January 1932, 245; Harry Houdini, *A Magician among the Spirits* (New York: Harper & Brothers, 1924), xvi.

146. For a comprehensive account of Spiritualists' legal arguments for religious expression and courts' gradual acceptance thereof after World War I, see Christine A. Corcos, "The Scrying Game: The First Amendment, State Regulation of the Crafty Sciences, and the Rise of Spiritualism, 1848–1944," *Whittier Law Review* 38 (2017), https://papers.ssrn.com/sol3/papers2.cfm?abstract_id=2071391##.

147. State v. De Laney, 122 Atl. 890 (N.J. 1923); Blewett Lee, "The Fortune-Teller Again," *Virginia Law Review* 16 (1929): 54.

148. N.Y. Cr. Code and Penal Law (Gilbert, 1929), 615, as quoted in Lee, "The Fortune-Teller Again," 55.

fortune-telling's legal framework by acknowledging that some occult forecasters were actually not frauds and were not pretending to foretell future events.

Yet even constitutional freedoms of religion and speech could not protect Spiritualists accused of making predictions with certainty, as the case of Alice Ashley revealed in 1918. Presenting herself as the president of and a minister in the Brooklyn Spiritualist Society, and not a fortune-teller, Ashley argued that she provided mere "advice," not prophecy, based on spirit communication. In this instance, the recipient of Ashley's advice was undercover police officer Margaret Seller, who, in March 1917, rang Ashley's bell, sat down at the table, and asked whether she would find a job and a husband. Seller and Ashley testified differently regarding the certainty of Ashley's predictions, with Seller recalling that Ashley stated, "The spirit of your mother comes to me and says that you are to do as you are prompted yourself. . . . Yes, you will get the position and you will marry the man and you will have a small family of two or three children." In Ashley's recounting, her answers were more ambiguous: "Anyone with your ability ought not to be without a position very long. . . . That remains for you to say whether you will [marry] or not. . . . It is always best for people to have a family because they seem to be happier with a family than without it." The court considered Ashley's answers, as she repeated them, neither "satisfying" nor "valuable," expressed skepticism that Seller would have paid a dollar for such a vague response, and ultimately discounted Ashley's testimony in favor of Seller's. The court also rejected the argument that a spirit medium's predictions were significantly different from a fortune-teller's. Ashley summoned as a witness Chicagoan Dr. George B. Warne, president of the National Association of Spiritualists, who sought to distinguish between prophecy (both religious and Spiritualist) and fortune-telling "for the sake of compensation or gain." However, the court's ruling, citing Old Testament denunciations of fortune-telling, associated fortune-tellers with "rogues and mountebanks and generally disreputable members of society to be summarily dealt with for the good of the community" and noted that antidivination law was "for the protection of the entire community, the credulous as well as the more seasoned members of the body politic." The court subordinated Ashley's appeal to religious freedom to its legal obligation to "prohibit acts and practices which are deemed detrimental to the community," deciding that Ashley's brand of Spiritualist advice was nothing more than "some old-time wrongdoing or indecency sought to be brought to life again" under the pretense of religious practice.[149]

149. People v. Ashley, 172 N.Y.S. 282 (N.Y. App. Div. 1918).

Similarly, Spiritualist minister John J. Hill, head of New York's Spiritualist Church of Advanced Thought, appealed his conviction of pretending to tell fortunes on the grounds that his constitutional freedom of speech had been violated. But the appellate court cited a list of specific pronouncements that Hill had made regarding the whereabouts of undercover officer Sullivan's lost jewelry and her grandfather. Hill predicted that Sullivan and her sister would soon celebrate their weddings and revealed that her "deceased brother, who in this life, had wanted to be a doctor, was . . . studying medicine in the spirit world and would soon graduate." Concluding that Hill intended to defraud Officer Sullivan, the court rejected his appeal and declared that "telling fortunes and pretending to tell fortunes are one and the same thing."[150]

The new legal framework for occult prediction in the World War I era depended on distinctions between fortune-telling, the science of astrology, and Spiritualism as religious practice, which law professor Blewett Lee attempted to sort out in a series of law review articles in the 1920s.[151] Lee began from the premise that forecasting in general was a ubiquitous and rational form of knowledge production in twentieth-century society:

> The familiar pragmatic definition of knowledge is that it is merely a mental arrangement of experiences which gives us the best available forecasts of future experience. All of us predict the future more or less, from the astronomer with his nautical almanac to the lawyer who foretells the outcome of a lawsuit, or the physician the result of an operation. The weather man makes a calling of predicting, and all who sit down and count the cost of a proposed undertaking are trying to unravel the future.[152]

Lee understood "this universal human practice of prediction" as one that would become more accurate with every step in the march of scientific progress. He noted that "the more the walls of ignorance are broken down the surer our predictions avail" and "only the poverty of our intellect and of information

150. "People v. Hill," *New York Law Journal* 66 (October 7, 1921). For a detailed account of *People v. Ashley* and *People v. Hill*, see Dyson, "Spiritualism and Crime," chap. 6.

151. Blewett Lee, "The Conjurer," *Virginia Law Review* 7 (1921): 370; Blewett Lee, "Psychic Phenomena and the Law," *Harvard Law Review* 34 (1921): 625; Blewett Lee, "Spiritualism and Crime," *Columbia Law Review* 22 (1922): 439; Blewett Lee, "The Fortune-Teller," *Virginia Law Review* 9 (1923): 249; Blewett Lee, "The Fortune-Teller Again," *Virginia Law Review* 16 (1929): 54.

152. Lee, "The Fortune-Teller," 249.

limits our prophetic powers," echoing the late nineteenth-century conviction that positivist science would yield certain knowledge of the future.[153]

But occult prediction, Lee reasoned, was not this kind of rational anticipation, and he contrasted the predictive calculations of the astronomer, lawyer, physician, and meteorologist with the predictions of fortune-tellers, "the whole tribe of future-knowers [who] are esteemed wonderful because they are believed to have sources of information not available to ordinary mortals." In Lee's assessment, fortune-tellers occupied an epistemic middle ground between rationalism and occultism, "a debatable land where science neglects to take hold, and superstition will not let go."[154]

Fortune-tellers were ubiquitous, Lee explained, given that "the need to know the future is as great as it ever was." In language reminiscent of Bellamy's "specter of Uncertainty," Lee imagined that foreknowledge would create a world free of "many of the most frightful afflictions." And in a passage similar to Bellamy's "Blindman's World" that depicted earthlings disadvantaged by lack of foresight, Lee observed that "the world is really a horribly dangerous place, where the most terrible misfortunes are liable to happen unexpectedly to perfectly innocent people, and if the mischiefs were foreseen many could be avoided. As it is, in our daily lives we are like blind men blundering over a battlefield."[155]

Congressional lawmakers similarly attempted to articulate the legal boundaries of fortune-telling in 1926 (albeit with less success), when New York congressman Sol Bloom cosponsored a bill to prohibit fortune-telling for a fee in the District of Columbia, where fortune-tellers were permitted to operate if they had paid twenty-five dollars for a license. Harry Houdini made a dramatic appearance before the House Subcommittee on Judiciary of the Committee of the District of Columbia, at hearings that took on a carnivalesque atmosphere, with frequent outbursts and pointed exchanges between lawmakers, Houdini, and the audience, which, according to the *New York Times*, "resembled a tabloid convention of Spiritualists and mediums."[156] Houdini was well known for his investigations of spirit mediums, whom he held responsible for defrauding gullible clients of "millions of dollars" each year, and his performance turned the hearings into a debate over whether Spiritualist practices should be pro-

153. Lee, "The Fortune-Teller," 250, 249.

154. Lee, "The Fortune-Teller," 251.

155. Lee, "The Fortune-Teller," 251, 252.

156. "Say Lawmakers Consult Mediums," *NYT*, February 27, 1926.

tected forms of religious expression.[157] Although Bloom's bill never made it out of committee, the hearings revealed the enduring instability of the category of *fortune-teller* and the persistent legal ambiguity of the phrase "pretending to tell fortunes."

Houdini's appearance before the committee was a performance of his masculine authority and expertise combined with a typical Spiritualist exposé.[158] After denouncing ordained Spiritualist mediums who hid behind "the guise of religion," Houdini unfurled a fifty-foot-long roll of "letters, ordinations, and certificates of churches" that served as mediums' credentials. Houdini called this piece of evidence a "sucker list" because it represented "people who are duped" and went on to describe one fortune-teller who had maintained a client list of forty thousand dupes. Houdini later insisted that no human was capable of predicting the future and that palmists and astrologists were frauds, declaring, "They can not tell from a chunk of mud millions of miles away what is going to happen to me." Houdini and the chairman went back and forth regarding whether palmistry was a science and whether Houdini's own stage magic was mere "trickery," as Houdini maintained, or "deliberate fraud." Houdini's authority and expertise hinged on his self-fashioning as both "psychic investigator" and "mysterious entertainer," both empiricist and trickster.[159]

Houdini's theatrics aside, the hearings displayed a lack of clarity and consensus regarding the definition of fortune-telling. Bloom, the bill's sponsor, defined telling fortunes as predicting the future but noted a clear distinction between giving advice and telling fortunes. H. P. Strack, secretary of the National Spiritualists' Association of America, argued that a fortune-teller "pretends" to foretell future events without any basis for doing so, whereas a medium may in the course of communicating with the dead foretell an event, with such prediction based in the occult knowledge of the spirit world. Spiritualist minister Jane B. Coates testified, "I am a premonitionist. . . . I would say I do

157. *Fortune Telling Hearings*, 9, 23. For a recent analysis of the fortune-telling hearings that emphasizes Spiritualists' ability to translate the traditional passivity of female mediumship into a form of cultural authority and political power that would be more fully articulated in late twentieth-century "pluralistic feminism," see Jeremy C. Young, "Empowering Passivity: Women Spiritualists, Houdini and the 1926 Fortune Telling Hearing," *Journal of Social History* 48, no. 2 (2014): 341–62, quotation on 341.

158. On Houdini as an exemplar of turn-of-the-century American masculinity, see John F. Kasson, *Houdini, Tarzan, and the Perfect Man: The White Male Body and the Challenge of Modernity in America* (New York: Hill and Wang, 2001).

159. *Fortune Telling Hearings*, 12, 13, 24, 28–29, 6.

not predict the future; I have premonitions," and Houdini offered, "I believe in premonition, absolutely, but not in spiritualism."[160]

Such disagreement about what constituted fortune-telling returned once again to the meaning of the phrase "pretend to tell fortunes." Former dean of the Washington School of Law C. Larimore Keeley, counsel for the palmists, emphasized the ambiguity of the word *pretending*, noting that the bill made no distinction between foretelling for play and for profit and thus "would apparently apply equally to the play of children, the amusement of adults, or the credulity of those who have learned to rely upon premonition and presentiment, of whom there are many." Were the word *pretending* applied indiscriminately, Keeley contended, then any account of futurity could be considered illegal, including "the prognosis of the physician, forecasts of the weather, predictions of the state of the markets, and whether or not she will be happily married." "But what is 'pretending,' standing alone and unless explained. Pretending what?" she asked. "If it relates to future matters, then the clause is repugnant, for it has long been established that one can not well be defrauded by statements as to things to happen in the future, since one man's guess is as good as another's."[161] Keeley's argument invoked not only the new legal framework for occult prediction that permitted the indeterminacy of fortune-telling's claims to foreknowledge but also the centrality of forecasting to economic life in the 1920s.

Although fortune-telling methods had not changed significantly, if at all, by the time Houdini and the Spiritualists faced off in 1926, legal and cultural understandings of fortune-tellers had been transformed. As the population of "future-knowers" expanded in growing late nineteenth-century cities, fortune-tellers, the vast majority of whom were women, fashioned professional identities that claimed scientific authority for their time-honored ways of knowing.[162] Fortune-tellers redefined their trade as the modern sciences of palmistry, phrenology, and astrology to appeal to turn-of-the-century consumers and to evade prosecution under antidivination laws that were selectively enforced in New York and other northeastern and midwestern cities and driven by moral reform and nativist ideologies. The case of New York in particular reveals how the urban policing of fortune-tellers hinged on and indeed perpetuated the classification of fortune-tellers as racial "others," a cultural status that was also commonly invoked in advertisements as part of fortune-tellers' predictive au-

160. *Fortune Telling Hearings*, 77, 3, 41, 73, 68.

161. "Officers Installed by Kappa Beta Pi," *Washington Post*, October 29, 1922; *Fortune Telling Hearings*, 118.

162. Lee, "The Fortune-Teller," 251.

thority. The popularity of fortune-telling as fashionable party entertainment in the 1890s and the meteoric rise of "society palmist" Evangeline Adams to fame and fortune in the 1910s suggest that the white, middle-class appropriation of fortune-telling was central to its assimilation into twentieth-century consumer culture. Indeed, a 1931 article observed that because of Adams, astrology had "seeped into the middle classes," having been transformed from "the romantic idiosyncrasy of the few to a commodity" with "no foreign nonsense."[163]

Legal debates over fortune-tellers' promises of love and money—just like debates over cotton estimates, the daily "Probabilities," and economic futures—revealed epistemic conflicts over whether it was possible to predict the future and with what degree of certainty. Although antidivination laws would remain on the books in major American cities—and still do in some places to this day—state courts in the 1910s articulated a new legal framework for occult prediction that shifted away from nineteenth-century moral critiques of fortune-telling as a fraud and a social threat in favor of a more circumscribed definition that permitted "pretending to tell fortunes" as long as fortune-tellers did not pretend to foretell with certainty. The ambiguity of promises of love and money kept twentieth-century fortune-tellers in business rather than landing them in court, and they would stay in business, as a *Harper's* essay noted in 1932: "Fortune-tellers flourish, and they will always flourish, for they offer what religion cannot promise nor science provide. Religion deals with futures too remote. Science, near to becoming a religion, predicts futures in the mass. What man wants is a glimpse of his individual future, not in heaven but on earth."[164] The commercial success and cultural authority of fortune-telling would endure throughout the twentieth century, as its practitioners sold visions of the future in an uncertain present.

163. Talmey, "Evangeline Adams and Her Stars," 258, 259.

164. Hoke, "Heyday of the Fortune-Tellers," 246.

EPILOGUE

Specters of Uncertainty

In 1896, Darwin P. Kingsley summoned Edward Bellamy's "specter of Uncertainty" to sell life insurance, confronting his audience with their own mortality so they would recognize a financial and moral imperative to protect their families from future ruin.[1] But by the 1920s, such a cautionary tale would have hardly been startling, given the proliferation of new types of insurance that signaled a cultural recognition of the unpredictability of economic life along with new attempts to manage it. The "specter of Uncertainty," not banished as Kingsley had promised, would take on new twentieth-century forms as a calculable risk.

Uncertainty and risk were famously redefined in 1921 by Chicago economist Frank H. Knight, whose treatise titled *Risk, Uncertainty and Profit* acknowledged an economic necessity and social propensity for prediction. Characterizing human thought as "forward-looking," Knight observed, "It is a world of change in which we live, and a world of uncertainty. We live only by knowing *something* about the future; while the problems of life, or of conduct at least, arise from the fact that we know so little. This is as true of business as of other spheres of activity." But in Knight's formulation, uncertainty was no longer a crisis but rather one half of a twentieth-century economy in which risk and uncertainty had diverged according to a new model of probabilistic judgment. Risk involved an unknown future with a calculable and thus insurable probability of occurrence, whereas uncertainty referred to an indeterminate future.

1. D. P. Kingsley, "The First Business of the World," pamphlet, 1896, folder 10, drawer 2, series 11, Edward Bellamy Memorial Archives, Edward Bellamy House, Chicopee, MA.

Knight's risk was manageable, its probability able to "be reduced to complete certainty" in aggregate, whereas "*true uncertainty*," which defied probability, offered economic reward to those entrepreneurs most able to reckon with a statistically unpredictable future.[2] Knight's theoretical reformulation emerged alongside new mechanisms for transforming everyday uncertainties into insurable risks.

After war broke out in Europe in 1914, American philosopher Josiah Royce observed the prevalence of "the insurance principle" that "transform[s] our modern social order" and "contributes to the sense of stability."[3] With US entry into the war came federal administration of war risk insurance between 1917 and 1921, "a milestone in the development of the American welfare state" that instituted an unprecedented system of social insurance for soldiers and their families.[4] The Bureau of War Risk Insurance, established in 1914 to mitigate "the unknown and uncertain new risks of naval submarine warfare" for Atlantic shipping, dramatically expanded its scope with the passage of the War Risk Insurance Act in November 1917. The act's three main provisions provided "allotments and allowances" equal to a portion of a soldier's pay plus a supplement based on the number of dependents, compensation for death or injury, and life and disability insurance.[5] As one journalist announced, the war had brought the arrival of "Uncle Sam, Underwriter."[6]

The War Risk Insurance Act sought to replace the uncertainties of Civil War pensions and charitable relief efforts with an objectively calculated and "predictable degree of protection."[7] The "state-socialistic program" of war risk insurance that hinged on the logic of the "government as insurance company"—along with the expansion of the wartime state and the temporary nationalization of the railroad industry—could have come from the pages of Bellamy's *Looking Backward*, as could one journalist's account of a dreamlike

2. Frank H. Knight, *Risk, Uncertainty and Profit* (Boston and New York: Houghton Mifflin, 1921), 199, 203, 232 (emphasis in original).

3. Josiah Royce, *War and Insurance* (New York: Macmillan, 1914), x.

4. K. Walter Hickel, "War, Region, and Social Welfare: Federal Aid to Servicemen's Dependents in the South, 1917–1921," *Journal of American History* 87, no. 4 (2001): 1362.

5. Thomas B. Love, "The Social Significance of War Risk Insurance," *Annals of the American Academy of Political and Social Science* 79 (September 1918): 46; Paul H. Douglas, "The War Risk Insurance Act," *Journal of Political Economy* 26, no. 5 (1918): 464–81.

6. William B. Shaw, "Uncle Sam, Underwriter," *American Review of Reviews* 60, no. 5 (1919): 508–14.

7. K. Walter Hickel, "Entitling Citizens: World War I, Progressivism, and the American Welfare State, 1917–1928" (PhD diss., Columbia University, 1999), 116.

vision of risk management: "John Doughboy himself had to rub his eyes when his Uncle Samuel appeared in the guise of an insurance agent, but in the end he succumbed—and with the best of grace."[8] Indeed, with the federal government selling war risk insurance as aggressively as it had sold war bonds and the war itself, close to 95 percent of those enlisted had purchased insurance by the end of 1918, with over 4.1 million policies totaling thirty-eight billion dollars' worth of coverage, more than all the private life insurance policies in the United States combined.[9] Although the volume of war risk insurance and the massive bureaucracy required to administer it were new, the program was modeled on the existing measures of Civil War pensions, workmen's compensation, and private life insurance. As the assistant secretary of the Treasury observed, "War risk insurance is nothing more than the extension by the government of the United States to some of the risks of war of business methods with which the people of the United States are already familiar in nearly every department of their life and ordinary daily business."[10]

This insurability of daily life was also evident in rising demand for weather insurance, which implicitly acknowledged the uncertainties of weather forecasts while relying on the knowledge infrastructure that produced them. In 1892, the *Harrisburg Patriot* joked about the lucrative possibility of a "weather insurance bureau" to aid planners of outdoor excursions.[11] Twenty years later, rain insurance, which had appeared in England by 1910, was well established in the United States, where the number of companies writing "pluvius policies" rose from three in 1921 to thirty in 1925, with premiums totaling ten million dollars in 1924. Although rain insurance was not initially profitable for American companies, they broke even by 1923 and, after raising premiums, lowered their loss ratios in 1924 and 1925.[12] The new legitimacy of weather

8. Douglas, "War Risk Insurance Act," 483; Shaw, "Uncle Sam, Underwriter," 509, 510. On the administrative capacity of the wartime state, see Barry D. Karl, *The Uneasy State: The United States from 1915 to 1945* (Chicago: University of Chicago Press, 1983), chap. 3.

9. Douglas, "War Risk Insurance Act," 479; Shaw, "Uncle Sam, Underwriter," 510.

10. Love, "Social Significance of War Risk Insurance," 46.

11. "Nuggets," *New York Times* (hereafter *NYT*), August 7, 1892.

12. Andrew H. Palmer, "Rain Insurance," *Tycos-Rochester* 14, no. 3 (July 1924): 16–17; G. Wright Hoffman, "Weather Forms of Insurance," *Annals of the American Academy of Political and Social Science* 130 (March 1927): 123–24; William Gardner Reed, "Weather Insurance," *Monthly Weather Review* 44, no. 10 (October 1916): 575–80. On the new demand for weather insurance, see, for example, "Wide Interest Felt in Rain Insurance," *NYT*, May 27, 1920; "Rain Insurance Grows Popular," *Los Angeles Times*, August 10, 1921; "Weather Insurance Takes Strong Hold," *NYT*, August 14, 1921.

insurance was recognized by the *Journal of Commerce*, which characterized it as "formerly regarded as a rather frivolous form of enterprise . . . now taking its place among the older and recognized forms of insurance."[13]

Insurance companies initially sold rain insurance for "outdoor events and businesses" like baseball and football games, state fairs, boating companies, and summer resorts, and subsequently for indoor events like department store sales and boxing matches.[14] Rain insurance would indemnify a policyholder against a minimum amount of rainfall (say 0.1 or 0.2 inches) in a time period ranging from three to twenty-four hours, with higher premiums for longer durations and lesser amounts of precipitation. According to the president of the Henry W. Ives Insurance Company, rain insurance represented "an entirely new class of risk."[15]

Insuring this new form of risk hinged on the relative accuracy of weather forecasts and the availability of climatological data. Clients had to buy "pluvius policies" at least a week before their scheduled event since short-term forecasts were considered reliable but weekly forecasts were not.[16] Companies writing rainfall policies were betting on the uncertainty of long-range forecasts, confident that no weather prophet—government forecaster or private citizen—could accurately predict the next week's weather. Insurance companies calculated their own risk using Weather Bureau data on the average number of days each month with at least 0.1 inch of rainfall in a particular location, which meant that in Philadelphia, for example, premiums were highest in March and lowest in September. As the *Philadelphia Evening Public Ledger* remarked, "Rain insurance sounds as though it were merely a gamble but, young as the idea is, the probabilities have been figured out scientifically."[17]

Yet weather insurance belonged to both worlds: probabilistic risk management and the uncertainty of speculation. A Weather Bureau employee turned crop and weather insurance expert described rain insurance as "a contribution to public service, as it tends to promote stability to business in general. . . . By removing an unavoidable hazard it relieves the promoter of such an event of

13. Advertisement, *NYT*, January 25, 1921, 20.

14. Palmer, "Rain Insurance," 16.

15. Palmer, "Rain Insurance," 16; "Much Weather Insurance Is Arranged For," *North Dakota Grand Forks Herald*, August 17, 1921; "You Can Now Insure Your July 4 Outing," *NYT*, June 17, 1920.

16. Kristine C. Harper, *Weather by the Numbers: The Genesis of Modern Meteorology* (Cambridge, MA: MIT Press, 2008), 19; "Insurance Against Rain," *Scientific American*, September 17, 1910, 214.

17. "Insuring Against Rain," *Philadelphia Evening Public Ledger*, August 24, 1922; Harper, *Weather by the Numbers*, 19–20.

a risk which is beyond human control."[18] Although insurance companies prevented overinsurance and "a semblance of gambling" (in which an individual would take out a policy for a location he did not plan to actually visit), one newspaper likened rain insurance to "playing a five-to-one shot" (with premiums between 2.5 and 25 percent of a policy's coverage).[19] Speculation, coupled with the possibility that a rain gauge might not accurately reflect precipitation (or lack thereof) a few miles away and thereby award or deny compensation erroneously, were among what *Scientific American* termed "the elements of uncertainty" in weather insurance itself.[20]

The uncertainties of "pluvius policies" paled in comparison to those of crop insurance, an even more complex mechanism that had to account for weather, climate, and commodity prices in addition to the unpredictability of the harvest. "The most important risk of the farmer has for its cause the uncertainties of weather and market conditions," wrote economist G. Wright Hoffman in 1925, analyzing recent experiments in private crop insurance in the 1910s and 1920s. Hail insurance first appeared in the United States in 1880 but took off in the 1910s, with premiums totaling forty million dollars nationwide in 1924. Farmers could also purchase tornado, livestock, and frost insurance, the latter of which became widely available in the United States after a 1920 industry survey revealed significant demand among citrus growers.[21]

In the early 1920s, falling agricultural commodity prices, along with newly available USDA data indicating an average of $2.6 billion in annual crop damage, led lawmakers to consider the topic of crop insurance for the first time.[22] A 1923 Senate investigation revealed the extent to which companies writing crop insurance, which was "still in the experimental stage" and not yet profitable, struggled to resolve several fundamental questions, the first of which was what exactly a policy should consist of: coverage against individual risks like hail or frost, or, as a few companies had introduced in 1917, an "all-risk"

18. "The Rise of Weather Insurance," *American Review of Reviews* 72 (July 1925): 105; quotation in Palmer, "Rain Insurance," 19.

19. Palmer, "Rain Insurance," 16; "Weather Insurance," *Scientific American*, July 15, 1911, 69; "Much Weather Insurance Is Arranged For," *North Dakota Grand Forks Herald*, August 17, 1921.

20. "Weather Insurance," *Scientific American*, July 15, 1911, 69.

21. G. Wright Hoffman, "Crop Insurance—Its Recent Accomplishments and Its Possibilities," *Annals of the American Academy of Political and Social Science* 117 (January 1925): 95, 110, 109; "The Rise of Weather Insurance," *American Review of Reviews* 72 (July 1925): 105.

22. Randall A. Kramer, "Federal Crop Insurance, 1938–1982," *Agricultural History* 57, no. 2 (1983): 181.

policy encompassing weather, plant and animal disease, and market volatility. Whether farmers could insure costs of production, either total or partial, and how much to underwrite each acre for had no clear answers either. Once the parameters of crop insurance had been determined, companies then had to calculate and price their risk, which proved difficult given the lack of local agricultural statistics more precise than existing state and county figures, as well as the same question of how to quantify a "normal" yield that had frustrated agricultural statisticians and crop reporters at the turn of the century.[23] Insurance experts testifying before Congress also worried about how to calculate risk given "the personal equation," which encompassed moral hazard, adverse selection (of "farmers who farm worse than the average in their community"), and the potential for farmers to speculate on crop insurance.[24] But the overarching question was whether "growing crops which represent goods in prospect rather than goods in existence" should be underwritten by private companies or the federal government, and there was no real momentum for the latter approach in the 1920s.[25]

Theodore H. Price, editor of *Commerce and Finance*, found the answer in war risk insurance, which he cited as a model and rationale for government-subsidized crop insurance that would convert the uncertainties of farming into risks. "One of the farmer's greatest difficulties," Price wrote, "is that he is compelled to be a speculator and to take risks that are not calculable. He is in fact a gambler against his will."[26] But the problem Price identified would not be addressed by the federal government until the Depression. As breadlines lengthened and dust storms intensified, US agricultural policy sought to stabilize commodity prices through the subsidized scarcity of the Agricultural Adjustment Act (1933) along with the creation of the Commodity Credit Corporation (1933) and the Federal Crop Insurance Corporation (1938). The FCIC began insuring wheat harvests in 1939, gradually expanding its scale and scope through the 1940s despite initially poor loss ratios in its few first years.[27]

23. *Hearing Before a Select Committee on Investigation of Crop Insurance*, 67th Cong., 4th sess. (1923), 109; Hoffman, "Crop Insurance," 96–98, 100, 117, 112.

24. *Hearing Before a Select Committee on Investigation of Crop Insurance*, 48, 107, 110, 16; Hoffman, "Crop Insurance," 118.

25. Quotation in "The Rise of Weather Insurance," *American Review of Reviews* 72 (July 1925): 105; Kramer, "Federal Crop Insurance," 183.

26. Theodore H. Price, "Crop Insurance—Is It Feasible?," *Commerce and Finance* 11, no. 4 (1922): 163.

27. Kramer, "Federal Crop Insurance," 181, 185–91.

The "peculiar hazards" of farming, as Secretary of Agriculture Henry C. Wallace had termed them in 1923, were now underwritten by the state.[28]

This array of new insurance mechanisms designed to protect Americans from the uncertain winds of war, weather, and agriculture marked both a confrontation with and an embrace of the commercialized "specter of Uncertainty," which would haunt capitalism and culture in different guises throughout the twentieth century.

*

Federal crop insurance was one among many new federal risk-management efforts during the Depression, which witnessed a more interventionist version of the infrastructural state power constituted by the government administration of agricultural statistics and weather forecasting that began in the late nineteenth century. Upon signing the Social Security Act into law in 1935—with Bellamyism among several socialist plans garnering national attention—Franklin D. Roosevelt described the welfare state as the solution to what William Dean Howells had characterized as the "economic-chance world" of the late nineteenth century. "We can never insure one hundred percent of the population against one hundred percent of the hazards and vicissitudes of life," FDR declared, "but we have tried to frame a law which will give some measure of protection to the average citizen and to his family against the loss of a job and against poverty-ridden old age."[29] The Dust Bowl also yielded new government interest in meteorological cycles, and World War II fueled research on long-range weather forecasting for the military (in which Henry Helm Clayton participated).[30] The crises of depression and war brought a renewed focus on the promise of long-range prediction that had generated such demand for the forecasts of late nineteenth-century prophets like weather forecaster W. T. Foster and commodity price forecaster Samuel Benner.

28. *Hearing Before a Select Committee on Investigation of Crop Insurance*, 6.

29. Franklin D. Roosevelt, Statement on Signing the Social Security Act, August 14, 1935, FDR Presidential Library and Museum, http://docs.fdrlibrary.marist.edu/odssast.html.

30. On the USDA's interest in the study of cycles in the 1930s, see Halbert P. Gillette, "Governmental Investigation of Weather Cycles and Their Causes," *Roads and Streets*, May 1936 (reprint), folder 8, box 4, Clayton Papers. On the Weather Bureau's expansion of long-range forecasting during World War II, see F. W. Reichelderfer to H. H. Clayton, September 28, 1942, folder 10, box 6, Clayton Papers.

The postwar world suffered unprecedented atomic anxieties about the future of humanity in a world of "mutual assured destruction," as policymakers and schoolchildren alike made preparations for nuclear war through scenario planning and "duck and cover" drills. American intelligence analysts regularly engaged in "prevision" regarding the likelihood of nuclear war with the Soviets, and readers of *Collier's* magazine read a timeline of "Principal Events of World War III," predicted to end in 1955 "as U.S.S.R. degenerates into a state of chaos and internal revolt."[31] The Cold War culture of prediction was indeed distinctive for its emphasis on forecasts of a global scope, whether in the context of geopolitics, population growth, or planetary ecology, but, as this book has demonstrated, it is not the case that "prediction was indisputably a Cold War product."[32] Although a new generation of future-oriented thinkers like Olaf Helmer of the RAND Corporation announced the arrival of "a new attitude toward the future" that produced "sober and craftsmanlike analysis of the opportunities the future has to offer," the postwar era was not the first time that groups of professionals imagined that their technocratic expertise would create a modern, scientific kind of forecasting integral to a capitalist system and notably distinct from "crystal-ball gazing."[33] Like the agricultural statisticians, weather forecasters, business-cycle adherents, and fortune-tellers of the late nineteenth and early twentieth centuries, the futurists of the 1960s and 1970s pointed to new technoscientific methods of prediction and new forms of expertise in claiming authority for their visions of the future. And just as had happened a century earlier, occult forecasting—this time in the form of

31. Matthew Connelly et al., "'General, I Have Fought Just as Many Nuclear Wars as You Have': Forecasts, Future Scenarios, and the Politics of Armageddon," *American Historical Review* 117, no. 5 (2012): 1431–60; "Principal Events of World War III," *Collier's*, October 27, 1951, 14.

32. Jenny Andersson, "The Great Future Debate and the Struggle for the World," *American Historical Review* 117, no. 5 (2012): 1411–30; Matthew Connelly, *Fatal Misconception: The Struggle to Control World Population* (Cambridge, MA: Harvard University Press, 2008); Paul Warde and Sverker Sörlin, "Expertise for the Future: The Emergence of Environmental Prediction c. 1920–1970," in *The Struggle for the Long-Term in Transnational Science and Politics: Forging the Future*, ed. Jenny Andersson and Eglė Rindzevičiūtė (New York: Routledge, 2015), 38–62. On forecasting as "a Cold War product," see Jenny Andersson and Eglė Rindzevičiūtė, "Introduction: Toward a New History of the Future," in Andersson and Rindzevičiūtė, *Struggle for the Long-Term*, 3.

33. Olaf Helmer-Hirschberg, "Analysis of the Future: The Delphi Method," RAND Paper P-3558 (Santa Monica, CA: RAND Corporation, 1967), 1–3, quotations on 1, 2. Available at http://www.rand.org/pubs/papers/P3558.html.

astrology—entered the mainstream of American culture, with best-seller lists and newspapers regularly featuring horoscopes in the 1970s.[34]

But the lessons of the late nineteenth-century history of forecasting were not lost on Helmer, author of *Looking Forward: A Guide to Futures Research.* In 1967, he translated the "specter of Uncertainty" into the language of systematic scenario planning shaped by Cold War social science and cybernetics, pointing to

> a growing awareness that a great deal can be said about future trends in terms of probability, and moreover that through proper planning we can exert considerable influence over these probabilities. . . . The future is no longer viewed as unique, unforeseeable, and inevitable; there are, instead, a multitude of possible futures, with associated probabilities that can be estimated and, to some extent, manipulated.[35]

The late nineteenth-century quest for certainty regarding a singular future had been replaced by an acknowledgment of multiple futures as well as the assumption—however hubristic—of the human capacity to engineer desired outcomes. Given the obvious role of calculation in producing probabilistic scenarios, it is not surprising that some historians have framed studies of forecasting around the postwar advent of digital computing.[36] But, as this book has revealed, the role of quantification in forecasting was evident in the late nineteenth and early twentieth centuries, in aggregations of climatological data, tables of cotton statistics, graphs of price cycles, and astrological charts, all of which demonstrate that postwar digital computational methods are part of a longer history of calculating the future.[37]

34. Matthew Connelly, "Future Shock: The End of the World as They Knew It," in *The Shock of the Global: The 1970s in Perspective*, ed. Niall Ferguson, Charles S. Maier, Erez Manela, and Daniel J. Sargent (Cambridge, MA: Belknap Press, 2010), 341–42.

35. Helmer-Hirschberg, "Analysis of the Future," 2. See also Kaya Tolon, "Futures Studies: A New Social Science Rooted in Cold War Strategic Thinking," in *Cold War Social Science: Knowledge Production, Liberal Democracy, and Human Nature*, ed. Mark Solovey and Hamilton Cravens (New York: Palgrave Macmillan, 2012), 45–62.

36. Matthias Heymann, Gabriele Gramelsberger, and Martin Mahony, eds., *Cultures of Prediction in Atmospheric and Climate Science: Epistemic and Cultural Shifts in Computer-Based Modelling and Simulation* (New York: Routledge, 2017); Harper, *Weather by the Numbers.*

37. On calculation and futurity in an earlier period, see, for example, William Deringer, "Pricing the Future in the Seventeenth Century: Calculating Technologies in Competition,"

In the late twentieth and early twenty-first centuries, our decision-making processes have become increasingly embedded in layers of predictive knowledge infrastructures that offer around-the-clock access to increasingly precise and local probabilistic forecasts. The most obvious example is government weather forecasts, which, although first introduced as the daily "Probabilities" in 1871, did not actually become probabilistic until 1965, when the US Weather Bureau began to issue on a regular basis the probability of precipitation forecasts we have today (e.g., 40 percent chance of rain).[38] Probabilistic forecasting, along with model-based computational meteorology, has given forecasters more complex methods for accomplishing the same task that bedeviled late nineteenth-century weather prophets: reckoning with and communicating uncertainty. In 1980, National Weather Service forecasters received an internal memorandum that emphasized, "REMEMBER—PROBABILITY IS ONLY A MEANS OF COMMUNICATION." The memo also acknowledged a broader context for predictive judgment in which individuals weigh the costs of taking precaution against any potential losses in order to determine their tolerance for risk: "The same principle is used when deciding whether or not to carry an umbrella or to buy life insurance. In fact every decision by each of us is made in the same way." This "cost-loss ratio" can be even more finely calibrated in the twenty-first century, when anyone with an Internet connection can access hourly, daily, and weekly weather forecasts produced by government and private weather services for locations around the globe, and the National Weather Service mobile phone app tells me—to the minute—what time a thunderstorm will pass through my town.[39]

Our present age of "big data" has accelerated the process of calculating and commercializing the future that was systematized on a large scale in the United States beginning in the late nineteenth century. Environmental forecasts have become both more global and more "hyper-local," as in the case of the Climate Corporation, which analyzes fifty terabytes of weather and climate data daily in order to sell "individualized" crop insurance along with digital monitoring applications that provide farmers with acre-by-acre analyses of

---

in "Forum: Paper Technologies of Capitalism," ed. Seth Rockman, *Technology & Culture* 58, no. 2 (2017): 506–28.

38. Allan H. Murphy and Robert L. Winkler, "Probability Forecasting in Meteorology," *Journal of the American Statistical Association* 79, no. 387 (1984): 492.

39. Lawrence A. Hughes, *Probability Forecasting—Reasons, Procedures, Problems* (Silver Spring, MD: National Oceanic and Atmospheric Administration, 1980), 1, 9; National Oceanic and Atmospheric Administration, "Weather.gov on Your Mobile Phone," http://www.nws.noaa.gov/com/weatherreadynation/mobilephone.html.

their fields to help them anticipate how best to adapt to climate change.[40] At the same time, weather derivatives trading and prediction markets like Betfair and PredictIt have emerged as new digital exchanges for speculating on and indeed commodifying the future.[41] Questions that one might have posed to a fortune-teller—in any century—can be answered by algorithms that identify the most compatible romantic partners (a scenario Edward Bellamy had once imagined), apps that predict fertility, and controversial personal genomics tests that may be used to indicate one's risk factors for disease.[42]

"Predictive analytics," now touted as an essential management tool for twenty-first-century global capitalism, is harnessed by corporations as well as government agencies to forecast human behavior. In 2013, online megaretailer Amazon patented a method for "anticipatory package shipping" in which data on customers' online activity, including which items they have purchased and which items they have merely looked at and for how long, allows Amazon to ship boxes to a nearby distribution center or put them on a delivery truck before a customer has even made a purchase.[43] And "precognition" of criminal

40. Michael Specter, "Climate by Numbers," *New Yorker*, November 11, 2013, http://www.newyorker.com/magazine/2013/11/11/climate-by-numbers.

41. Samuel Randalls, "Weather Profits: Weather Derivatives and the Commercialization of Meteorology," *Social Studies of Science* 40, no. 5 (2010): 705–30; Leighton Vaughn Williams, "Betting the Future Online," *Significance*, March 2006, 10–12. Betfair (https://www.betfair.com), a London-based company founded in 2000 as one of the first online sports betting sites, became an industry giant before agreeing to a merger with a Dublin company in 2015. Operated by the Victoria University of Wellington, PredictIt (https://www.predictit.org) is a political prediction market in which users buy and sell "yes" or "no" shares (for between one between ninety-nine cents per share) in the likely outcome of future events. PredictIt is a research project that makes its data freely available to academic researchers working in fields including microeconomics, political science, and game theory.

42. John Tierney, "Hitting It Off, Thanks to Algorithms of Love," *NYT*, January 29, 2008; Janet Burns, "A Higgs Boson Scientist Created a Fertility App That's as Effective as the Pill," *Forbes.com*, October 17, 2016, http://www.forbes.com/sites/janetwburns/2016/10/17/a-higgs-boson-scientist-created-a-fertility-app-thats-as-effective-as-the-pill/#7ac916277690; Antonio Regalado, "For $999, Veritas Genetics Will Put Your Genome on a Smartphone App," *MIT Technology Review*, March 4, 2016, https://www.technologyreview.com/s/600950/for-999-veritas-genetics-will-put-your-genome-on-a-smartphone-app/. For a futurist's optimistic account of the potentially liberating effects of predictive analytics despite a decrease in privacy in the age of big data, see Patrick Tucker, *The Naked Future: What Happens in a World That Anticipates Your Every Move?* (New York: Current, 2014).

43. Victoria Wagner Ross, "Amazon Knows What You Want and Has It Ready to Ship to You," *San Diego Examiner*, January 21, 2014.

activity has been taken from the pages of Philip K. Dick's 1956 science-fiction short story "Minority Report" and implemented in the "crime forecasting" currently under way in state prison systems that use algorithms to predict inmates' recidivism rates, which some have hailed for reducing crime and others have criticized for perpetuating the systemic racial bias at the core of the American carceral state.[44] Ostensibly objective methods of calculating the future—whether cotton forecasts during the American Civil War, Cold War scenarios of a Communist collapse, or forecasts of criminal behavior today—are shaped by, and indeed perpetuate, the ideologies, politics, and cultural ideals of their times.

Yet probabilities are not certainties, regardless of how precisely they may be calculated. For months leading up to November 8, 2016, the presidential election in the United States was simulated by statistical models thousands of times each day as poll aggregators like Nate Silver's FiveThirtyEight and Sam Wang's Princeton Election Consortium calculated the likelihood of Democratic or Republican victory. In the fall, psychologists and psychiatrists reported widespread electoral anxiety and "hypervigilance" on the part of their patients who checked FiveThirtyEight multiple times each day to see if the forecast had changed.[45] On the eve of the election, a broad consensus existed among pollsters and journalists that former First Lady, US senator, and secretary of state Hillary Rodham Clinton would defeat real-estate magnate and former television reality-show star Donald J. Trump, with the major poll aggregators calculating the probability of a Clinton victory between 70 and 99 percent (and President Obama's 2008 campaign manager David Plouffe coming in at "one hundred percent").[46]

44. Philip K. Dick, "The Minority Report," in *The Minority Report and Other Classic Stories* (1987; repr., New York: Citadel Press, 2002), 71–102; Christopher I. Haugh, "Prison by Algorithm," *Atlantic*, June 26, 2016; Kelly Lytle Hernández, Khalil Gibran Muhammad, and Heather Ann Thompson, "Introduction: Constructing the Carceral State," *Journal of American History* 102, no. 1 (2015): 18–24.

45. Lesley Alderman, "Talking to Your Therapist about Election Anxiety," *NYT*, October 20, 2016.

46. Steve Lohr and Natasha Singer, "How Data Failed Us in Calling an Election," *NYT*, November 10, 2016, http://www.nytimes.com/2016/11/10/technology/the-data-said-clinton-would-win-why-you-shouldnt-have-believed-it.html?_r=0; John Cassidy, "Media Culpa? The Press and the Election Result," *New Yorker*, November 11, 2016; quotation in David Remnick, "Obama Reckons with a Trump Presidency," *New Yorker*, November 28, 2016, http://www.newyorker.com/magazine/2016/11/28/obama-reckons-with-a-trump-presidency.

In the ensuing days, journalists and pundits offered different explanations for why pollsters had failed to predict that Hillary Clinton would win the popular vote while narrowly losing the Democratic strongholds of Pennsylvania, Wisconsin, and Michigan, and thus the electoral college, and many, recalling the surprise Brexit vote in June, questioned the value of the polling industry itself.[47] A handful of forecasters—deemed outliers during the polling season—had correctly predicted the outcome, among them political scientist Helmut Norpoth, whose statistical modeling of the primary elections and analysis of electoral cycles (that traditionally have not favored two-term incumbent parties) predicted a 97 percent chance of a Trump victory. Historian Alan Lichtman, who had correctly predicted the winner of the popular vote in every presidential election since 1984 using his "Keys to the White House," thirteen true/false statements on the structural conditions of the economy, foreign policy, and party politics, declared that "very, very narrowly, the keys point to a Trump victory . . . [and] to a generic Republican victory."[48] A Chinese monkey known as the "king of the prophets" and a Russian polar bear were also hailed for their accurate, if less technical, prognostications.[49]

47. Vann R. Newkirk II, "What Went Wrong with the Polls?," *Atlantic*, November 9, 2016, http://www.theatlantic.com/politics/archive/2016/11/what-went-wrong-polling-clinton-trump/507188/?utm_source=atltw; Steven Shepard, "Pollsters Tackle What Went Wrong," *Politico.com*, http://www.politico.com/story/2016/11/polling-election-what-went-wrong-231207; Matt Sedensky, Associated Press, "'Big Data' Questioned in Wake of Trump's Surprising Victory," *Boston.com*, http://www.boston.com/news/politics/2016/11/11/big-data-questioned-in-wake-of-trumps-surprising-victory. On the eve of the election, Nate Silver's FiveThirtyEight model indicated a 10.5 percent chance that Clinton would win the popular vote but lose the electoral college.

48. Christopher Cameron, "Political Science Professor Forecasts Trump as General Election Winner," *Stony Brook Statesman*, February 23, 2016, https://www.sbstatesman.com/2016/02/23/political-science-professor-forecasts-trump-as-general-election-winner/; Peter W. Stevenson, "Trump Is Headed for a Win, Says Professor Who Has Predicted 30 Years of Presidential Outcomes Correctly," *Washington Post*, September 23, 2016, https://www.washingtonpost.com/news/the-fix/wp/2016/09/23/trump-is-headed-for-a-win-says-professor-whos-predicted-30-years-of-presidential-outcomes-correctly/.

49. Peter Holley, "This Mystical Chinese Monkey Has Figured Out Who Will Win the U.S. Election," *Washington Post*, November 5, 2015, https://www.washingtonpost.com/news/worldviews/wp/2016/11/05/this-mystical-chinese-monkey-has-figured-out-who-will-win-the-u-s-election/; Damien Sharkov, "2016 U.S. Presidential Election: Russian Polar Bear Picks Trump," *Newsweek*, November 8, 2016, http://www.newsweek.com/presidential-election-russian-polar-bear-trump-518404.

In the week following the election, media coverage of the overall failure of the polling industry and poll aggregators to predict the election of a "Black Swan president" revealed a struggle with the limitations of predictive analytics in the age of big data.[50] The *New York Times* examined "How Data Failed Us in Calling an Election." But it was not only data that had failed. Panicked post-mortems by Democrats and rapid-fire finger pointing on cable news, Twitter, and Facebook revealed fundamental questions about the nature of forecasting itself. Was the failure in the state-level polling methods or the statistical models of the poll aggregators? Did algorithms fail to adequately account for uncertainty, or did the relatively high number of undecided voters in 2016 increase the likelihood of state-level polling error, as Nate Silver has argued?[51] Did polls fail to capture "shy voters," "likely voters," or voters with no landlines?[52] Or did some elements of voters' decision making simply defy prediction? Was the greatest failure ultimately in our subjective interpretation of probabilities as certainties? As writers for the *New York Times* data analytics blog *The Upshot* remarked after the election, "we humans are bad at understanding uncertainty."[53]

As journalists, pollsters, and pundits struggled to account for the failure of expert predictions, new uncertainties about the future emerged in the weeks after the election. Many economists and market watchers had predicted that a Trump victory would send markets into steep decline, among them Justin Wolfers and Eric Zitzewitz, who forecast an 8 to 10 percent drop in the stock market. That prediction seemed right on target until midnight on election night, when the S&P 500 futures index began to rebound, and major markets in the United States drove stock prices higher until the Dow Jones Industrial Average posted its biggest weekly gain in five years, defying analysts'

50. *Politico* magazine, "The Black Swan President," *Politico.com*, November 12, 2016, http://www.politico.com/magazine/story/2016/11/donald-trump-president-white-house-policy-black-swan-214450.

51. Nate Silver, "Election Update: Where Are the Undecided Voters?," *FiveThirtyEight.com*, October 25, 2016, http://fivethirtyeight.com/features/election-update-where-are-the-undecided-voters/.

52. Steven Shepard, "Pollsters Tackle What Went Wrong," *Politico.com*, November 10, 2016, http://www.politico.com/story/2016/11/polling-election-what-went-wrong-231207.

53. Amanda Cox and Josh Katz, "Presidential Forecast Post-Mortem," *The Upshot* (blog), *NYT*, November 15, 2016, http://www.nytimes.com/2016/11/16/upshot/presidential-forecast-postmortem.html?smid=tw-upshotnyt&smtyp=cur&_r=0.

expectations.[54] Uncertainties regarding social policy led to a surge in applications for health insurance under the Affordable Care Act, which the president-elect and congressional Republicans have vowed to repeal, and women rushed to investigate options for long-term birth control, fearing that access would be curtailed.[55] World leaders and millions of Americans wondered whether the president-elect, a foreign-policy "enigma," would dramatically alter the role of the United States on the world stage, with respect to the NATO alliance, the Middle East, and the 2015 nuclear arms agreement with Iran.[56] Reacting to the president-elect's promises to withdraw the United States from the historic 2015 Paris climate agreement and to cripple the Environmental Protection Agency, as well as his widely cited statement on Twitter that climate change was a hoax invented by the Chinese, climate scientists and environmentalists envisioned an even more dire planetary future.[57]

In the immediate aftermath of the election, millions of Americans tried to anticipate how the nativist rhetoric of the Trump campaign would translate into action or policy. Six days after the election, the *Washington Post* asked, "The Postelection Hate Spike: How Long Will It Last?"[58] As of Novem-

54. Justin Wolfers, "Markets Sent a Strong Signal on Trump . . . Then Changed Their Minds," *The Upshot* (blog), *NYT*, November 18, 2016, http://www.nytimes.com/2016/11/19/upshot/markets-sent-a-strong-signal-on-trump-then-changed-their-minds.html; "Dow Industrials Hit Fresh Record in Best Week Since 2011," *Wall Street Journal*, November 11, 2016.

55. Ana B. Ibarra, "Uncertain about Obamacare's Future, Millions Ask if They Should Enroll," *PBS Newshour*, November 17, 2016, http://www.pbs.org/newshour/rundown/obamacares-uncertain-future-millions-ask-enroll/; Erin Ross, "Women Rush to Get Long-Acting Birth Control after Trump Wins," *Shots: Health News from NPR*, November 11, 2016, http://www.npr.org/sections/health-shots/2016/11/11/501611813/women-rush-to-get-long-acting-birth-control-after-trump-wins.

56. James Kitfield, "The Knowns and Unknowns of Donald Trump's Foreign Policy," *Atlantic*, November 19, 2016, https://www.theatlantic.com/international/archive/2016/11/trump-foreign-policy-flynn-sessions-obama-isis-iraq-muslim/508196/; Carol Morello, "Iran Nuclear Deal Could Collapse Under Trump," *Washington Post*, November 9, 2016, https://www.washingtonpost.com/world/national-security/iran-nuclear-deal-could-collapse-under-trump/2016/11/09/f2d2bd02-a68c-11e6-ba59-a7d93165c6d4_story.html.

57. Coral Davenport, "Donald Trump Could Put Climate Change on Course for 'Danger Zone,'" *NYT*, November 10, 2016, http://www.nytimes.com/2016/11/11/us/politics/donald-trump-climate-change.html.

58. Samantha Schmidt and Jasper Scherer, "The Postelection Hate Spike: How Long Will It Last?," *Washington Post*, November 14, 2016, https://www.washingtonpost.com/news/morning-mix/wp/2016/11/14/making-sense-of-the-post-election-spike-in-harassment-and-intimidation-how-much-how-long/?tid=sm_tw.

ber 19, 2016, the Southern Poverty Law Center had tallied 701 reports of harassment and intimidation targeting immigrants, African Americans, Muslims, Jews, women, and the LGBTQ community, sometimes invoking the name of the president-elect alongside the language of white supremacy, including the spray-painted words "Make America White Again" in Wellsville, New York and "swastikas and the words 'Go Trump' . . . discovered in graffiti" on a Brooklyn playground.[59] In the week after the election, as tens of thousands of protestors filled the streets in cities across the country, college students organized rallies calling for the protection of the millions of undocumented residents facing threat of deportation, and civil rights activists and Muslim advocacy groups denounced Islamophobia and sought to reassure schoolchildren who feared for their safety.[60] As a *Toronto Star* headline announced, "Racist Incidents, Policy Uncertainty Dominate News Since Trump's Victory."[61]

Despite the myriad uncertainties unleashed by the election of 2016, the failure to predict its result is but the most recent reminder of the epistemic struggles that characterized the late nineteenth- and early twentieth-century culture of prediction. Beginning in the mid-1860s, Americans became newly reliant on systematized, ubiquitous, and often controversial forecasts while at the same time confronting the limitations of new forms of knowledge that aspired to but could not achieve predictive certainty. The emergence, by the 1920s, of an acknowledgment of uncertainty as an inescapable feature of economic life, and indeed forecasting itself, in turn led to new attempts throughout the twentieth

59. Southern Poverty Law Center, "Update: Incidents of Hateful Harassment Since Election Day Now Number 701," *SPLC Hatewatch*, November 18, 2016, https://www.splcenter.org/hatewatch/2016/11/18/update-incidents-hateful-harassment-election-day-now-number-701; Ben Tsujimoto and Ellen Przepasniak, "Disturbing Acts in Wellsville, at Canisius College Follow Election Day," *BuffaloNews.com*, November 10, 2016, http://buffalonews.com/2016/11/09/disturbing-acts-in-wellsville-at-canisius-college-follow-election-day/; Sarah Maslin Nir and Ashley Southall, "Rally in Brooklyn Park Condemns Swastikas and 'Go Trump' Graffiti," *NYT*, November 20, 2016.

60. Alan Taylor, "'Not My President': Thousands March in Protest," *Atlantic*, November 10, 2016, http://www.theatlantic.com/photo/2016/11/not-my-president-thousands-march-in-protest/507248/?utm_source=atltw; Jonah Engel Bromwich, "Trump Camp's Talk of Registry and Japanese Internment Raises Muslims' Fears," *NYT*, November 17, 2016, http://www.nytimes.com/2016/11/18/us/politics/japanese-internment-muslim-registry.html.

61. Daniel Dale, "Racist Incidents, Policy Uncertainty Dominate News Since Trump's Victory," *Toronto Star*, November 11, 2016, https://www.thestar.com/news/world/2016/11/11/racist-incidents-policy-uncertainty-dominate-news-since-trumps-victory.html.

century to tame uncertainty and manage risk with increasingly complex tools for calculating the daily probabilities.

The allure of objectivity remains powerful when it comes to anticipating the future. It is comforting, perhaps, to have a probabilistic glimpse of our own future, or at least to believe that we can comprehend it with professional expertise and methods of rational calculation. Yet despite the array of forecasts that we have at our fingertips today, the "specter of Uncertainty" haunts us still, evident in the very extent to which we are beholden to the routinized forecasts of everyday life. Although we cannot definitively know what may come, we can—if we look closely—see how our convictions about prediction and uncertainty influence our decision making today. Looking forward gives us the clearest view not of the future, but of ourselves, with all our aspirations and anxieties, in the present.

ARCHIVAL COLLECTIONS

Edward Atkinson Papers, Massachusetts Historical Society

Edward Bellamy Compositions, 1860–1939, Houghton Library, Harvard University

Edward Bellamy Correspondence, 1850–1898, Houghton Library, Harvard University

Edward Bellamy Memorial Archives, Edward Bellamy Homestead, Chicopee, Massachusetts

Blue Hills Observatory Papers, Milton Historical Society, Milton, Massachusetts

Henry Helm Clayton Papers, 1877–1949 and undated, Smithsonian Institution Archives, Washington, DC

Commercial Telegraph Company Archives, Clara Waldron Historical Room, Tecumseh District Library, Tecumseh, Michigan

Dudley M. Hughes Papers, Hargrett Library Broadside Collection, 1900–1919, Hargrett Rare Book and Manuscript Library, University of Georgia Libraries, http://broadsides.libs.uga.edu/hbro/search

Records of the National Agricultural Statistics Service (Record Group 355), Crop Report and Historical Materials, 1914–1918, National Archives College Park, College Park, Maryland

Records of the Supreme Court: 1848–1960, Division of Old Records, New York County Clerk's Office, New York City Municipal Archives

Records of the United States House of Representatives (Record Group 233), Records of the Committee on Agriculture, National Archives Building, Washington, DC

Records of the United States Weather Bureau (Record Group 27), National Archives College Park, College Park, Maryland

- A1–148, Office of the Chief Signal Officer: Administrative Orders, Circulars, Instructions, and Memorandums, 1871–1891
- NC3–18, Meteorological Correspondence of the Signal Office: Letters Sent (Agricultural Series), 1873–1878
- NC3–19, Meteorological Correspondence of the Signal Office: Letters Received (Agricultural Series), 1873–1877

NC3-33, Weather Bureau: Letters Sent by the Chief, 1891–1895, 1897–1911
NC3-35, Forecast Division: Letters Received, 1893–1894
NC3-46, Records Relating to the Transfer of Meteorological Functions from the War Department to the Agriculture Department, 1887
NC3-47, General Correspondence of the Weather Bureau, 1870–1912, 1894–1942, Letters Received, 1896–1912
NC3-50, General Correspondence, 1912–1965, 1912–1942
UD-20, Reports on Value of Climatological Publications, 1903–1908
UD-25, Records Relating to Opinions Concerning Emerson Hough's Article on the Weather Bureau, 1909
UD-87, Surface Land Observations: Newspaper Meteorological Reports, September 1871–February 1873

Abbot Lawrence Rotch Papers, Houghton Library, Harvard University

PRIMARY SOURCE DATABASES

American Periodicals Series Online
America's Historical Newspapers
British Newspapers, 1800–1900 (now available through 19th Century British Library Newspapers Database)
Brooklyn Daily Eagle Online (now available through Newspapers.com)
Cornell University Library Making of America Collection
Google Books
HathiTrust Digital Library
Internet Archive
JSTOR
Library of Congress Chronicling America: Historic American Newspapers
New York Public Library Digital Collections
Newspapers.com
Nineteenth Century US Newspapers
NOAA Central Library Digital Collections (Reports of the Chief of the Weather Bureau, US Army Signal Corps/Weather Bureau Annual Reports, 1861–1942)
ProQuest Congressional Publications
ProQuest Historical Newspapers (*Atlanta Constitution*, *Boston Globe*, *Chicago Defender*, *Chicago Tribune*, *Los Angeles Times*, *New York Times*, *Wall Street Journal*, *Washington Post*)
Readers' Guide Retrospective: 1890–1982
Utah Digital Newspapers

INDEX

Note: Page numbers in italics indicate figures.